Jan Myrick 8/08

ALSO BY GWYNETH CRAVENS

Gates of Paradise

Heart's Desire

Love and Work

Speed of Light

The Black Death

POWER TO SAVE THE WORLD

POWER
TO SAVE THE
WORLD

The Truth About Nuclear Energy

Gwyneth Cravens

ALFRED A. KNOPF, NEW YORK 2007

Grateful acknowledgment is made to the following for permission to reprint
previously published material: *The Independent*: Excerpt from "Nuclear Power is
the Only Green Solution" by James Lovelock (*The Independent*, May 24, 2004),
copyright © 2004 by *The Independent*. Reprinted by permission of
The Independent. • *The New York Times*: Excerpt from "Nuclear Risk and
Reality" by Herschel Specter (*The New York Times*, May 20, 2003), copyright
© 2003 by *The New York Times*; and "The Low-Risk Pool" by Herschel Specter
(*The New York Times*, October 17, 2004), copyright © 2004 by *The New York
Times*. Reprinted by permission of *The New York Times*. Richard Rhodes:
Excerpt from "The Genie is Out of the Bottle" by Richard Rhodes (*The
Guardian*, August 6, 2002), copyright © 2002 by Richard Rhodes.
Reprinted by permission of Richard Rhodes.

Library of Congress Cataloging-in-Publication Data

Cravens, Gwyneth.
Power to save the world / Gwyneth Cravens.

p.   cm.

Includes bibliographical references and index.

ISBN 978-0-307-26656-9

1. Nuclear engineering—United States. 2. Nuclear power plants—United States.
I. Title.

TK9146.C65 2007   333.792'4—dc22   2007017611

Manufactured in the United States of America
First Edition

For
Marcia Beauregard Fernández

Lili del Castillo Critchfield and Lewis Critchfield

And in memory of my father, Robert F. Jones, 1910–2006

¡Ay, ay, ay, ay!
Canta y no llores,
porque cantando se alegran,
cielito lindo,
los corazones.

Most of us were taught that the goal of science is power over nature, as if science and power were one thing and nature quite another. Niels Bohr observed to the contrary that the more modest but relentless goal of science is, in his words, "the gradual removal of prejudices." By "prejudice," Bohr meant belief unsupported by evidence.

— RICHARD RHODES

A NOTE TO THE READER

A full glossary of terms commonly used regarding
nuclear energy begins on p. 395.

CONTENTS

INTRODUCTION: GWYNETH'S PILGRIMAGE

Richard Rhodes

GWYNETH CRAVENS EVOKES AN old tradition in this very modern book: seeking understanding by going on a journey. In John Bunyan's *Pilgrim's Progress,* published in the 1660s during the English Restoration, a seeker named Christian encounters the Slough of Despond, the House Beautiful, the Valley of the Shadow of Death, Vanity Fair, and Doubting Castle, among other challenges, before he finally achieves the Celestial City. Cravens encounters similarly colorful places: abandoned mines that breathe radon; laboratories where nuclear reactors were deliberately melted down; a hellish place of heat, coal dust, and shattering noise; quietly vigilant nuclear power plants; vast humming turbines; a giant crystal called WIPP; Yucca Mountain. She accumulates knowledge as she goes, guided by her own Virgil, a steadfast scientist named Dr. Rip Anderson. She achieves greater understanding of the deep things of the world, as her predecessors did, and as they also did, she shares it generously.

Unlike the chronicles of those ancient worthies, however, Gwyneth's narrative is factual, not allegorical, a tour from discovery to mining to waste disposal of a modern technology, nuclear power, that harnesses the first major source of energy that does not depend, directly or indirectly, on the sun. Nor did she begin her journey, as Bunyan did, with an ax to grind. If anything, she was antinuclear. Her journey to knowledge changed her perspective.

I had a similar experience when I began writing about energy issues for national magazines back in the 1970s, in the years of the Arab oil embargo and the ensuing energy crisis. I was as reflexively antinuclear in those days as too many of my colleagues continue to be today. Then I researched and wrote my book *The Making of the Atomic Bomb.* In the course of doing so, I got to know the extraordinary men and women who developed the science of nuclear physics, many Nobel laureates among them, and learned to my surprise that they looked at nuclear energy very differently from their perspective of firsthand knowledge than I did from my perspective of secondhand ignorance. I asked them questions, I read their work, I visited laboratories and power plants, I researched and wrote a book-length history of the development of nuclear power in the United States, and what I learned changed my mind. I came to understand that nuclear power is one of the best solutions to environmental public health problems, as well as a necessary and probably central part of any effort to reduce global warming.

Energy transitions take time. One example is the historic substitution of coal for wood, which was fundamental to the Industrial Revolution. Coal had been known and used for three thousand years, but only marginally. The inhabitants of sixteenth-century London, however, suffered from a problem familiar to conurbations in developing countries today: as the city grew, a greater and greater area around it became deforested, and as transportation distances increased, wood became more expensive. The poor had to switch to coal; the rich followed later.

A second major energy transition originated in the United States. In the mid-nineteenth century, petroleum was first used as a substitute for whale oil for illumination in the form of kerosene. At the beginning of the twentieth century, coal still accounted for more than 93 percent of all mineral fuels consumed in the United States, and electric light was rapidly displacing the kerosene lantern in urban America, with eighteen million lightbulbs in use by 1902. Large oil fields were discovered in Texas and California early in the century. Railroads in the West and Southwest almost immediately converted to oil burning because local oil was cheaper than distant coal when transportation costs were figured in. That conversion prepared the way for the use of gasoline in automobiles.

The emergence of nuclear power in the twentieth century has been haunted by preadaptation. In the United States, the Soviet Union, Britain, France, and China, nuclear reactors were developed first of all to breed plutonium for nuclear weapons. The development of power reactors was delayed in the years immediately after the Second World War because everyone involved in the new atomic energy enterprise believed that high-quality uranium ore was too rare in the world to be diverted from weapons production. Early in the 1950s the U.S. Atomic Energy Commission even considered extracting uranium from coal fly ash, where burning had concentrated coal's natural complement of uranium. Well into that decade, almost the entire U.S. production of uranium and plutonium was dedicated to nuclear weapons. Finally the federal government offered bonuses to uranium prospectors for high-quality finds and the prospectors unearthed the extensive uranium resources of the Colorado Plateau.

Another delay arose from concerns for secrecy. The Atomic Energy Act of 1946 made atomic energy an absolute monopoly of the federal government. All discoveries were to be considered "born" secret—treated as secret until formally declassified—and the penalty for divulging atomic secrets was life imprisonment or death. All uranium and plutonium became the property of the government, as beached whales once became the property of kings. No one could build or operate a nuclear reactor except under government contract, nor could one be privately owned. These restrictions and mind-sets had to be revised before utilities could own or build nuclear power stations.

It's clear in hindsight that the careful evolutionary development of nuclear power in the United States, including the types of reactors developed and the nurturing of a solid political constituency, was a casualty of the cold war. Early in the 1950s, the Soviet Union announced a power reactor program, and by then the British were developing a power reactor fueled with natural uranium that countries without enrichment facilities might want to buy. In both cases Congress feared the United States might be left behind. It amended the Atomic Energy Act in 1954 to allow private industry to own and operate reactors, and government-subsidized construction began on a 60-megawatt demonstration plant at Shippingport, Pennsylvania, the same year. The reactor was a modified Westinghouse Large Ship Reactor, a system under development at the time to power aircraft carriers. Naval reactors use highly enriched, weapons-grade uranium for fuel and water as a heat-transfer agent; Admiral Hyman Rickover, who shepherded the early development of navy and civilian power reactors, sensibly but daringly decided to use low-enriched uranium oxide in the civilian reactor to lessen the risk that the fuel could be diverted and used to make bombs.

The configuration met the needs of the U.S. Navy, but it was less than ideal for commercial power. Sodium might have been a better heat transfer agent than water, but water was more familiar to civilian engineers. Uranium oxide, which became the standard fuel for commercial reactors, is less dense than uranium metal and conducts heat much less efficiently. Since the pressurized-water reactor isn't a breeder, it wastes most of its fuel; that in turn increases the volume of long-lived radioactive waste. To make their compromise competitive in a field dominated by relatively cheap fossil fuels, reactor manufacturers pushed design limits, maximizing temperatures, pressures, and power densities. Tighter design limits led to more frequent shutdowns and increased the risk of breakdowns, which in turn required more complex safety systems.

More crucially, manufacturers began pursuing economies of scale by selling larger and larger reactors, without fully addressing the changing cost and safety issues such reactors raised. "The largest commercial facility operating in 1963," two policy analysts wrote, "had a capacity of 200 megawatts; only four years later, utilities were ordering reactors of 1,200 megawatts." But the safety equipment that government regulators judged sufficient at 200 megawatts they no longer judged sufficient at 1,200 megawatts. So they began requiring add-on safety systems, escalating engineering and construction costs.

Construction time increased from seven years in 1971 to twelve years in 1980, roughly doubling the cost of the plants and raising the price of the resulting electricity. Nuclear regulatory commissioner Peter Bradford would write later that "an entire generation of large plants was designed and built with no relevant operating experience, almost as if the airline industry had gone from Piper Cubs to jumbo jets in about fifteen years." Because of the scale-up in size and the correspondingly larger inventory of fuel, "engineered safety" replaced "defense in depth" as a design philosophy, and it became impossible to demonstrate that large U.S. power reactors were acceptably safe. Nor was a safety culture developed and maintained among the operating teams at private utilities, which lacked experience in nuclear power operations.

It was these problems, not antinuclear activism, that led to the cancellation of orders and the halt in construction that followed the Arab oil embargo of 1973–74. Orders for some one hundred U.S. nuclear power plants were canceled, but so were orders for eighty-two coal power plants, because the Arab oil embargo stimulated dramatic improvements in energy conservation in the United States and these stalled a long-standing trend of increasing demand. "Who . . . would have predicted," Oak Ridge National Laboratory director Alvin Weinberg would write, "that the total amount of energy used in 1986 would be only 74 quads, the same as in 1973?" Today, with demand once again increasing, U.S. nuclear power is thriving: existing plants are being relicensed to extend their operating life; plants left unfinished will probably be finished and licensed; and new reactor construction utilizing newer, safer, and more efficient designs is pending.

France has a different story, with 80 percent nuclear electricity *today*, energy security in consequence, and air pollution reduced fivefold. Japan is pursuing a highly successful nuclear power enterprise as well. Secrecy extended to nuclear power generation reached a deadly extreme in the Soviet Union. Operators were forbidden to share information about problems and accidents from one plant to the next, making systemic improvements impossible, and information about the developing Chernobyl disaster was delayed for crucial days while Moscow hunkered down. The worst large-scale consequence of Chernobyl has been thyroid cancer in Ukrainian and Belarusian children. I am told by Dr. Stanislav Shushkevich, the first Belarusian head of state and a nuclear physicist, that every Soviet fallout shelter held a supply of potassium iodide that would have protected the children by saturating their thyroid glands, preventing the uptake of radioactive iodine-131, but that Moscow refused to allow the tablets to be distributed until it was too late and the children had already been exposed. Chernobyl was a failure not of nuclear power but of the Soviet political system.

A word about public opinion and antinuclear activism. Weinberg has argued that nuclear power faltered "because nuclear optimists ignored social, political, and economic realities." I would emphasize the economic part. Almost every nuclear power plant built in the United States was designed for its site rather than prefabricated. These large, expensive plants needed a construction license to build but then had to stand idle while an operating license was negotiated, often with considerable and expensive delays. That mistake has been corrected; one license at the outset now legally covers both conditions. Resistance to nuclear power, such as it was—every U.S. nuclear power plant that was completed was licensed—was less concerned with risk and safety than it was with nuclear power's associations with nuclear weapons, highly centralized political and economic systems, and technological elitism. In other words, antinuclearism was primarily a political movement, and proponents of nuclear power missed the point when they defended nuclear power on technical grounds alone.

But enough. Let's join Gwyneth's pilgrimage. When I set out on it I thought I knew a lot about nuclear power. She and Dr. Rip taught me something new on every page. I enjoyed the journey, and I think you will, too.

Part 1 **ORIGINS**

Look always at the whole.

—MARCUS AURELIUS

1 SURVIVAL

IMAGES OF EXPLODING NUCLEAR BOMBS gave me nightmares when I was a child. And I would often be startled awake by the roar of fighter jets splitting the sky. Were the Soviets finally attacking Albuquerque, my hometown? I'd once watched a television show about what would happen to America if the Russians bombed us—a crowd of upturned faces, a blast of white light incinerating them—then had run outside in search of cover.

Our school faced the nearby Sandia Mountains, their tall facade like a cutout made of blue construction paper. One day at recess as my best friend, Peggy Smith, and I were talking about how we loved them, she told me that she'd overheard her father, an engineer at Sandia National Laboratories who made frequent trips to the atomic test site in Nevada, tell her mother top-secret information: one of those mountains had been hollowed out and filled with thousands of atomic weapons. I pictured a vaulted cavern filled with row after row of big, white, egg-shaped bombs, looking just like the replicas of the ones dropped on Japan that were always displayed at Kirtland Air Force Base on Armed Forces Day along with a battery of real missiles pointed skyward.

When I was about eleven, I compared notes on the nuclear threat with Janet Johnson, my neighbor and pal. She had heard from some Los Alamos kids that Albuquerque was the Soviet Union's number one target in the Western Hemisphere. That had to be because of the hollow mountain full of nuclear weapons. As every schoolchild in New Mexico knew, the first atomic bomb in the world had been tested in the southern part of the state. Janet and I paged through *Life* magazine's photo spreads of the ruins of two Japanese cities, of multicolored mushroom clouds, of life-size dolls flung around in dust storms inside living rooms as houses imploded during atomic tests.

We estimated that, living as we did on the edge of town, far from the base and the labs, and on the last street before the open mesa, we'd probably survive the explosion and could save our poor parents, who downplayed the impending catastrophe to us and were not prepared for it. But *we* made a plan, and the preliminaries for it consumed one summer. Carrying shovels, we set out barefoot across the mesa, with its fragrant sagebrush, rippling grama grass, and cholla and prickly pear cactus, to a nearby arroyo. There we began scooping out a refuge in pinkish gravel and clay among the roots of a clump of rabbitbrush. Our mothers wouldn't let us bring blankets or canned goods to this hole in the ground, but we did store bottles of water there as well as cloth bags we sewed and stuffed with matches, Band-Aids, packets of rice, tapioca, and tea, and knives that we carved out of bamboo that grew in my yard.

On a hot day we'd lie in the cool hollow and gaze up at the sky, with its circling hawks, big thunderheads, and fighter jets. Besides plotting our survival strategy, we read science fiction. For us the future was divided into two visions, and in discussions of them, we often used the word *conceivably*. We saw ourselves thriving as scientists in a prosperous utopia with atoms as a source of energy that would power our airborne cars. Wearing beautiful dresses made of miracle fabrics, we'd fly to the moon and Mars in atomic rockets piloted by our handsome husbands. At the same time we worried that World War III might reduce our country, starting with Albuquerque, to a smoldering, radioactive wasteland. Then we'd have to subsist the way humans did during the Stone Age.

If and when we felt the initial shock wave from an exploding warhead, we planned to grab our family members and pets, and run as fast as possible to our shelter. The glowing cloud would whoosh up over the city with a rumble. When the smoke and dust cleared, only a flickering crater would remain. What would be radioactive? The food in the supermarket? Maybe we'd have to take our archery sets up to the mountains and bring back game for our mothers to roast.

By the end of the summer, surveyors' stakes had appeared on the mesa, heralding a housing development. A flash flood had swept away our hideout, and when we returned to school, boys began to loom as a much bigger concern than nuclear holocaust. Still, for many years the image of the hollow mountain visited my thoughts and appeared in my dreams.

I now live in New York, but I often fly back to New Mexico. Smog and clouds usually obscure the eastern half of the journey. When the plane angles toward the Southwest, the sky opens up and light pours down. The smooth, cultivated expanses of the Great Plains begin to give way to wrinkled tracts with occasional irrigated squares or circles of green and then fade into browns and mauves. Towering, sun-glazed thunderheads send violet shadows racing on the ground below. The landscape crumples and splits as the jetliner arcs over the vast Chihuahuan Desert. Its parched, undulating surface, barren stony ridges, salt lakes and salt pans, and meandering, branching dry riverbeds remind me that in the West, destiny is determined by geology and climate.

Soon the Sandia mountain range rears up, a giant hinge opening westward to an elevation nearly two miles above sea level, the first obvious sign that the land is tilting toward the highlands of the Continental Divide. On the approach to Albuquerque, the airliner sometimes glides over the vertical pink slabs and steep canyons that make up the western face of the Sandias. Next come foothills cut through with arroyos and dotted with scrubby piñon trees where, as a girl, I went riding on a ranch that has been transformed into a subdivision with a country club, its golf course shouldering up against the fifty thousand mostly empty acres of the military base and Sandia National Laboratories.

Access to them had always been restricted, and so had their airspace. But when I flew in one day in the mid-1990s, some time after the Berlin Wall came down, the flight path went over the base instead of skirting it. Below, for the first time, I saw what had been hidden there: widely spaced clusters of buildings, towers, ramps, and earthworks, and something unexpected: a small mountain—a foothill, really—girdled by a railroad track and ringed by three tall, concentric fences. Planed nearly clean of vegetation, shaped like a pyramid, and with several giant doors set into its flanks, this odd apparition seemed to have been transported here from Egypt's Valley of the Kings. I realized I'd glimpsed it years earlier from horseback, my gaze dazzled by grids of steel mesh. Until now, I'd never connected Peggy's secret with the mysterious facility I'd seen—how would a mere kid ever come upon a mountain full of nuclear weapons? Or was that just a myth?

During that particular visit to Albuquerque, I went to a gathering at the home of my oldest friends, Lili del Castillo and her husband Lewis Critchfield; they're flamenco artists. Lew is the son of Charles Critchfield, a Los Alamos theoretical physicist who had participated in the Manhattan Project. Lew and Lili's close friends, Rip Anderson, a scientist at Sandia National Laboratories, and his wife, Marcia Fernández, an educator and a singer, were there. I knew them but not well. While others danced, sang, and played flamenco music, Rip—nobody calls him D. Richard Anderson, Ph.D., except professional journals—stayed in the background, as usual. He has curly, receding gray hair, a full mustache, and blue eyes. Wiry of build and quick of step, he favors yoked gingham or flannel shirts with mother-of-pearl snaps, faded jeans, a Stetson hat, and cowboy boots. He always seemed relaxed, but I'd found out that he was extremely observant. Of someone at a reception we'd attended, he said, "That guy spoke for about ten minutes without using any pronouns." When Rip was four, his father, a rancher, began to teach him to be a tracker. Once as we drove at sixty miles an hour across the high desert, in answer to my questions, Rip had rapidly identified wildlife moving through his peripheral vision. He had a reputation for being able to fix anything. Lew and Lili would describe how Rip and Marcia always showed up to help whenever there was a problem, and how, during breaks from mysterious Sandia-related trips to places like Korea, Finland, and the National Academy of Sciences in Washington, D.C., he would tackle house construction, wiring, plumbing, welding, car repair, farming, and livestock wrangling. He didn't care to talk about himself, and at parties I would try to draw him out. That's how I found out that in his youth he'd been a rodeo contestant. At a brunch at Rip and Marcia's I'd happened to glance out the window just as he was vaulting onto the bare back of a runaway mare. One time I was present when he and the Southwestern mystery writer Tony Hillerman compared their boyhood experiences of hand-plowing on foot behind a mule. With some wonderment at the thought, Rip realized that the two of them were part of a tiny number of Americans who still knew how to hitch up a team of horses and put them to work. Rip and Marcia are grassroots activists, tenacious crusaders for clean air and water and preservation of open land. Together with a group of volunteers, they established and maintain a wildlife sanctuary in Albuquerque.

Although I'm primarily a novelist, I've also written articles on scientific topics. I'd once asked Rip how the atomic clock at the National Bureau of Standards operated, but I was careful never to make inquiries about his job. In New Mexico, it's bad manners to ask federal employees about their work or about information that might be secret. In fact, for years I wasn't sure what my father did after he left the Forest Service to become a security-cleared civilian employed by the air force. (It turned out he inspected construction projects and surveyed the site of a new runway.) Still, I was hoping that if commercial flights could now pass through formerly restricted airspace, maybe scientists such as Rip could speak more freely.

I told him what I had seen from the air and, emboldened by the careful way he listened, asked him whether thousands of nuclear weapons were hidden in that mountain.

To my astonishment, he replied, "Sure. Bombs and warheads used to be stored there." His manner of speaking, courteous and laconic, with the occasional archaic word thrown in, reminded me of the Old West.

"What happened to them?"

He replied, "We're getting rid of 'em. Because of the Strategic Arms Limitation Treaties."

"Well, I'm all for that."

"Me, too," he said.

"You are?"

"Heck yah. I think most of us in the nuclear end of things are."

That gave me pause. "But how do you get rid of a nuclear warhead?"

"Take it apart."

Even though I'd participated in ban-the-bomb rallies in Greenwich Village after I moved East to study literature in graduate school and rejoiced when the United States and the USSR called off the arms race, I'd never considered the fate of a retired weapon. "And what happens to the pieces?"

"The pits—the uranium and plutonium components—will one day be processed into fuel. The rest is carefully disposed of."

"Bombs are being turned into fuel?" I was incredulous. "For what?"

"To make electricity."

I don't know which surprised me more—confirmation that the mountain had really contained bombs or the news that weapons could provide constructive power.

That was the beginning of a long dialogue and, although I could not then have imagined it, of an unexpected journey through the nuclear world with Rip as my Virgil. I came to understand that he had a mission. Trained as a chemist and oceanographer, he grasped the interconnections among the abyssal bed of the ocean, the depths of the earth, the fluctuations in global climate, the behavior of particles and energy, the inside of a nuclear reactor, and the meanderings of public policy. Conversant with deep time—the remote past and the planet ten thousand years from now—and with the universe of risk and consequence, he had acquired an international reputation in the fields of probabilistic risk assessment, environmental health, and nuclear safety. His colleagues told me that he was highly respected for his original thinking and the comprehensive sensibility he displayed when managing large programs. To this day he encourages those who work for him to do their best to make the hypothesis under scrutiny fail. Those who have worked for him tell me that this maverick imperative has made for the best science they've ever done—it has held up decades later—and they speak of his contribution to Big Science, his uniquely comprehensive approach to finding solutions, his unwavering respect for scientific objectivity, his status in his fields of expertise, his willingness to take unpopular stands in the scientific and technical communities, and his refusal to compromise with prevailing notions unsupported by science.

"OK, the bombs that used to scare me are going to be used for electricity," I said that night. "But aren't nuclear plants deadly? And can't reactors be used to make bombs?" Worried that my daughter would be harmed, that evacuation would be impossible in case of an accident, and that my organic garden on Long Island would be contaminated by cooling tower emissions, in the 1980s I'd joined a successful campaign to prevent the opening of a new nuclear plant thirty miles from my home.

Rip quietly replied that no other technology to produce energy steadily on a large scale had a better safety record than nuclear power and emphasized that few people understood that you could make a nuclear weapon without having a reactor. "In the countries that have the bomb, civilian nuclear power plants have not been used in weapons programs. It's not an efficient or easy way to get weapons-grade material. What you need to do that is uranium ore and a uranium enrichment plant. It's expensive, hard to build, and it's big, and it's really messy. Or, if you have spent fuel from a reactor, you can reprocess it to get out the uranium and plutonium—that's another big, dirty, complicated job that requires special equipment, and the yield from low-enriched commercial nuclear fuel is so small that it's not worth the effort. Anyway, banning nuclear power plants will never ban bomb production. To do that, you'd have to somehow make all the technical knowledge disappear from the earth. Political restraint is what keeps countries from using bombs, not banning nuclear plants. Nuclear is the best option on the scale we need."

"What do you mean? Why would we need it?"

He began speaking about how for decades he had been watching a very alarming trend, particularly as it related to one of his specialties, the ocean. He began speaking about a topic that was somewhat familiar, mainly from occasional reports in the science pages of newspapers and from a seminal book, *The End of Nature*, by Bill McKibben.

"The earth is warmer today than it has been in thousands of years," Rip said, and described the remarkably rapid increase in certain gases in the upper atmosphere that trap heat from sunlight, preventing it from radiating back into space. The result: conditions like those inside a greenhouse. These gases are mostly carbon dioxide and methane that are released when we burn coal, oil, and natural gas, grow rice, and raise cattle. If we keep producing them at the present breakneck rate, the consequences to the intricate web of life on the planet will be grave. And even if we could stop these emissions right now, it would take centuries for the effects to dissipate. However, he told me, we have solutions. We can eliminate this deadly waste from the systems we use to make energy. "Nuclear power doesn't emit greenhouse gases when it makes electricity and is the largest displacer of greenhouse gases on the planet," he said. "It's safe, reliable, and clean."

"Clean?" I bristled. "And the *waste*?"

"Every energy resource has its risks and benefits," he said. "With nuclear, the benefits outweigh the risks by far." (I didn't know that he was working hard to get certification for the world's first long-term, deep geologic repository for nuclear waste.) "Look at the big picture. If you got all of your electricity for your lifetime from nuclear power, your total share of the waste would weigh two pounds and fit into one Coke can. Of that, only a trace is long-lived. And all that waste is kept out of the biosphere. We get half of our electricity from burning coal, and its huge amount of waste all goes into the biosphere."

The contrast is staggering, as I later learned. The annual solid residues of coal combustion come to eight hundred ninety pounds per American. Think of the total amount as fitting in one million railroad cars that would form a train extending nine thousand six hundred miles—three and a half times longer than the distance between New York City and Los Angeles. If an American got all his or her electricity from coal over a lifespan of seventy-seven years, that person's mountain of solid waste would weigh 68.5 tons. Picture a soda can next to that.

Without a thought about where electricity comes from, we expect it to serve us whenever we flip a switch. Every year, we burn more fossil fuels, obliviously pumping thousands of tons of greenhouse gases and particulates into the environment. According to the United States Energy Information Administration, from 1990 to 2005 our output of these gases rose by 17 percent, with most coming from residential sources. On average, an individual American's contribution is 24.5 metric tons of greenhouse gases a year and growing. Combined, they account for 25 to 30 percent of the world's total even though we make up only 5 percent of the world's population. For each mile driven, every car releases about a pound of carbon. Making electricity contributes 40 percent of world carbon dioxide emissions. To keep a 100-watt bulb burning for ten hours you have to burn a pound of coal. The electricity use of an average American home results in eighteen thousand pounds of carbon dioxide a year released into the atmosphere. Practically everything we do, wherever and however we live, results in dispatching more and more carbon from deep in the earth into the atmosphere. The consequences for the natural world and humanity are huge and disastrous, and they are already occurring.

If you picture energy generation on our planet as a house with five rooms, fossil fuels occupy the three biggest ones. Coal combustion generates 39 percent of the world's electricity, followed by natural gas combustion (19 percent) and oil combustion (7 percent). Soot, smoke, toxins, radionuclides, carbon dioxide, methane, and noxious gases from those rooms spread throughout the whole house, spattering it with greasy black particulates, and causing the temperature to rise. Nuclear (about 16 percent) occupies a cellar.

Renewables, those resources used to generate electricity that are replaced naturally, are not necessarily free or limitless and, unfortunately, they're not always available. They include hydroelectric, solar, wind, and geothermal technologies, methane gas from landfills, and energy produced by burning municipal waste or biomass—wood chips, dung, or agricultural waste. In our planetary house, renewables (except for hydroelectric dams) get only a closet (1 percent). In this closet is a dollhouse with a miniature cabinet, and there, in a drawer so tiny you have to open it with a pair of tweezers, we find wind and solar power (0.02 percent). In the United States, the total contribution of renewable energy has remained at 6 percent since 1970, and hydroelectric and biomass still account for almost all of that percentage, with solar providing 0.2 percent and wind 0.13 percent. Some blame can go to the erratic federal funding of research and development; however, in recent years renewables have received considerable government and corporate funding. All of our energy is subsidized, with those who produce electricity from fossil fuels getting the biggest handouts by far.

Rip believed that humanity would be at higher risk if we did *not* increase our reliance on nuclear power. Otherwise our civilization would rely mainly on fossil fuels. Their pending depletion and the impact of their waste was already causing a global crisis. He knew intimately what daily life without electricity meant. Before transmission lines reached the Idaho ranch where he'd grown up, all the water for the household and the livestock had to be pumped by hand. Counting the time to pump and haul the water, chop the wood for the kitchen range, and heat the water, his mother had to spend four hours to bathe her four children. The family's lives were ruled by the sun. Candles and coal-oil lanterns, burdensome to maintain, supplied a dim glow—not enough to read by at night. "It would take a hundred of those damned lanterns to put out as much light as a single 100-watt bulb," Rip said. The family rejoiced when electricity finally arrived. Some sixty years later, the average hourly demand per American for electricity comes to 1,400 watts around the clock.

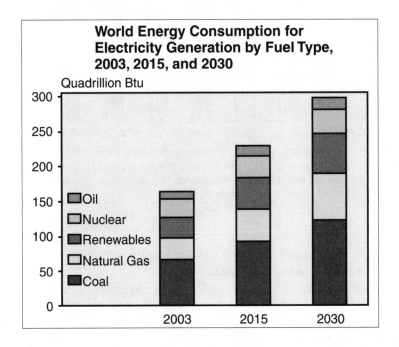

World electricity use is expected to more than double by 2030, with Americans and Chinese as the biggest consumers; the fastest growth will occur in developing countries. In the United States, demand will probably increase 45 percent.

Despite energy-saving appliances, which became available in the 1980s and have improved in efficiency over the years by half, and despite other conservation measures, such as switching to fluorescent lighting, Americans have been exuberantly bumping up consumption. New houses are bigger than the ones our grandparents owned; they have to be, to contain all the giant appliances, Jacuzzis, entertainment centers, and climate control gadgetry. The digital revolution has also devoured a lot of wattage. To feed this growing appetite for electricity, national and international agencies predict, much more fossil fuel will be burned.

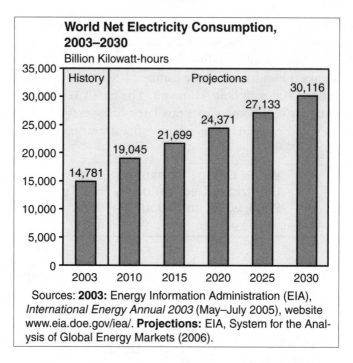

World Net Electricity Consumption, 2003–2030

Billion Kilowatt-hours

History | Projections

- 2003: 14,781
- 2010: 19,045
- 2015: 21,699
- 2020: 24,371
- 2025: 27,133
- 2030: 30,116

Sources: **2003:** Energy Information Administration (EIA), *International Energy Annual 2003* (May–July 2005), website www.eia.doe.gov/iea/. **Projections:** EIA, System for the Analysis of Global Energy Markets (2006).

That evening, Rip and I went on to have a polite conversation. He didn't say much unless I asked a question.

"If nuclear power is so great, why are so many people afraid of it?"

He replied that shortsightedness, stupidity, and poor design had caused some accidents. Lack of transparency about them had produced fear and suspicion in the public mind. He said that these incidents had been entirely preventable and that carelessness in the chemical and fossil fuel industries had led to far worse disasters. Thorough training, backup safety mechanisms, and foresight could reduce nuclear risks and already had to a great extent. He mentioned the nuclear navy, which has demonstrated the successful operation of reactors for over fifty years without a harmful release of radioactivity, and he pointed out that in over forty years of operation no deaths attributable to nuclear power had occurred in the United States.

I didn't know then that he was a leading expert on risk assessment or that the major program he was then managing—it addressed the long-term disposal of nuclear waste—required him to assess forecasts of climate change, energy use, and societal transformations for the next 10,000 years.

Rip told me that we needed to increase nuclear power because it very efficiently and predictably produced electricity on a large scale to meet growing demand and, in doing so, emitted almost no carbon dioxide or other pollution for the simple reason that it didn't burn anything. "If we don't use more nuclear, then we are for sure going to get more coal-fired plants. That's already happening."

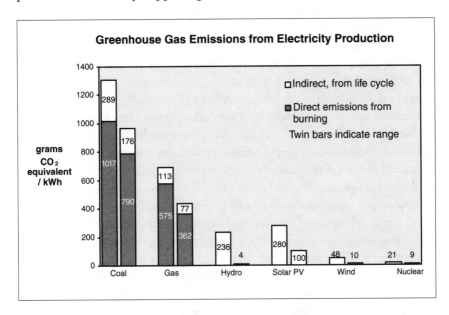

Greenhouse Gas Emissions from Electricity Production

I believed that wind and solar power and conservation were the best solutions, that nothing was more unnatural and scary than nuclear power, and that people who supported it must hate nature. True, I'd already noticed Rip's profound attunement to the natural world and had always found him to be honest and direct. I respected his goodwill and intellect. Still, I was annoyed, and despite the arresting points he was making, I could scarcely take his position seriously, so contrary was it to my worldview.

When I returned to New York I looked for ammunition that would prove Rip's claims wrong. Instead, I found that they had considerable scientific support. I'd written articles about epidemiology with John S. Marr, M.D., former director of New York City's Bureau of Infectious Disease Control, so I was aware of the gulf between what scientists understood about risk and the public's perception, which could be exaggerated in one direction or the other. For example, in the event of an epidemic people might—because of poor communication, wrong assumptions, or lack of accurate information—tend to respond with anxiety despite the low probability of contracting the disease in question, or those at high risk might fail to act in their own best interest. Others might not be bothered by familiar dangers but instead would worry about the exotic. A pack-a-day smoker might fear getting West Nile virus, not realizing how small the odds were compared to the toll taken by tobacco-related illnesses, which kill half a million Americans every year.

When I turned to the statements of antinuclear groups, I naturally found that they echoed many of my own assumptions: that uranium mining and processing, depleted uranium, nuclear accidents, and nuclear waste had killed or would one day kill huge numbers of people, caused mutations and birth defects, and turned pristine places, usually home to Native Americans, into radioactive wastelands; that all man-made radioactive material was lethal and we lacked any natural defenses against it; that all radiation was bad; that cancer clusters occurred around nuclear facilities; that to stop proliferation of nuclear weapons all reactors had to be shut down; that terrorists could easily overwhelm a nuclear plant or waste dump; that such facilities could explode atomically; that there was no safe place to put our towering heaps of nuclear waste, which would remain harmful for millions of years; that reliance on uranium was futile because soon we'd be running out of it, just as we were running out of oil; that nuclear power put almost as much carbon into the environment as coal, gas, and oil; and that instead of using fossil fuels and nuclear power we could instead practice conservation and obtain all our energy from wind, sunlight, tides, and geysers.

On subsequent visits to Albuquerque, I'd confront Rip with what I had dug up in the interim. He responded with explanations about laws of nature and showed me peer-reviewed, rhetoric-free documents that contained information widely agreed upon in the scientific community. And he revealed how he thought about risk.

"People demand zero risk, but that's impossible," he said. "Just living has its risks. Every energy resource does too. But if you look at the science, nuclear always comes out ahead. The nuclear world today has a culture of safety that's drilled into everybody. We all have to attend yearly classes and review radiation protection. But the public still has an exaggerated perception of risk from nuclear power.

"Nuclear plants in Western Europe and the United States have never killed anyone. Chernobyl failed because the reactor was badly designed and badly managed—during an equipment test the operators disconnected the safety backup systems! Contamination escaped because there was no containment building. Every reactor in America has to have one, and an independent body, the Nuclear Regulatory Commission, strictly supervises every plant. Eventually, we'll replace fossil fuels with nuclear, and we'll probably use nuclear plants to make hydrogen or synthetic fuels for our transportation. Every energy industry takes at least fifty years from the time it goes from a few tenths of one percent of what people use up to ten percent to fifteen percent. Same with nuclear. It's been making electricity for about half a century and today about twenty percent of all electricity generated in the United States comes from nuclear."

"Right, and we're getting all that radioactivity," I said.

"Actually, we're not. If you went and sat on a fence of a coal-fired plant you'd likely get a much bigger radiation dose than you would if it was a nuclear plant. A nuclear plant that emitted as much radiation as a coal-fired plant would be shut down."

During these dialogues, Rip graciously refrained from referring to holes in my knowledge that I was to discover on my own. I lacked a clear sense of what radiation actually was, didn't know much about its sources, didn't distinguish between low-dose radiation and high-dose radiation, and was foggy about the difference between exposure and dose and about radioactive decay. I had no idea how a nuclear plant worked. He merely pointed out that there were real but controllable risks attached to the nuclear fuel cycle and there were also highly improbable ones.

"If you take a panoramic, long-term view about the rise in global temperature and the increasing demand for electricity," Rip said, "and you look at all our resources, then you immediately ask how we can decarbonize our energy systems on a global scale. Are the risks of mining uranium greater than the risks of mining coal? Are the risks from nuclear waste greater than the risks from waste from fossil-fuel combustion? Can we replace coal and nuclear with something better in the next few decades? As far as I know, nobody has ever written anything accurate for the public about the risks and benefits of uranium throughout its fuel cycle—its trip from cradle to grave—and how they compare to other energy sources. The Sierra Club used to promote nuclear energy as the best choice for protecting habitat and clean air and clean water."

"No!"

"In the late 1960s, the Sierra Club in California did not want big hydroelectric projects messing up wilderness areas. The environmentalists back then knew that a nuclear plant has a much smaller footprint and would save thousands of acres of habitat from the destruction that occurs when you build a big dam and flood a canyon. And when dams break, lots of people can die. I know scientists who were members until the Sierra Club turned against nuclear power purely for political reasons— not because of the science."

I later found that in California this leading environmental organization once had as a slogan "Atoms not Dams," and that its board voted in favor of nuclear plants—the one at Diablo Canyon, for example. In 1970 the organization wrote a "crisis report" asking irrigation districts to build and manage a nuclear plant for the city of San Francisco instead of constructing yet another hydroelectric dam. Five years later, after an internal upheaval, the Sierra Club reversed its policy and began opposing nuclear plants and seeking to shut down Diablo Canyon. Today, throughout the United States, the official Sierra Club policy is strictly antinuclear.

"I've looked at the future of energy use from every angle," Rip said. "If there's a workable alternative out there, I want to know all about it. We need everything—more conservation, wind, solar, and cleaner fossil-fuel technologies. But for at least the next fifty years or probably longer, there's nothing else, not a hint of anything on the horizon that can be harnessed on the scale we require, and our needs are going to keep growing. We have no choice but to use fossil fuels, nuclear, and hydroelectric power until something like fusion comes along, which we may finally master many decades from now." (Nuclear fusion occurs when two atomic nuclei combine, as when the sun fuses the nuclei of hydrogen atoms into helium atoms, releasing an enormous amount of energy.)

The U.S. Energy Information Agency and the International Energy Agency estimate that to meet our growing needs we will require a much bigger supply of baseload electricity. "Baseload" refers to the minimum amount of proven, consistent, around-the-clock, rain-or-shine power that utilities must supply to meet the demands of their millions of customers. That power comes from energy stored in coal, natural gas, oil, uranium, and inherent in falling water. Wind and sunshine are weak, intermittent, and less efficient than other sources. They can't provide baseload now, because the technology for storing energy remains in its infancy. Photovoltaic panels, which are semiconductors that use photons from sunlight to make electricity, and turbines, which are turned by wind, also require a lot of space. A nuclear plant producing 1,000 megawatts takes up a third of a square mile. A wind farm would have to cover over two hundred square miles to obtain the same result, and a solar array over fifty square miles. Many people, including some environmentalists, do not want a sprawling heavy industrial site near their homes. It's likely, however, that these considerably subsidized renewable resources will continue to grow and perhaps in twenty-five years will supply 10 to 20 percent of our power.

Today, about three-quarters of our baseload electricity comes from burning precious natural resources created a quarter of a billion years ago. The resulting pollution, mostly exempt from regulation, makes the United States the foremost contributor of heat-trapping gases. Without a drastic intervention, the earth's burden of these gases will probably rise by nearly 60 percent in the next twenty years. I knew coal pollution was bad but had no idea of just how bad. As they make 51 percent of our electricity, coal-fired plants cause the premature deaths of twenty-four thousand Americans every year as well as hundreds of thousands of cases of lung and cardiovascular diseases. China's coal pollution, so immense that it creates smog on the West Coast of the United States, kills four hundred thousand Chinese each year.

Hydroelectric plants, a cleaner resource, provide around one twentieth of our baseload power, but drought has been on the rise, lowering output. In any case, water is a finite resource: most American rivers that can be dammed already have been. Dam failures are the direct cause of the greatest number of deaths of any energy source. In one disaster in China in 1975, flooding from a typhoon led to multiple breaks in a dam system that killed 26,000; another 145,000 died during subsequent epidemics and famine, and about six million buildings collapsed. In the United States dams killed over 1,000 in the last century.

As I researched energy options, I began to understand why a survey taken in 1980 of randomly selected scientists listed in *American Men and Women of Science* found that nearly 90 percent favored expansion of nuclear power. Most scientists polled today agree. About 73 percent of emissions-free electricity in the United States comes from nuclear plants. An inclusive analysis of the life cycle of nuclear power—the extraction of uranium and its transformation into fuel, the construction of plants, the decommissioning of reactors, and the disposal of waste—shows that throughout the process, nuclear power emits about the same amount of carbon or slightly less than is produced during the typical life cycle of wind turbines and solar panels. Radiation from nuclear plants, strictly regulated, is so insignificant that it's difficult to distinguish it from natural background radiation. Even when American plants have accidentally released radioactive materials, the actual exposure of humans has been minuscule compared with what we receive daily from natural sources.

This information, so contrary to my suppositions, was to lead me to examine my own prejudices and their origins. Nevertheless, how could I overlook the threat of more accidents like those at Three Mile Island and Chernobyl, the terrorist scenarios, the health risks associated with uranium, and the legacy of long-lived nuclear waste?

After quietly listening to my objections on several occasions, Rip asked, "Why don't you see for yourself what goes on in the nuclear world?"

"You're pulling my leg," I said. "No ordinary mortal could gain access."

"If you're really curious, I can get you in," he said.

The truth is, I was fascinated by the possibility of gaining entrance to the secret domain I'd wondered about since childhood. But I was also cautious. "I wouldn't want to get exposed to plutonium. One speck of that stuff and you're dead."

"Some air bubbles injected in a vein and you're dead," he said. "An injection of two grams of potassium cyanide and you're dead. If the contents of a hardware store or a garden center were dispersed freely in a populated place, some of those chemicals and pesticides would kill or poison people. But we don't worry about that because the solvents, lye, chlorine, arsenic, and all the rest of the nasty stuff stay contained. Same with plutonium. It's always contained. It takes an internal dose of plutonium to cause health effects, and even lab workers who have accidentally gotten internal doses are alive forty years later."

In a radiation safety class, Rip had once held a lump of plutonium the size of a robin's egg on his palm, on a piece of paper. "It felt a little warm, but the radioactive particles it gives off didn't touch me, because they can't even penetrate a piece of paper, or skin. Anyway, no one leaves plutonium lying around where you can get at it. Never. It costs five times more than gold. And nuclear facilities are strictly regulated. They're very clean and very safe. They have to be. Otherwise you durstn't set foot in one. I wouldn't, and nobody else I know would either."

I wanted to sort out which of my long-held convictions were correct and which were without scientific basis. After some reflection, I took Rip up on his offer and went with Marcia and him on what her son, Miguel Fernández, dubbed "the Nuclear America Tour" to see the places where uranium was born, did its work, and was entombed. On the journey, I learned about risks at each stage and the specifics about why scientists do not share the public's nervousness about nuclear power. I discovered that various practices I'd assumed were going on had long since been discontinued but indeed lived on in myths that were still circulating. I was surprised to hear investigators who witnessed the worst that radiation exposure can do to large populations endorse nuclear power as safer and healthier than fossil-fuel combustion. Ultimately, I came to understand uranium's benefits and their far-reaching implications for the fate of humanity and the planet.

In recent years, news coverage about global warming has moved from the science pages to the front page. The United Nations Intergovernmental Panel on Climate Change (IPCC), the National Academy of Sciences, the American Geophysical Union, the American Meteorology Society, the American Association for the Advancement of Science, the National Aeronautics and Space Administration, the National Oceanographic and Atmospheric Administration, and academic institutions around the world are now issuing increasingly urgent warnings. We are facing a catastrophe that could kill up to a billion people, mostly in poor countries, extinguish thousands of species, drown our coasts, create deserts, wreck our agriculture and our economy, and otherwise radically affect our lives. The 2007 UN panel's report on climate change, prepared by over 1,000 scientists, states with a confidence level of more than 90 percent that heat-trapping gases produced by human activities have been the chief cause of global warming since 1950. Relying on an enormous collection of data and on supercomputer simulations, the IPCC predicts a temperature increase of as much as ten degrees Fahrenheit by 2100 and a rise in sea levels of as much as two feet. Since the panel didn't take into consideration the melting of land ice, many scientists believe that the levels will be much higher, perhaps as much as twenty feet. But even if global warming skeptics should turn out to be right, the health impacts from combustion are so well documented and so great that we cannot afford to ignore alternatives.

Alvin Weinberg, a physicist, was for many years director of the Oak Ridge National Laboratory. Under his leadership, the laboratory researched, in the 1950s, radiation and human genetics; in the 1960s, the environmental impact of various energy systems, including the effects of increased carbon dioxide from fossil fuel combustion on global temperature; and in the 1970s, enhanced reactor safety and energy conservation. After his retirement from Oak Ridge he proposed the establishment of what was to become the National Renewable Energy Laboratory. In 1994, he wrote:

> Carbon dioxide poses a dilemma for the radical environmentalists. Since nuclear reactors emit almost no carbon dioxide, how can one be *against*

nuclear energy if one is concerned about carbon dioxide? To my utter dismay, indeed disgust, this is exactly the position of some of the environmentalists. Their argument is that extreme conservation, and a shift to renewables—that is, solar energy—is the only environmentally correct approach to reducing carbon dioxide. When I point out to them that conservation might be feasible in industrialized countries, but that it is hardly a choice for India and China, they seem to ignore the point. Or when I argue that solar energy is hardly a choice at this time, or even for the next century, my environmental critics simply disagree: spend on solar energy what has been spent on nuclear energy, and solar energy will be cheap. But we have yet to discover a technical breakthrough—the solar equivalent of fission—and unless we do, rejection of fission energy condemns the world to a future of very expensive energy. . . . And when I point out that France has reduced its carbon dioxide emission by a good 20% in the past decade by aggressive deployment of fission reactors, I am greeted by silence.

In 1996, former vice president Al Gore said, "Nuclear power, designed well, regulated properly, cared for meticulously, has a place in the world's energy supply." Ten years or so later, Patrick Moore, a founder of Greenpeace; Hugh Montefiore, a former board chairman of Friends of the Earth; James Lovelock, author of *The Ages of Gaia;* and Stewart Brand, of *Whole Earth Catalogue* fame, began to echo the observations of Gore and Weinberg. Major environmental organizations are also taking a fresh look at nuclear power, and so are leading politicians of both parties. Several countries, including the United States, are planning construction of more nuclear plants—just as Rip had predicted.

But let's go back to the beginning.

2 ALWAYS LOOK AT THE WHOLE

ALBUQUERQUE, FOUNDED BY SPANISH CONQUISTADORS three centuries ago, had around 180,000 citizens during my childhood and now has over 700,000. The open tableland and arroyos I knew have mostly vanished, replaced by suburbs surrounding a few high-rises. The city now stretches from the manor-encrusted foothills of the Sandia Mountains down to the valley where the Rio Grande flows through the bosque, a narrow strip of woods and farms—about the only verdure in the asphalt-grouted, aquifer-draining mosaic of military facilities, laboratory and technology complexes, malls, car lots, housing tracts, and crowded freeways. The formerly luminous air has been replaced on most days by a pall composed of tailpipe exhaust, water vapor, dust, and, in winter, wood smoke.

On the other side of the river, the city resumes its expansion, straining toward a row of volcanoes and cinder cones that rise from the valley's western escarpment. There, pale brown cross-hatchings in what was once an ancient seabed delineate future roads and subdivisions. Each new household and its cars will be adding around sixty thousand pounds a year to the world's burden of greenhouse gases, which we humans collectively increase daily by eighty million tons, and yearly by twenty-eight billion tons.

Albuquerque, like its rapidly growing counterparts throughout the Southwest, seems mostly oblivious of its setting, which has a grandeur that can make a person feel small and mortal. Such a setting lends itself to contemplation of the framework underlying humanity's possibly transitory existence: the mountains of billion-year-old weathered granite; the volcanic mementos of past cataclysms and of the grinding tectonic plates and magma infernos lurking beneath the earth's crust. The turquoise dome above can contain several different weathers simultaneously; the play of high-altitude light seems to have an intelligence of its own (here sunsets unfold like a series of long thoughts). At night, thousands of stars—at least where the glare of artificial light and pollution have not hidden them—can inspire contemplation about where we came from, what part we have to play during our brief sojourn, and where we're going.

One fall day a couple of years after the turn of the century, I emerged from the Albuquerque airport and found Rip Anderson waiting at the wheel of a blue, green, red, and white two-ton pickup that he bought secondhand years ago for his stepsons and rebuilt so frequently with spare parts that the only original components left were the chassis and the hood.

We headed toward his home, soon crossing over the Rio Grande—a mere trickle because of a prolonged drought—and into the South Valley. Poor neighborhoods of tar-paper shacks and run-down brick or frame houses abutted stuccoed cinder-block subdivisions and low-rise apartment complexes, and a few miles away new Southwest-style mansions were springing up on big lots. There was a sewage treatment plant, as well as junkyards, wrecking lots, and a nearby petroleum tank farm that leaked. An Environmental Protection Agency cleanup took care of a metal-working site where decades ago nuclear weapons components and later jet engines had been fabricated. We wound through a patchwork of small farms with fields of alfalfa and chile, and clusters of adobe houses. Some were old haciendas; still others were crowded family compounds with omnipresent double-wides—ranch-style prefabs.

At the end of a cul-de-sac was a chain-link steel gate with a twenty-foot-high metal arch that said, in cutout letters, THE FERNANDERSONS. A racing-mule ranch, an irrigation ditch, a field, and some pitch-roofed adobe houses with small yards bordered the spread. Ancient cottonwoods, bright yellow in fall, stood out against the dark blue sky and the cobalt-hued Sandias. A high fence enclosed an alfalfa field, a big organic vegetable garden, a cluster of beehives, and some corrals containing cows and their calves.

The driveway and an area around a garage were occupied by an ancient wooden hay wagon with iron-bound wheels from the Idaho ranch, a battered van, an antique school bus Rip had turned into a camping vehicle, various tools and farm equipment, and a metal-working shop with sheets of corrugated metal, lengths of pipe, and coils of wire. The place had the feel of a working ranch where everything is salvaged, saved, recycled, or composted. Tall rose and forsythia bushes, shade trees, a small orchard, and a very green lawn fertilized by wild and domestic ducks surrounded a traditional one-story, thick-walled adobe with a pitched, galvanized roof and a new passive-solar addition.

We stepped into a living room with a brick floor and a wood-burning stove, long couches, and wide windowsills with greenery spilling from pots. The aroma of baking bread emanated from the kitchen, a long room with an oak dining table and shelves of trophies—children's academic and music awards. The walls, covered with family photos from the nineteenth century to the present, included portraits of Marcia's maternal grandmother, of Mayan and Mexican ancestry, and Cajuns from her father's side, and of Rip's paternal grandfather, who was born in Sweden, and of Rip's seven grandchildren. Tucked into a corner and hidden by philodendron tendrils were some plaques and awards with Rip's name on them. Missing was a photo that Marcia had once shown me of Rip in full scuba gear diving down through the clear blue coolant water inside a nuclear reactor at Sandia Labs.

Before he retired, in January 2002, he was director of the Nuclear Waste Management Programs Center there. Among his other duties, he was responsible for counterterrorism planning in regard to nuclear and chemical materials. He holds several patents, continues to work on inventions, and has served as the editor of two international journals of environmental remediation and waste management. He's sought after as a consultant on a variety of projects. "Rip sees the obvious problem that everyone else overlooks, and he stays focused on it, and he tries to solve it by considering all the angles, pro and con," a longtime colleague told me.

Marcia has a mezzo soprano voice, a ready grin, lustrous, dark eyes, and a cloud of long, black, curly hair. She's a community leader with a history of activism in education as well as progressive political and environmental issues. Today she was on the phone trying to shut down a sloppily run toxic landfill and to whip up support for an agricultural campus in the South Valley on land that would otherwise become a strip mall.

Rip sat me down at the dining table, a pad of graph paper and a pencil at hand, and began to prepare me for the first leg of our journey tracing the arc of uranium from its emergence from the earth to its ultimate burial underground. Tomorrow we would visit New Mexico's uranium belt, which I took to be the element's birthplace.

"Not exactly," he said. "Uranium was born in exploding stars long before there was an earth."

Most cosmologists assume the universe to have sprung forth around fourteen billion years ago from a void of unknown origins with the manifestation of a single point, in a primal fire hotter and denser than anything since. As that fire slowed down and cooled, some of the energy

hardened into particles, and a fundamental force, one of attraction between negatives and positives, bonded the particles into increasingly
more elaborate structures: the first atoms. Within them was caged a vast
quantity of energy. An atom, the smallest unit of a chemical element,
consists of a single positively charged nucleus encircled by much smaller,
negatively charged particles—electrons. The first atom, hydrogen, had a
nucleus composed of one proton (a positively charged particle) orbited
by a single electron. Attraction and residual heat drew particles together
to form the first elements. Their binding energy was enormous. Eventually, as snowflakes condense out of a cloud of water vapor, the first stars
began to shine.

"A star cooks hydrogen and helium by shoving their nuclei together so
that they get hot enough to transform matter back into energy," Rip said.
The waste products of that fusion process become new elements. "Now,
here's the reason nuclear energy is different from the kind we're more
used to. Burning coal or wood or gas only affects electrons in the outermost shell of the atom—that's a chemical reaction. 'Burning' uranium—
speeding up its decay—affects the nucleus of the atom. When wood
burns, a chemical reaction releases energy in the outermost shell of electrons, and the waste is ash. As the star burns hydrogen, a reaction going
on in its atomic nuclei creates as its ash a second element, helium. Helium
has two neutrons, two protons, and two electrons." (Neutrons, which
lack an electrical charge, help make up the nucleus of every atom.)

When the hydrogen is all used up, the star, now more compressed,
fuels itself with helium, and a new element, carbon, is given off as waste.
Once the star burns all its helium, it collapses further. After billions of
years, resources bankrupted, it shrinks into a cold, dark cinder. That's the
fate of most stars, but for some, another process occurs, which is, from a
strictly material point of view, why we exist.

Though threatened by gravitational collapse because of its own
increasing density, a star of a certain size can maintain its existence by
continuing to unleash energy from the nuclei of atoms, creating one element after another—the oxygen our lungs are taking in, the calcium
in our bones, the iron in our blood. We are in effect made of celestial
nuclear waste. But this star's attempts to stave off gravitational collapse
are also ultimately defeated. Fuel exhausted, the energy released by the
nuclear reaction can no longer prevent the powerful attraction that
causes the star's substance to rush toward a common center. Its heat and
matter concentrate into a single, profoundly dense object that explodes
in a supernova, a display of light matching that of a hundred billion suns
and hotter than anything since the birth of the universe. In such fur-

naces, all the rest of the elements on the periodic table were forged when simpler elements were transmuted into the most complex ones, like uranium.

Rip led me to a breezeway off the kitchen, to a wall chart of the periodic table of the elements. If he has a sacred text, a representation of the mystery of the universe, this is it.

"When a star dies and explodes, it blasts matter in big arcs throughout space, seeding galactic clouds with hydrogen and all the elements, including the 'ums,' " Rip said, pointing to the far right-hand side of the chart, "uranium, thorium, plutonium, and lots of other unstable elements. Their atoms give off particles to become more stable. That's called radioactive decay. The 'ums' with heavier and more unstable nuclei than uranium are higher on the chart than uranium and are called transuranics. These elements, like plutonium, californium, einsteinium, and americium, decay relatively quickly and so they don't exist naturally on earth now. They're made in reactors. But they were here way back when we solidified out of molten whatever.

"The scattered elements start clumping together to give birth to new stars. Then some of those stars exploded. And the process started all over again. We've been processed at least once—we know that. And probably the stars have been recycled a number of times. Otherwise we wouldn't be here. Our atoms had to be made from earlier stars that became supernovas. Our solar system is composed of all the sweeping and gathering up of debris from previous suns. Luckily, in our case, the accumulation of elements occurred in a bath of hydrogen, and we wound up with a sun and planets, made from dust and hot gas that cooled down.

"Our sun, a nuclear reactor, is cooking hydrogen right now. We receive the sun's energy in waves, from infrared radiation through X-rays. Most of the energy that's used in chemical reactions on earth comes from the sun. Sunlight—ultraviolet and visible radiation—bombards the earth daily, along with microwaves left over from the big bang. As the earth turns, the part facing the sun gets bathed with warming radiation, and the white ice caps and glaciers reflect a lot of that thermal energy back into space—unless certain gases and water vapor in the upper atmosphere are there in a big enough quantity to trap the heat."

He went on to say that most elements are made up of stable atoms, but radionuclides, the unstable atoms in radioactive materials, give off invisible particles and rays that are detectable with instruments. He drew a uranium atom. It had a nucleus that looked like a sphere made of basketballs encircled by little Ping-Pong balls for electrons, and he sketched in bristling arrows to indicate particles being emitted from the nucleus.

"Unstable atoms can occur naturally—or we can coax uranium into making them."

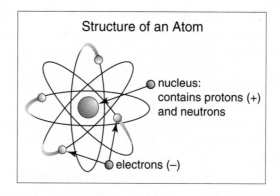

"And the man-made ones give off harmful radiation," I said.

"To the body, there are only particles and rays—whether they come from cosmic radiation or from a watch that has a glowing dial or from uranium that's been in a reactor."

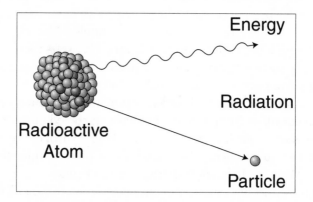

We get exposed every day to damaging radiation from nature—in fact our greatest dose by far is usually natural—but our bodies know how to deal with it. "How many times have you gotten a sunburn and recovered from it?" he asked. "That's because the body is equipped to heal itself after exposure to low-dose radiation. Our cells don't discriminate between natural and man-made radiation. You doubled your exposure just by coming from New York to New Mexico. You got an increased dose in the plane because you were above most of the atmosphere—it acts as a shield against cosmic radiation. Then here at this altitude we have less atmosphere than you do at sea level. And we have a lot more uranium in the rocks and soil than you do in New York, and its decay products give us a dose. But all of this is low-dose radiation, and we're exposed to it all the time. The only time you or just about anyone else will be exposed to high-dose radiation is if you have radiation therapy or medical diagnostic radiation. But we'll get into all that another time. Right now we'll just go over the basics."

Rip was born in Idaho's Upper Snake River Valley. His father, a rancher, imparted to his children a firm sense of the importance of leading an ethical life as a part of a greater whole, and he championed doing work that conformed to one's conscience and helped the family and the community. Rip remembered times when he was young and would adopt an absolute stance about something. "Dad would politely point out that there are multiple points of view on any subject."

Marcia hadn't expected Rip, being a scientist, to believe in God. "Now I see that he has strongly held spiritual beliefs that seem to be his moral compass," she said. "They guide quite a lot of what he does in the scientific world. He doesn't seem to have any problem reconciling the two."

Rip remarked that religions have stayed exactly the same for thousands of years. "But the science of today is built on the research carried out by people like Madame Curie, Rutherford, Einstein, Bohr, and lots of other thinkers in all disciplines. Their results have to stand the test of empirical proof and peer review. So science keeps evolving. Why doesn't the equivalent of that exist for religion? If you look at science, it's not doing so well either at putting together the big picture. But the discovery aspect is just wonderful."

Rip always liked learning how things worked, and as a small boy he'd been interested in nature and its processes and how they were interwoven. He observed the seasonal changes in a pond on the ranch, studied the best ways to grow vegetables, and, after watching a mechanic who came to fix equipment, took over the job. "On a ranch you have to pay attention to the real world and you have to be pragmatic—it's one-trial learning. You better get it right the first time. I guess that helped prepare me for the scientific method. But it hadn't dawned on me until a particular moment in the college chem lab when I was a freshman at Idaho State College that just about anything that happens anywhere in the entire universe is a chemical reaction, that is, the combining of the outermost shell of electrons of one atom with the outermost shell of electrons of another atom. All of a sudden, this world that I thought was a certain way, that I was used to and looked out at, turned out to be totally foreign and unknown."

Rip went around making endless discoveries in this brilliant new light. "Until then I'd been like someone who owns a car, and all he knows is that you put gasoline in a hole in the side and you drive it. Then one day you lift up the hood, and you find the most amazing thing there that makes the car go in a very elegant way. I realized that the controlling factor, the key to the door of nature, was the periodic table of the elements. I spent one weekend memorizing it. Chemistry was fun and felt good. It helped me understand how air and water and rocks worked, how metals came to be the way they are. I had to learn more about how things come together and come apart and why they do what they do. I never saw matter and its processes in the same way again."

He then took organic chemistry. "I fell in love with carbon, because it was the basis of all life. Carbon combines with *everything*. Organic chemistry explained to me still other things: how the body works, how the gut digests food, how wood in the fireplace burns. Analytical chemistry taught me to be accurate and to understand how, for instance, you can't ever have a one hundred percent yield. If you transform one chemical using another, there's always going to be some residue. You can't get all of the cake batter out of the bowl you mixed it in. You can't do a hundred percent of anything. This is a common misunderstanding nonscientists have about science: they expect it to take care of a problem a hundred percent." Physics and math added another dimension for him, explaining processes that were happening all around but were not chemical reactions.

Rip's passion led him to acquire a doctorate in theoretical organic chemistry and chemical oceanography at Oregon State University. While working as a salvage diver, lab assistant, and chemistry instructor, he created an organic hydrocarbon molecule and experimented on it; he also researched the path of radioactive particles in the Columbia River to learn where its current went when it entered the Pacific.

"The discipline that can teach you the most about the world is chemistry, followed by physics. Chemistry tells you what things are made of. Physics tells you what they do. Mathematics describes their underlying order. That combination explains just about everything, like how the water gets up to the leaves at the top of a tree. You could say chemistry occurs out in the countryside, in that outermost electron shell. But I kept moving and went into the suburbs: nuclear chemistry. Then I fell in love with that, because I started to understand how the world worked at the atomic level—that is, with the electrons in the innermost shells of the atom. Knocking those electrons around forms X-rays—the kind your dentist uses. Then I went to the city, the nucleus. Once you're inside that, you start determining what the different buildings are—neutron apartment buildings, proton office buildings. And that takes you into physics and quantum mechanics. But the key to it all is the periodic table."

As we stood before the chart in the breezeway, Rip told me that even if all present knowledge of chemistry were to vanish along with the rest of civilization, sooner or later, as people began to wonder what made the stars shine and how elements were made and how chemistry worked, the periodic table would be reconstructed. It was first elaborated in the nineteenth century by a German chemist and a Russian chemist independently of each other. "The knowledge of nuclear energy would inevitably be reconstructed too," he said, "as soon as the periodic table was laid out."

Rip began to trace the harmonious choreography of the periodic table. Though static in representation, the periodic table shows that some elements are being continuously and predictably transformed in a manner that obeys definite laws. "That transformation takes place in the 'city'— the atomic nucleus. Like other elements, the element of uranium has a family of isotopes."

All of an element's isotopes have the same number of protons in their nuclei (think of the element as the family name) but each isotope has a different number of neutrons in its nucleus (think of that number as a given name, like uranium-235 or uranium-238). Isotopes with an odd number of neutrons are more unstable, and they keep ejecting particles and rays (electromagnetic pulses) in order to become balanced. The more radioactive the isotopes of an element such as uranium are, the more likely they are to wind up—as the nucleus sloughs off parts of itself over time—as some other element.

"Mother Nature is always trying to stabilize things," Rip explained. "Unstable uranium isotopes, like U-235, try to balance the number of neutrons and protons in the nucleus." He compared this situation to the dances he used to go to at the Grange as a boy. If more males showed up than females, there would be a limited number of stable couples, and the extra males would go outside and drink and fight. The nucleus with an odd number of neutrons sheds rays and particles, decaying into a whole series of other elements until it reaches lead, which is stable.

When time began, the big bang's energy burst forth and poured itself into the form of elements of increasing complexity—the elements we would be made of—and flowed through stars, which are still freezing energy into matter and thawing matter into energy. Life itself is energy, brought into being by increasingly complex manifestations of energy, and life turns energy into matter and matter into energy, and conducts it, and stores it. One of the best containers of energy on our particular planet happens to be matter in the form of uranium.

"We talk about particles and such," Rip remarked as we stepped outside for some air. The stars were beginning to come out. "But really, there's nothing in the universe but energy."

3 AMBROSIA LAKE

BY 7:00 A.M., MARCIA WAS AT THE WHEEL, and we were driving out of Albuquerque, passing pylons bearing skeins of high-tension wires.

New Mexico gets about 63 percent of its electricity from coal-fired and natural-gas plants, 16 percent from uranium in a nuclear plant in Arizona, 8 percent from wind, and the rest as purchases from various sources. The state extracts plenty of fossil fuels and emits twice the amount of greenhouse gases per capita as the national average. In Connecticut, New Jersey, Illinois, and South Carolina, about half the electricity comes from uranium, and in Vermont it's three-quarters. Nuclear plants make about a third of the electricity lighting up New York State.

Electricity generation is miraculous. Those of us lucky enough to enjoy it mostly ignore it. To paraphrase the Gnostic Gospel of Thomas, the kingdom of electricity is spread out before us but we do not see it. This silent energy coursing through our walls, along our streets, and across our landscape, instantly and constantly ready to serve, is a boon created when the universe began. Since then, no new energy has been made; it's just gotten rearranged among various containers. "Uranium, thorium, wood, water behind a dam, and hydrocarbons like oil, gas, and coal are all ways of storing energy for long periods," Rip said. "Most people don't think of these as energy storage mechanisms like a battery, but they should."

Many countries, having recognized their lack of indigenous energy resources as well as the problems of dependency on imported fuels, started making long-range plans decades ago that included nuclear power. A number of European nations presently get from half to three quarters of their power from nuclear plants. Worldwide, about 440 reactors in thirty countries supply about 16 percent of the electricity; over a billion people rely on them for some or all of their power. Several countries are building new reactors to meet increased demand for electricity and to phase out fossil-fuel combustion.

A thousand additional reactors worldwide in place of fossil-fuel plants would reduce civilization's greenhouse gases by 25 percent, according to

The Future of Nuclear Power, an analysis written by a panel of professors from the Massachusetts Institute of Technology and Harvard University. Although no new reactors have been ordered in the United States since 1979, aging equipment is routinely replaced with more efficient components so that the oldest plants retain few of their original parts; the effect is one of rejuvenation. Over the years reactors have increased their output significantly and on average run at over 90 percent of their capacity. The Nuclear Energy Institute, the chief industry organization, wants to expand nuclear's 20 percent share of electricity generation to 50 percent by 2050. This would require replacing many of today's reactors, adding new plants to replace large numbers of fossil-fuel plants, and increasing uranium production.

We climbed out of the Rio Grande Valley and began to cross the Zuñi Uplift, with its desiccated sagebrush plains bordered by striated hills and mesas and indigo ranges and punctuated with eroded red formations— the petrified bodies of ancient monsters, according to the Diné (as the Navajo people call themselves). As we passed through sovereign lands belonging to Native Americans, we began to see heaps of black, broken lava, and the occasional lone volcanic plug or cinder cone, the lesser kin of Mount Taylor. This immense volcano rises over eleven thousand feet above sea level at the southeastern edge of the Colorado Plateau, which is neither a plateau nor confined to Colorado but rather a vast ellipsoidal feature, mostly flat or concave, a mile high on average, its basin populated by many plateaus and encircled by highlands. The Colorado Plateau covers about 130,000 square miles of New Mexico, Colorado, Arizona, and Utah and contains some of the most stupendous scenery in the world: the Grand Canyon and Glen Canyon; petrified forests; sandstone and limestone sediments intricately carved into spires and towers and natural bridges and labyrinthine chasms painted in vermillion, ochre, lavender, verdigris, and black. At times, seas have covered the Colorado Plateau. It has been thrust up by the jostle of tectonic plates and pushed down by the weight of sediments, only to emerge again during epochs of mountain building, which, on its borders, heaved up and stretched the surrounding terrain. Meanwhile, the formation, extremely old, has survived intact as a distinct piece of the continent's crust, an island serenely adrift, it is thought, on a sea of molten rock.

Around the edges of the Colorado Plateau, the contents of that fiery sea spurted upward and eruptions of boiling water and steam showered the surface with heavy minerals from deep within the planet's interior.

These washed into valleys and were buried under dirt, ash, and rock until new cataclysms exposed them or humans dug them up.

"When the earth condensed out of gas and dust, the heavier elements sank toward the center," Rip said. "They congregated and stayed hot, because of pressure and thermal energy released by fission—the spontaneous splitting of atoms—and by radioactive decay. This energy keeps the iron core molten. Electrons cause that iron to produce magnetism and the electromagnetic field—it's essential for all of life."

The heat drove convection currents in the liquid iron, with some of them looping upward toward the crust, bringing small amounts of heavy elements that to this day continue to be dispersed during eruptions. "Most of the radioactive heavy elements are still down toward the center in the earth. We got a lot of them—uranium, protactinium, californium, et cetera. Volcanism on ancient Mars and currently on some of Jupiter's moons means they have or once had a radioactive core like we do. To get one, you need uranium, thorium—those kinds of heavy, unstable elements. Radioactivity gives Earth its dynamism. The heat flow out of our core makes everything here on the surface possible. Otherwise we sure as heck wouldn't be here. Earth would be a frozen rock. We lucked out when collecting elements from an earlier sun."

Today the air was so clear that it gave the illusion that distant mountains were a short hike away—except to the west, toward coal and oil country, where brown haze persisted. The closer we drew to Mount Taylor, its broken peak snowy, the dirtier the horizon became and the more tortured the terrain, called by the Spaniards *El Malpais*, "The Badland": home to dozens of volcanic craters as well as numerous lava beds, flows, and tubes. Mount Taylor last erupted two and a half million years ago; it was pushed up ten million years ago when a huge block of the earth's crust dropped down, causing the Rio Grande Rift. It's among the newest rift valleys in the world, and runs from southern Colorado to northern Mexico. To compensate for the rift, on either side of it mountain ranges like the Sandias were thrust upward, along with big volcanoes.

As a mapmaker for the U.S. Forest Service, my father had surveyed parts of the Colorado Plateau. He hiked through the boot-sole-shredding lava beds with his equipment, and he brought home bits of pottery, arrowheads, lumps of copper and silver ore, fool's gold, pebbles of turquoise and quartz, and a small, heavy, gray-brown rock encrusted with yellow crystals. That one was especially good for show-and-tell: it would make a Geiger counter click more rapidly than any other rocks did.

It wasn't unusual for geologists to show up in a classroom or at a Girl Scout meeting to teach us how to use a Geiger counter. Uranium was a popular topic, and radioactivity seemed an interesting phenomenon, like gravity, or God: invisible, powerful, omnipresent. It was wonderful, except in bombs, when it was terrible. From exhibits and programs staged by Sandia National Laboratories as part of the Atoms for Peace initiative launched by President Eisenhower, and from educational films such as *Our Friend the Atom*, I learned how uranium made my hometown special. As a grade-schooler, as a Scout, as a member in my adolescence of a rocket club and of Future Scientists of America, I was allowed on special occasions to visit the other side of the chain-link fence. Our group would be shepherded past guard booths manned by military police and then through a building containing display cases of minerals, some streaked with yellow, some glassy, like the lumps of trinitite—fused sand from the ranch down south where the first atomic bomb had been tested in 1945. We used manipulators to play with materials, isolated in transparent boxes, that a Geiger counter verified in a metallic clicking language as being "hot."

Those crackling packets of energy promised a future of limitless abundance: they would prevent the tyrants behind the Iron Curtain from succeeding in their plans for world domination, they would power our cities, and they would eventually transport us beyond the earth. When the rocket club made annual pilgrimages to the White Sands Missile Range in southern New Mexico to see test firings of the vehicles that might one day deliver satellites into earth orbit, or men to the moon, or bombs to the cities of our enemies, we heard about a scheme for a nuclear-powered rocket for space travel.

The whole state, meanwhile, was in a fever about the element that would provide all this energy. A few years after the end of World War II, a uranium rush had begun on the Colorado Plateau. Grants, a town near Mount Taylor and an hour west of Albuquerque that used to proclaim itself Carrot Capital of the World, now dubbed itself the Uranium Capital of the World—a title also adopted by other uranium boomtowns. They all began to hold Miss Uranium and Miss Atomic Age contests. Winners, garbed in evening gowns, would receive truckloads of the surprisingly plentiful ore as prizes. People used to say that if the price of uranium got high enough you could dig up your backyard in Albuquerque and sell it to the mill. At the state fair, promotional jars of yellowcake—refined ore—were handed out. In Grants, the Uranium Café still serves its Yellowcake Breakfast and Uranium Burger Lunch.

Marcia, Rip, and I dropped by the New Mexico Mining Museum,

which we had mostly to ourselves. Wandering past exhibits of local history, we looked at ancient Native American artifacts and peered into a glass display case at an apple-sized chunk of uranium ore. It contained billions of years of history that had begun within exploding stars and was encrusted with glittering bright-yellow crystals.

Pure uranium is a silvery-white heavy metal with a bluish hue, prone to rapid tarnishing when exposed to air and, if turned to fine dust, spontaneously bursting into flame; it's malleable and ductile, a little softer than steel, nearly twice as dense as lead, and nearly twenty times heavier than water. Three tablespoons of uranium would weigh two pounds; a pint-sized kitchen measuring cup of it would weigh almost twenty pounds. Uranium doesn't exist in nature as a metal but rather in mineral compounds, and the mineral compounds are everywhere. From our daily food and water we take in about 5 micrograms of natural uranium—the equivalent of a few grains of salt—and we excrete almost all of it within days. Traces may be laid down in bones and kidneys. In some ore-rich areas, uranium in water and soil can damage the kidneys: its chemical toxicity is about the same as lead's. (Mercury is ten times more chemically toxic.) Living near any kind of mine or a coal-fired plant, or working in a phosphate fertilizer factory, can expose you to uranium. However, according to the Centers for Disease Control even high-level exposure to natural uranium or to depleted uranium (DU) has never caused a case of cancer in humans or animals, and uranium isn't considered radiologically dangerous by any government agency. Nevertheless, misinformation abounds about radiation effects from DU, which is incorporated into tank armor and used for armor-piercing bullets as well as shielding for medical radiation equipment and to supply weight in sailboat keels and golf clubs. Most of DU's fissile isotopes have been removed and so its radioactivity is one-third that of natural uranium's. Someone smoking three packs of cigarettes a day gets the same radiation dose to the lung as an individual inhaling about a *pound* of uranium would receive in a year.

Going back to antiquity, the sole applications of uranium were aesthetic—to impart lustrous hues to glassware and pottery glazes. In 1789, Martin Heinrich Klaproth, a German scientist, discovered that pitchblende, a sticky, black ore from the mountains of central Europe, contained an unknown element. It was isolated in 1841, by Eugène-Melchior Péligot, a French chemist who named it after the newly discovered planet Uranus.

In 1896, in France, Antoine-Henri Becquerel saw that uranium minerals had the same effect on a photographic plate as exposure to the sun, and he realized that the phenomenon was caused by radiation: the spontaneous emission of particles and energy. Scientists then found that the uranium atom, whose nucleus has the greatest mass of all the natural elements, not only gave off radiation, it had another exceptional attribute: it fundamentally changed after relieving itself of surplus energy by expelling some of its matter, and it would keep expelling bits of itself in a series of disintegrations until it became stable.

I had assumed that a long half-life was proof of uranium's enduringly lethal nature—that something that could remain radioactive for ages must be extremely hot to begin with. As we looked at that chunk of ore in the glass case, Rip corrected my impression.

"A long half-life means that it takes a long time for half the number of atoms in a particular sample to decay," he said. "To put it another way, the half-life tells you how rapidly the nucleus is breaking apart unassisted, due to insufficient binding energy to keep the protons and neutrons together. The longer the half-life, the less radioactive the sample, because the rate of disintegrations is slower—the fewer the number of rays and particles emitted per minute or hour. So the lower the level of radiation, the less risk to humans. An individual atom of uranium can decay at almost any time, from a microsecond to billions of years from now. The more rapid the decay—that is, the shorter the half-life—the greater the number of radiations as more rays and particles are emitted. A person exposed to a radionuclide with a short half-life would get more of that energy deposited in cells."

But for that to happen, the person must be close to the radioactive source. The distance the rays and particles can travel varies from nanometers to hundreds of meters, depending on which kind they are—alpha, beta, or gamma—and what medium they're penetrating. Concrete or rock or other dense material several feet thick can stop them. Uranium can decay over long periods into other radionuclides—"daughters"—that can cause cancer if you are exposed to enough of them for a long enough period of time.

"Uranium goes *kling-kling-kling* down the decay series like a pinball machine," Rip continued. The isotopes of the uranium family—the isotopes of any element, actually—are chemically identical, but their possessing different numbers of neutrons in their nuclei makes them dissimilar in their ability to undergo fission. In the fission process, a heavy, unstable atomic nucleus, such as uranium-235's or plutonium-239's, splits into two daughter nuclei, releasing neutrons and a large amount of energy. About 99 percent of natural uranium is made up of the more sta-

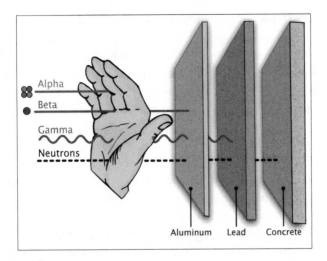

Alpha
Beta
Gamma
Neutrons

Aluminum Lead Concrete

ble and therefore more slowly decaying isotope uranium-238. An atom of U-238 has 92 protons and 146 neutrons in its nucleus—add the two numbers and you get 238. In natural uranium, the proportion of U-235, which has only 143 neutrons and which is the sine qua non of the nuclear chain reaction, comes to 0.7 percent. In large amounts, about half the uranium-238 in a given sample will decay in 4.5 billion years (about the age of our planet) into daughter products, among them radium-226, which in turn takes 1,602 years to decay into radon-222, a gas that takes 3.8 days to decay into polonium-210. After 138 days, polonium-210—a radioactively hot particle present in tobacco smoke—has given up enough energy to become stable, as lead-206. So half of the uranium atoms in the chunk of ore my father brought home will, in 4.5 billion years, be gone, turned into other elements such as lead.

Marie Curie was a heroine of mine—the only woman scientist I ever heard about as a child. She and her husband, Pierre, painstakingly separated a tiny bit of radium from a lot of pitchblende in 1898 and began discovering this new element's surprising effects: it could glow in the dark, burn skin and flesh, and it had the extraordinary power to stop the growth of malignancies. As the Curies published their findings, radium became a sensational and costly commodity. Prospectors fanned out across the continents looking for the element in various ores. The Belgian Congo turned out to be a good source of high-grade pitchblende, and the Colorado Plateau was found to be abundant in radium-rich carnotite, a compound also containing uranium. Carnotite mining proceeded in Colorado and Utah, where refineries extracted the precious radium and discarded the rest.

In the early 1930s, an enterprising fellow noticed that the uranium in the mine tailings was high-grade and sold it to pottery and glass factories back east, where it was used to make many different colors, including the flaming orange glaze for Fiestaware, a popular crockery manufactured in the 1930s and 1940s. I ate oatmeal daily out of those bowls, ignorant of the fact that their glaze offered up stored energy from an ancient supernova.

On September 12, 1933, a stream of associations—about a science fiction novel by H. G. Wells on the far-fetched subject of "atomic" war, about an uncharacteristically pompous remark made by Sir Ernest Rutherford, the father of nuclear physics, dismissing any notion of the practical application of the energy in the atomic nucleus as "moonshine"—flowed playfully through the brain of a man named Leo Szilard as he walked through central London. As he waited at a traffic light, the idea came to him that the right sort of atom could be split, and from a small quantity of matter a great deal of energy could be released and set to work. The thought triggered a series of actions, reactions, and consequences that were to alter the world as significantly as the development of any other invention in the entire course of human history. Szilard, whose insight was derived from his familiarity with the experiments of Rutherford, Marie Curie, and others, came to envision the nuclear chain reaction: a neutron emitted at very high speed from a nucleus by fission would smash into another nucleus, releasing more neutrons that would repeat the process.

But which element possessed the nucleus with the necessary properties? It would have to be energy dense and unstable. In response to Szilard's published hypothesis, the Curies' daughter, Irène, and her husband, Jean-Frédéric Joliot, determined that uranium was ideal for obtaining a chain reaction. Physicists and chemists exchanging information about these findings began to understand that uranium's energy, properly released, could be used for the good of humankind; such enormous energy in so tiny a package promised cheap and abundant power for all. However, World War II intruded on the scientists' utopian dream. The Germans and the Japanese, each possessing some of the requisite technology along with access to uranium, could well make an atomic weapon instead.

This concern led physicists in the United States—many of them refugees from Hitler—to apply their knowledge of the chain reaction to building an atomic bomb. The government launched a secret initiative, the Manhattan Engineer District—called the Manhattan Project for short—and sent its agents to the Colorado Plateau to buy up that mostly worthless by-product of radium production, uranium. Some mining towns became guarded enclaves, though their workers didn't know how this metal fit into the war effort. The Manhattan Project also acquired uranium from Africa and Canada. Richard Rhodes describes in *The Making of the Atomic Bomb*, the epic and definitive history of that world-changing enterprise, how its director, General Leslie Groves, believed that high-grade uranium ore (that is, relatively rich in fissile uranium-235) was scarce and attempted to buy up the world supply. At lower grades, the element has turned out to be as common in the earth's crust as tin and more abundant than gold and silver; it occurs in various minerals, rocks, and sands (and in building materials made from them), as well as in seawater. No good refining technology for low-grade ore existed during the Manhattan Project, but by the 1950s one was in place.

A card in the museum display case next to the chunk of uranium ore identified it as being the same size as the famous one that a Diné rancher named Paddy Martínez had found. One day in 1950, while nuclear physicists were in their labs pondering the rarity of uranium, Martínez's wife sent him to the village of Bluewater to fetch baking powder. As with all founding legends, versions about what happened next vary, but here's a representative one: a rainstorm hit and he took refuge under a dusty rock ledge. When he resumed his trip, his pants were covered with yellow powder. When he appeared in a bar frequented by prospectors, they quizzed him about the yellow dirt. Very few people knew that the atomic bomb's secret ingredient was uranium. But word had spread in the Southwest that the federal government would pay good money for the ore, and prospectors acquired Geiger counters. In one version of the legend, his pants set them off and the prospectors began to question him. Where had he been? Had he seen any yellow rocks in the area? Was the vegetation of a purplish hue? (Uranium near the surface can affect plant pigmentation, turning it purple.) Martínez went back the way he had come, picked up some yellow rocks on Haystack Mountain, and presented them to the mayor of Grants. Soon the first big uranium mine in the region opened, and Martínez was given a lifetime job scouting minerals. He prospered to the tune of $400 a month while his discovery earned millions for others.

The biggest uranium rush in history had begun. Near cliff faces, helicopters hovered as geologists in harness dangled down with their Geiger counters. Mine shafts were sunk and ore mills built. The Atomic Energy Commission (AEC), America's sole purchaser of all uranium, wanted a lot of it in a hurry. The public soon learned that it was needed to make weapons for the arms race with the Soviet Union. That boom began to fade when the AEC, having accumulated sufficient reserves, cut back its purchases in the early 1960s and by 1970 stopped them altogether. An economic depression in the region followed. About five years later—as an energy crisis arose, new nuclear plants came online and the government permitted commercial sale of uranium—its price rose and a short-lived second boom ensued.

Workers from coalfields and hard-rock mines around the country streamed to the Colorado Plateau during both booms to make their fortunes in big corporate mines or in one-man operations. As independent contractors, miners were paid by the number of pounds of uranium they delivered, earning as much as $100,000 a year. They used their own equipment, bought their own supporting posts and explosives, and mined by themselves, except when they hired a trusted relative as a helper. Lone prospectors, some of them Native Americans on the reservations, would burrow on their own into an outcropping and bring out ore to sell to middlemen.

In the museum, we met a shy, weathered old-timer in Western dress who had worked in all the big mines in the area. I asked him to recommend one for us to visit.

"You cain't, ma'am. They're all closed. Since Three Mile Island in 1979 nuclear power hasn't been the same."

Nuclear plants already under way were completed, but plans for additional ones were scrapped. The price of yellowcake, which had been in oversupply, plunged. Mines and refineries began shutting down. By the turn of the century, no conventional mining of uranium was going on in the Southwest or anywhere else in the United States. In fact, almost all types of hard-rock mining had been outsourced.

The old-timer went on to say that the only uranium extraction going on was some in-situ leaching, also called solution mining. "You pump natural groundwater in a uranium-rich area to the surface and filter out the uranium," he said. The water is then injected back into the ground. Since the removal rate is slightly greater than the injection rate, groundwater flows inward from the site's perimeter, providing a natural containment of the mining operation. As a precaution, an outer ring of wells monitors any leakage of water. When the leaching is finished, the company by law must completely restore the site to its original condition. Little water loss occurs, because it's recycled.

"This way, you don't need miners and there are no tailings," the man said. "It's a time-tested method. Some ore bodies can only be reached by conventional mining. But here, if uranium operations in New Mexico ever start up again, you can bet they'll mostly use in-situ leaching. You won't see mines like the ones we had during the boom ever again. Even all the head frames are tore down but one at the Chevron mine at Mount Taylor." A head frame is a wooden structure, about three stories high, on the surface above the shaft; it has pulleys to operate the lifts in the mine. "But right below we have the only underground uranium mine museum in the world."

Down we went. The basement simulation was stocked with items from real mines: silvery timbers, canteens, 380-pound pneumatic drills. Although Rip had visited a working uranium mine near Grants in the 1960s, this was the closest I would ever be able to get to the experience, at least in the United States. We followed a tunnel with concrete walls made to look like rock. Once Geiger counters had been used to locate ore and assess its grade, a pod of uranium would be assigned to a miner. "He'd drill out holes and place sticks of dynamite in them," Rip said. "After the blasting, men would drag the big chunks of ore out in buckets that were hauled up to the surface on ropes."

In a mock-up of a lunchroom stood a battered wooden table, and on the wall hung personal dosimeters, still containing the film that registered the wearer's total radiation dose from natural and man-made sources.

That dose came chiefly from radon. A decay product of uranium and thorium, this chemically inert, dense, invisible, odorless gas has a half-life of about four days, after which it decays to radioactive particles—radon daughters that can cause lung cancer. Any underground mine or tailings pile may harbor radon, but in uranium and radium operations the exposure can be higher. After the discovery of the association between radon and lung cancer and the establishment of safety regulations in 1972, every miner had to wear a dosimeter, and operators installed huge fans that practically turned the drifts into wind tunnels. With the reduction of radon levels, the incidence of cancer among those working in company mines—some of which were on reservations and were required by the tribes to hire a high percentage of Native Americans—dropped to that of the normal population.

However, a uranium mine could consist of simply a pick, a shovel, a wheelbarrow, and a path down a canyon to an exposed vein. No one monitored health conditions in these small mines and so radon exposure remained a risk. Such one-man enterprises, called "dog holes," could be lucrative; they flourished throughout the Colorado Plateau, including on the reservations; the total number is unknown. By tunneling a hundred yards into an ore vein in the side of a cliff, a miner might receive in one month as much exposure as anyone in a big, ventilated mine would get in several years.

The Navajo Nation, which covers twenty-seven thousand square miles in the Four Corners area, where New Mexico, Arizona, Colorado, and Utah come together on the Colorado Plateau, has an epidemiological office that has made detailed maps of over a thousand small uranium-mining sites and has been restoring each one to its previous condition—sealing the tunnels, grading the area, and planting native vegetation. Some Diné once used mine tailings to build houses. These have mostly been identified, destroyed, and replaced.

Contrary to the assumption of antinuclear groups, most of the thousands of miners came from out of state, and only a small percentage of the total number of employees in the uranium industry has been or is Native American. The maximum possible percentage who were uranium miners would have come to about 3.6 percent, according to the University of New Mexico Bureau of Business and Economic Research. Because of the informal nature of small mines in remote areas, getting an accurate count of the individuals who worked there is difficult. The 1980 census in New Mexico showed that of the twelve thousand or so people employed in metal mining, 14.7 percent were Native Americans. I spoke with the epidemiologist John D. Boice, Jr., of the International Commission on Radiological Protection, which sets standards worldwide. He formerly worked for the National Institutes of Health and is now at the International Epidemiological Institute in Rockville, Maryland. His specialty is the impact of uranium mining on health. "A study of over four thousand uranium miners from 1950 to 1964 that has gone on for over forty years includes about twenty percent Native Americans, primarily Navajo," he said. "In another study, 5.3 percent of 3,469 underground miners in New Mexico were Native American." He said that in the overall population of uranium miners on the Colorado Plateau, Native Americans were a minority.

Studies of a small cohort of Native American miners who got lung cancer suggest that this misfortune may well be the result of their time in underground rather than surface mines. Lee Marmon of Laguna Pueblo told me he worked from 1951 to 1981 in a big open-pit mine on the reservation, where 90 percent of the miners were Lagunas. "Lagunas are superstitious about going down into the earth," he said. "The under-

ground work was done by other groups. Those people got the problems—we didn't." In New Mexico, overall cancer incidence rates are
lower for Native Americans than for Anglos. A comparison study showed
that Native Americans living in Alaska have a cancer rate 2.5 times
greater than that of New Mexican Native Americans.

Stories abound about exposure to contaminated water and radioactivity from abandoned mine sites throughout Navajo lands, and political
wrangles between tribal officials and the federal government have complicated cleanup and compensation. On the one hand, mining companies, in the absence of strict regulation, failed over the years to control
waste adequately; on the other hand, people on the reservation, unaware
of the risks, did not know to protect themselves and so went on living in
places that may have caused health problems. In some old mining areas,
the water is pure, but in others it is tainted not only by uranium but also
by arsenic and lead, all toxic heavy metals common to the soil of the
region. Mineral extraction of all kinds—lead, gold, silver, copper, coal—
has led to health and environmental damage at former mining sites
around the country.

The Navajo Nation, which sits on about one hundred million pounds
of known reserves of uranium ore, voted in 2005 to ban uranium extraction and refinement. Terry Fletcher, a former chairman of the New Mexico Mining Commission, expressed regret, saying that the decision was
based on incorrect information. "A vast resource for the Native Americans has been lost, at least temporarily," he told me.

Those who smoked in the mines got lung cancer at a much higher rate
than nonsmokers. According to Leo Gómez, a radiation biologist who
conducted radon research in uranium mines in the early 1970s, NO
SMOKING signs were posted everywhere. "But I saw cigarette butts all
over the place," he said. For smokers, injury to the lungs was so severe
that cancer in those cases became very aggressive.

The mining companies had their employees get medical exams before
they started work and during their employment, because those who had
done mining elsewhere, particularly of coal and toxic heavy metals,
tended already to have medical problems. Dr. A. A. Valdivia, working at a
clinic in Grants, examined, tested, and took health surveys of tens of
thousands of miners. Mine inspectors and unions also kept records,
checking dosimeter film and noting how much radiation an individual
was exposed to in a year or cumulatively. Though people were not being

monitored as they would be in a scientific experiment, and record keep-ing was uneven and underwent considerable transition over the years, the data has created a picture of what radon can do and the role played by tobacco.

Most of our exposure to natural radiation comes from radon that rocks and soil give off as they weather and erode, and for people who spend time in enclosed subterranean places, that exposure is usually higher. We inhale and exhale low levels of radon gas all the time, and none of it remains in the body. But radon decays by emitting radionuclides that can hitch a ride on grains from smoke or dust. The nose and upper airways filter out most of these radionuclides; however, if they're fine enough they may get into the lungs and concentrate on the inner surface. People smoking in the presence of excessive radon inhale radionuclides that a strong drag on a cigarette can suck deep into the tissues. Some of these emit alpha particles. The alpha particle has a high mass and very little penetrating power, but if it's emitted close to the surface of a lung cell it can produce damage to it and to nearby cells by depositing enough energy to break molecular bonds. Sometimes when that occurs the can-cer process is initiated.

Every radiation expert I met on the nuclear tour brought up smoking. An analysis of nearly five decades of data by the Radiation Research Effects Foundation, a cooperative organization jointly sponsored by the United States and Japan, found that lung cancer is one of the higher rel-ative risks associated with radiation exposure and that when combined with smoking the risk increases. In a comparison study with Japanese atomic bomb survivors, researchers investigating lung cancer concluded that "the exposures of the cases selected from the two populations were such that far more of the uranium miner cases than A-bomb survivor cases were likely to have been caused by radiation exposure."

After the collapse of the American uranium mining industry, re-searchers at the Department of Medicine at the University of New Mex-ico Medical Center (UNMMC) in Albuquerque compiled records about miners and assembled the cohort of 3,469 of them. These men, who had worked underground one complete year or more in the Grants area and who are now in their seventies, continue to be followed. One of the researchers is Professor Charles Key, M.D., Ph.D., an epidemiologist and pathologist at UNMMC who served for over thirty years as princi-pal investigator for the National Cancer Institute Surveillance, Epidemi-ology, and End Results program, in addition to being medical director at the New Mexico Tumor Registry, a population-based project that tracks incidences of cancer in the state. "Some of those miners have died," he told me. "About a quarter of the general population gets cancer, but in this group there's an increased rate of lung cancer—more than you see in other residents in the state. The increase is related to estimates we've made about the doses they received. The range of exposure is pretty broad, depending on where miners had worked previously or whether they worked in unventilated mines. The greater the miners' exposure to radon, the likelier they've been to develop lung cancer. Most cases have tended to be smokers. Are you at greater risk being a nonsmoking uranium miner or being a smoking nonminer? It's a toss-up. If you're in a mine that's ventilated and monitored, the risk is lower. In the Navajo mines in the 1940s and 1950s, smoking was not common, but there are a number of cases of lung carcinoma among Navajo uranium miners."

Until 1984, experts thought that radon was a problem only for miners or for people with houses built with or near uranium tailings. Household radon was first discovered in 1966 in Grand Junction, Colorado, in homes that had indeed been built on such tailings. Radon at high levels turned up in houses in other states where builders had used tailings or sited houses on them. Tailings from the mining of phosphate and the refining of radium that have been used in construction materials also emit radon. Remediation of tailings, removal of soil and radioactive building materials, and razing of houses followed. But then one day a nuclear worker set off radiation detectors as he entered the Limerick nuclear power plant, near Philadelphia. The plant had not begun operating. There were no uranium tailings in the area. Radiation protection specialists, unable to locate any source of contamination, went to his home. In a woodworking shop in the basement, detectors picked up radiation levels hundreds of times higher than natural background radiation. Radon was emanating from a crack in the concrete slab, which rested on the Reading Prong, a geological formation rich in thorium—it decays into radium, which in turn decays into radon. The gas had in turn decayed into radioactive particles that attached electrostatically to polyester in the worker's clothing. The Reading Prong, which runs from Pennsylvania into New Jersey and New York, emits more radon than a uranium mine. The Environmental Protection Agency (EPA) has estimated that, after tobacco, household radon is the second leading cause of lung cancer, accounting for over 13 percent of deaths from the disease. Despite a campaign to have people check their basements and ventilate them if radon is found to be above a certain threshold, few people do it.

"The radon dose that miners got after the mines were ventilated was not any more than farmers tilling the soil get," Rip remarked as we left the museum. "Probably less. Farmers who plow their fields inhale a lot of radon. Someone up high on his tractor would get less, but people who are down turning over the soil get an appreciable dose."

Since 1990, the Radiation Exposure Compensation Act, in part the result of lawsuits brought by Native American and environmental groups, has guaranteed a payout to anyone contracting a uranium-related disease. Additional laws Congress has passed sometimes compensate workers who received negligible exposures and developed cancers not usually considered radiologically induced. Meanwhile, some workers who, based on the science, deserve compensation do not receive it, because of the way risks are calculated.

"The best report on the health problems of uranium miners was assembled and published by the National Academy of Sciences," Dr. Fred A. Mettler Jr. told me. Formerly chairman of the Department of Radiology at the University of New Mexico, he has worked for many years for national and international agencies in the field of radiation protection and written books about medical radiation. "There have been eleven major cohort studies. There's no evidence that uranium mining caused cancers except lung cancer from radon, and some from exposure to arsenic in the mines." Mettler mentioned Boice's study of miners and the general effects of uranium mining in a Texas county; Boice found no increase in cancer there at all compared with the incidence in surrounding counties.

Mettler recounted testimony he gave in a lawsuit. "Look, uranium miners are the worst-case scenario," he said. "They worked down in a mine where the ventilation was poor for twenty years and we don't find anything in them but cancer from radon plus smoking. So how are people who live ten or twenty miles away from a mine or a refinery going to claim their health has been affected?" Mettler turned to the metabolism of uranium. "The breast doesn't concentrate it. Large studies about very long exposure show no increase in lymphoma. The federal government just added prostate cancer to the list of diseases a uranium miner can be compensated for, although there's no evidence of any relationship between prostate cancer and radiation exposure."

We continued westward, turning north before reaching the Continental Divide and a town called Thoreau—pronounced "Thuh-ROO" around here. A prehistoric sea that had once covered the region eventually shrank until it encompassed only the Colorado Plateau and environs, pooling and concentrating heavy metals that had eroded out of rock. The sea finally disappeared, and its organisms, which had transformed solar radiation—and one another—into energy, became buried over time. Under the pressure of accumulations of rock and soil weighing down upon them for hundreds of millions of years, these plants and animals turned into coal, oil, and gas enclosed in impermeable formations in what is now the San Juan Basin, farther west. But long before that, water was already dissolving uranium out of sea sediments and, on this side of the divide, depositing it in a two-mile-wide band of amoeba-shaped pods that extends across a wide, shallow basin known as Ambrosia Lake and continues eastward through the flanks of Mount Taylor and toward the Indian reservations near Albuquerque. The ore bodies lie deep underground except for some outcroppings here and there. One is close to Santa Fe, where adobe walls are speckled with enough uranium to be used to calibrate radiation detectors.

The name Ambrosia Lake may conjure up an image of a lush oasis, but the vegetation is mainly that of the high desert: purple sage, yellow-flowering chamisa, and gray-green grama grass. When Spaniards settled here, they dubbed a small, ephemeral body of water La Laguna del Difunto Ambrosio after the arrow-pierced corpse of one Ambrosio was found floating there. By the twentieth century, the lake had dried up, and anglicization of the name had erased the defunct Ambrosio from the picture altogether. In 1925, during a national oil boom, a young Los Angeles businesswoman named Stella Dysart bought Ambrosia Lake, then a big sheep ranch, and she became a wildcatter, drilling everywhere—she kept logs of her attempts—and coming up dry every time. Eventually she subdivided most of her property into thousands of tiny lots that she sold to small investors.

As we drove mile after mile on a two-lane road across the Dysart ranch, we passed no cars or residences. We saw a lone saloon with peeling paint, and a few cows and horses grazing against a backdrop of red bluffs—some with round-mouthed caves gouged in their vertical faces, the remaining evidence of the radon-rich "dog holes." And there were big, contoured earthworks, some resembling dams and others small mesas. Unlike natural features, these were dark gray instead of reddish brown, and their sides were symmetrical, regular, smooth, and barren. They and the dog holes would be the only clues to the uninitiated visitor that something big and unusual had happened here.

After Paddy Martínez set off the 1950 uranium rush, Dysart showed her drilling logs to a geologist, who used them to pinpoint a thick seam of uranium. Further exploration revealed that the Ambrosia Lake uranium district, which includes Mount Taylor, contained 70 percent of the nation's uranium reserves in deep, large ore-body clusters of high grade—3 percent uranium oxide or more. Hundreds of shafts were sunk, a forest of head frames sprang up, along with several refineries, and Ambrosia Lake became the biggest producer in the Southwest, contributing nearly 20 percent to the entire national production of uranium between 1956 and 1988.

Following the old-timer's directions, we reached the edge of the basin bounded by the slopes of Mount Taylor, forested with piñon pines, and we found the deserted Chevron mine and the last remaining head frame. Nearby were rusting tanks and water towers and big mounds of tailings. We were stopped by a high chain-link fence with a NO TRESPASSING sign. Somewhere deep in the flanks of this extinct volcano lay more than one hundred million pounds of uranium oxide embedded in a six-mile-long ore trend, perhaps the biggest in the United States. The Department of Energy (DOE) estimates that national reserves of uranium of good quality and within reach of present mining technology add up to about eleven hundred million pounds, or about half a million metric tons. The Grants mining district, which encompasses Mount Taylor, alone harbors a total of about half a billion pounds of uranium. Only Wyoming has more abundant known reserves in the United States.

Uranium oxide is usually scattered through silica—sand particles glued together to make a sedimentary rock. Whether the ore is commercially mineable is dictated by the concentration of uranium oxide. If that's less than half a percent, you don't bother. In Africa there are very high-grade deposits—from 20 percent to 50 percent. The annual requirement of all the operating reactors in the world comes to about seventy-seven thousand metric tons of uranium oxide concentrate. Today most of that is supplied by Australia, Canada, and South Africa and the remainder by stockpiles owned by utilities or various governments; by reenriched depleted uranium; by weapons-grade military uranium from dismantled warheads and bombs; and by recycled uranium from spent nuclear fuel. Owing to political decisions, the United States no longer reprocesses spent fuel, and consequently about thirty years' worth of uranium lies in spent fuel pools awaiting interment. As I was to learn, that may change.

From the abandoned mine, we followed the route that the trucks carrying ore would have taken. After several miles we arrived at a rise where a collection of tall prefab buildings and big metal frameworks stood in disrepair—the Rio Algom uranium mill, where we were scheduled to take a tour. We passed a sun-bleached yellow sign with a magenta trefoil and the words:

RADIATION
CAUTION
RADIATION AREA

Up to this point, I'd been having fun. And I was trying to keep an open mind. But I was about to step into the place where the ore had been transformed into yellowcake. How much exposure to radiation would we be receiving? Had we in fact entered a "radioactive wasteland"?

I reminded myself that Rip was an expert in nuclear environmental safety and that he had to know what he was doing. Still, I had to ask. "Do you really think it's safe to go on?"

"Yep," he said. "We'll be exposed to less radiation here than we were at Mount Taylor. You'll see why."

We entered a modest one-story building—the offices of what was for years the biggest uranium mill in the United States and possibly in the world. In a lobby were glass cases containing dinosaur bones, crystals, ore samples, and a vial of powder resembling fine cornmeal: uranium oxide, or yellowcake, the mill's end product and the basis of fuel for nuclear plants and also for nuclear weapons.

This innocuous-looking stuff became famous when President George W. Bush in his 2003 State of the Union speech claimed—erroneously, the administration later acknowledged—that Saddam Hussein had been attempting to buy African yellowcake. The implication was that Iraqis could rapidly and covertly turn it into bombs to be used against Americans. Whenever reports about Iraq's nuclear capability aired, I'd phone Rip, and he kept assuring me that they were hogwash. To turn uranium oxide into weapons-grade material would require large facilities and materials that Iraq, after years of sanctions and surveillance, did not have. Scientists from the DOE's national laboratories had been accompanying international inspection teams on visits to Iraqi nuclear facilities since 1978. Iraq had built a reactor that was supposed to be for peaceful uses and to be monitored by the International Atomic Energy Agency (IAEA); the Israelis destroyed it in 1981, before it began operating. There remained in the country, under seal of the IAEA, enough yellowcake to make hundreds of atomic bombs. Rip kept emphasizing that making a bomb always requires a uranium enrichment plant or a reprocessing facility. A reprocessing plant—a large, highly technological facility equipped to handle extremely radioactive material—separates out uranium and plutonium from waste products in uranium fuel that has been cycled through a reactor. "To extract them is a nasty process," he said. "You've got a whole lot of dangerous material and you're trying to extract a little product out of it. You're in a very dangerous situation that requires a huge, expensive infrastructure." The plutonium isotope, plutonium-240, which is produced in power plant reactors, is very poor bomb material; it is thermally very hot and much more radioactive than the preferred isotope for a nuclear weapon, plutonium-239, which is made in a special kind of reactor, called a production reactor. That is the reason no nation makes bombs with power-plant-generated plutonium.

"What could terrorists do with yellowcake?" I asked Rip.

"Not much in the way of causing any real harm. They could wrap conventional explosives around yellowcake and disperse it. That would make a Geiger counter tick a little faster and scare people who didn't know much about yellowcake, but uranium particles are very heavy and would tend to fall right back to earth, unlike airborne chemical agents that can travel distances. If you were near the explosion you could take a shower as a precaution—to remove any radionuclides that might have collected on your body. The yellowcake could be cleaned up. But at this point in its life cycle uranium is only mildly radioactive. When I visited Ambrosia Lake back in the 1960s, there were lots of barrels of yellowcake sitting around. You could safely put your hand into it. It was OK as long as you didn't ingest it."

And why was that?

"Uranium oxide emits mostly alpha particles, and they can be stopped by your skin. But if they get inside the body they can do damage."

And what did Rip see when he looked at the yellowcake?

"I visualize atoms of uranium suspended in the crystal lattice and oxygen atoms and uranium atoms sharing electrons in their outermost shells. I see the uranium nuclei as wobbling, unsteady balls of neutrons and protons held together by the binding energy of the nucleus. They contain subatomic particles popping in and out of existence. Most of the atoms would have 238 neutrons and protons in their nuclei, some would have 233, and a very few in the yellowcake would have 235, which is the most fissile isotope because it's the most unstable. Only seven-tenths of a percent of the yellowcake is U-235. It fissions when its nucleus becomes wobbly enough to break apart spontaneously—think of a spinning raindrop—or when a neutron from outside bombards a U-235 nucleus. If you imagine the moon plunging into the earth, you can understand the impact of a neutron on a nucleus. It's only in reactors and bombs that you need to bombard nuclei with additional, external neutrons to have a controlled chain reaction or an explosive chain reaction. When a neutron splits an atom, some of its neutrons fly out and those neutrons hit other nuclei, and on it goes, as long as the conditions are right. In a chain reaction in a power-plant reactor, you want to control the neutrons to maintain a steady release of thermal energy, so you siphon them off by using neutron-absorbing material—the kind contained in control rods." Control rods in most reactors are set up to drop among the fuel rods to absorb neutrons and thus guide or shut down the chain reaction.

"Bombs don't have control rods," Rip continued. "The enrichment process involves increasing the proportion of a particular isotope in, usually, uranium by turning the uranium oxide into uranium hexafluoride gas and then patiently, over a long period, separating out by weight some of the heavier U-238. In this way, you increase the proportion of U-235 from seven-tenths of a percent in natural uranium to three to five percent. This gives you low-enriched uranium to sustain a controlled chain reaction that makes thermal energy to run a power plant. If the uranium is enriched to twenty percent and the chain reaction is uncontrolled, you could get an explosion—not a bomb. By explosion I mean that the uranium would be scattered around, but there is only a small amount of the chain reaction occurring, and you'd need many tons of uranium. If the uranium is enriched to a much higher percentage, like ninety percent, and the chain reaction is uncontrolled, you could get a bomb. And for that you need only a few kilograms of highly enriched uranium, but to make it takes a lot of time, effort, and equipment."

In the display case next to the vial was an indented metallic rod three-eighths of an inch in diameter and about half an inch long, roughly the

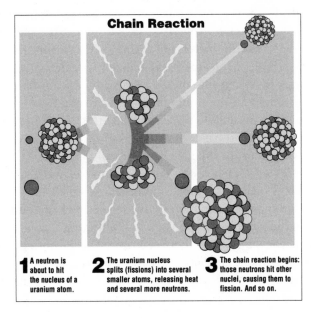

Chain Reaction

1 A neutron is about to hit the nucleus of a uranium atom.

2 The uranium nucleus splits (fissions) into several smaller atoms, releasing heat and several more neutrons.

3 The chain reaction begins: those neutrons hit other nuclei, causing them to fission. And so on.

size of two pencil erasers placed end to end. I was to think of that pellet many times during the nuclear tour. It takes a ton of ore to make four to six pounds of yellowcake, which, after going through enrichment and fabrication processes, becomes a pellet of low-enriched uranium-oxide fuel weighing 0.24 ounces—about seven grams, or slightly less than the weight of three pennies. "There's no other energy resource as dense as uranium," Rip said. "That's a major plus."

We were greeted in the lobby by Terry Fletcher, a tall, husky, sunburned man in a yellow hard hat. In addition to serving at the time as chairman of the New Mexico Mining Commission, he was the head of operations at the mill. He told us that nobody in the United States was producing yellowcake anymore. "We were the last licensed mill in the country," he said. "Demolition started here last week."

"Where will nuclear plants get fuel?" I asked.

He replied that domestic production of uranium oxide makes up less than 10 percent of the fifty-three million pounds a year needed to fuel American nuclear plants. "We import uranium from Canada, Australia, South Africa. And the U.S. government is now buying Russian uranium from dismantled Soviet bombs. It's being given to power plants at prices we can't compete with. Right now, the burn-up rate of uranium world-wide is twice the production rate."

"We'll run out of uranium then?" I asked.

"No—nobody wants to have our lifestyle disrupted because of insuffi-cient electricity," Fletcher replied. "So we'll import it from places that can mine and process it more cheaply. About sixty percent of the world's uranium now comes from Australia. Canada has high-grade ore that's mined robotically. When the United States runs out of the uranium we have stockpiled and all the uranium from dismantled nuclear weapons, we'll have another mining boom here to fuel the thousand new reactors we're going to need."

"In fifty years this mill will have to be rebuilt," Rip said.

Fletcher nodded. "That sounds about right. Today, the United States has forty years' worth of uranium to burn before we'll need to go back to mining it. I know where the best ore around is, and I'm going to stake a claim here in the heart of it, for my grandchildren."

"How can you be so sure there will never be a shortage of reactor fuel?" I asked, thinking of claims by antinuclear-power groups that only sixty-five years of uranium reserves remain, or less if more nuclear plants are built.

"We have abundant known reserves in the United States," Fletcher said. "And, if the demand goes up, I'm confident much more uranium will be found. Nobody's been prospecting, because of the low price of yellowcake and the low demand. There won't be any more uranium mills in this country unless it becomes government policy again to have them. When this place was built in the 1950s, the cost was staggering. But the cold war was on, and the Atomic Energy Commission needed uranium production right away. Today, you have to be a huge economic entity to go into mining and refining of any kind. It's a big investment to make before you get a return. We have plenty of uranium here that we would mine for twenty dollars a pound if there were a market. If the price went up to forty-five dollars a pound, well, then the supply is unlimited."

At the time of our visit to Ambrosia Lake in 2003, the price of uranium oxide was $7 a pound. As utilities here and abroad began to plan new reactors, and the program to recycle former Soviet warheads into fuel was approaching an end, the price shot to over $135 by mid-2007 and, boosted by uranium fever among investors, was expected to keep going up. On the Colorado Plateau, claims rose ninefold, and prospecting resumed along with schemes to reopen mines and to sink new shafts, and especially to do more in-situ leaching. Fletcher predicts this method will become the biggest source of American uranium. Certainly mining techniques and mine safety have improved since the 1950s and environmental laws have changed; any new operations would have to meet far more stringent health, safety, and environmental requirements.

Fletcher introduced us to our guide, Alberto Delgado, a slender, graying, soft-spoken man in blue coveralls. He had come here from southern New Mexico in 1981 as a chemist and became mill operation supervisor a year before the refinery started winding down.

He gave us forms to fill out and a pamphlet about risks and hazards we might be about to face:

One of these may be the possibility of individuals being exposed to radiation or radioactive materials. However, due to the type, location, and duration of activities that you will be involved in while on the premises, radiation exposure is expected to be negligible. Nevertheless, several simple procedures should be followed.

We were instructed to obey all posted warning signs; to wash hands prior to eating, drinking, or smoking; and to eat, drink, and smoke only in designated areas—no smoking allowed in any of the buildings. We signed liability forms. After donning hard hats and protective plastic glasses, we climbed into a van. Before Delgado turned on the ignition, he looked solemnly at each of us. "Our company is very safety conscious," he said. "Please fasten your seat belts." Such is the culture of safety in the nuclear world today, and it's thanks to a mantra, ALARA—As Low As Reasonably Achievable—first uttered decades ago. That means that workers are encouraged to keep the level of risk way down by heightening their attention and strictly following every procedure as if danger were present. Everyone in the nuclear world is supposed to abide by the principle that unless there is a good reason to be exposed in the line of duty, you should do all you can to avoid additional exposures.

Our first stop was a few hundred yards past the buildings, by the side of a two-lane road. A solitary tree here and there only emphasized the desolation. In somewhat melancholy tones, Delgado spoke about the bustle that was no more. "You can't tell it now, but we had everything here. About sixteen hundred people in the mines and the refinery. That's just this one company. There were other companies around also. Early in the morning in 1981, semis would be backed up here for a quarter of a mile, waiting to go to the weighing station and then unload ore. Each day eight thousand tons of ore were processed, and from 1958 on the mill ran twenty-four hours a day, seven days a week." Nearby, a boomtown had sprung up, with housing and a school. Now those buildings are part of a ranch.

He pointed to an area the size of a baseball field that had been scraped and smoothed over and was now covered with grama grass. A few months earlier, a mine shaft and a head frame had occupied the spot. Now the only evidence of a mining operation was a black hose that came out of the ground and extended all the way to the refinery complex. "The hose is from when we were injecting water into old mine workings here and pumping out the solution. It contained uranium. We shipped that concentrated slurry to Wyoming for processing into yellowcake. That's over now, and we're just cleaning the water we use. The mine shaft has been covered, and natural vegetation is taking over. The shaft will be filled in and capped off. Around the state you hear about kids falling into old gold, silver, copper mines. We'll make it impossible for somebody to break into the shaft for the next two hundred years."

The atmosphere of the mill buildings verged on the surreal. We passed a structure whose walls had been removed. On an upper floor a lightbulb burned and the blades of a small fan fluttered against a backdrop of sky, mountains, and the broad, shallow valley with its sculpted mounds.

After passing a crew in white decontamination suits carrying crowbars and other tools of demolition, we trooped through cavernous rooms where the ore had been crushed and then pulverized into sand before water was added and the slurry pumped to leaching tanks. In big vats, resembling those in a winery, chemicals were added to dissolve out the uranium. The sand was repeatedly rinsed to obtain any residual uranium, and then the waste was piped away to a pond bed to dry. The uranium, a yellow, wet powder, was pumped to a dryer and, finally, packed into fifty-five-gallon drums. It took twenty-four hours to turn ore into yellowcake. "We were set up here to make seven to eight million pounds of yellowcake a year," Delgado said. "It's sad to hear only echoes here now."

He brightened when he showed us a pond outside where fish darted and the ion-exchange building, which contained a filtering system he'd designed to get every trace of uranium out of the water. "All the extractions in any mining industry follow the same pattern: you hit it once, and you hit it again and again. Our waste water still contains some impurities, but so few that they're measured in parts per million. For practical purposes, it's clean water. We pump it into the pond, and the fish are fine."

Now I understood why at the mill radiation exposure was lower than in the surrounding area: the uranium oxide was just too valuable after all that effort not to be collected and sold.

As we left the assortment of buildings and other structures for transporting slurry and treating it, Delgado remarked on their imminent disappearance. The decommissioning of the mill has to meet the standards of the EPA and the Nuclear Regulatory Commission (NRC), which had approved the environmental impact statement. "Everything will be buried here on site. Lead paint was used back in the old days. We didn't know then that it was toxic. That's the environmental impact of the mill, not radioactivity."

"Yep," Rip said. "It's usually chemical or toxic heavy metal waste that's the problem."

Although artists have created large-scale earthworks in the Southwest, the ones in Ambrosia Lake were composed by cleanup teams remediating slurry tailings left after all the uranium oxide had been extracted. Because of a slurry dam failure in 1979 at a uranium mill in Church Rock, New Mexico, which had exposed people and livestock to contaminated water, and because of the closing of the mines, public attention turned to the tailings and their potential risk, and the New Mexico Mining Act was passed. It's retroactive and calls for the statewide restoration of grounds with surface damages caused by any type of mining. A big project is returning tailings piles to a natural state and planting them with native vegetation.

The NRC, which now regulates uranium-processing facilities, approves and monitors all their reclamation. The NRC's rules are based upon conservative—that is, worst-case—assumptions, and therefore the mill has been ordered to secure the tailings for a millennium and to make sure that a "maximum probable event" does not disperse them. The site must be protected from erosion caused by extremely heavy rainstorms of the kind that occur only once in a thousand years, so Rio Algom is building gigantic rock diversion channels.

The five uranium mills in Ambrosia Lake all had tailings piles, but Rio Algom's, which covered four hundred acres and towered ninety feet, was probably the biggest in the world, and Fletcher said it had been visible from space. No one had ever reclaimed a tailings pile before—Rio Algom was a pioneer, and people from other countries with tailings to remediate would visit the site to get pointers.

"The tailings from uranium milling make up the largest volume of nuclear waste in the United States," Rip said. "But the waste is classified as low level. The slurry residue is mildly hazardous—not from the uranium, because it's been removed, but because of the daughter products that were *not* removed, and because of all the thorium and its daughters."

Each of thirteen daughter products into which uranium successively decays is radioactive until the decay chain terminates with stable lead. Those daughters remain in the tailings. Normally, in thorium and uranium formations radon percolates upward to a level about a meter or so underground. But mill tailings tend to put thorium, and thus radon, directly into the biosphere, where weathering and wind can release it. If present conservative estimates about exposure to low-dose radiation are correct, radon could conceivably produce health problems if in the remote future people decided to build houses on or near the earthworks. Since the EPA decrees that the odds of getting cancer from them must be no more than one in ten thousand, they must be covered.

Delgado drove us around the project. Gigantic earthmovers clawed open burial pits. A track hoe moved slowly up and down a pond bed where slurry had once been spread out to dry and which had been cleaned out and lined with crushed rock. Bulldozers had already scraped up windblown material from adjacent properties and had sheathed the tailings with a layer of thick clay that kept out water. They were then covered with a foot of flat river rocks to stabilize them.

"The radionuclides in the piles emit radiation," Rip said. "It's important to immobilize radionuclides in nuclear waste. Clay sorbs radionuclides."

"Sorb" is a chemical term meaning "bind." I was often to hear it. Nature has its own way of controlling a chain reaction and of preventing radionuclides from bouncing around in the environment as they depart from their homes in atoms. Sorbers attract radionuclides the way polyester fabric attracts cat hairs. Positively charged radionuclides adhere to "active sites," where they are immobilized by negatively charged particles of earth and rock. Once a radionuclide falls into the embrace of an active site, it goes no further. One of the best sorbers around happens to be clay, and it's used to keep nuclear waste from migrating.

"So—how radioactive are these tailings?" I asked.

"If you've taken out all of the uranium, and capped the tailings with clay and rock, where's the problem?" Delgado asked. "Our recovery in this mill is the most efficient of all the mills. The uranium was extracted as completely as possible. If you go outside of our property, the natural background radiation is higher than it is right here. We don't calibrate our detection equipment from the tailings because the uranium has all been removed."

"If there's any danger," Rip interjected, "it's not from uranium. It's from the terribly fine dust from the tailings and the daughters left in them. When the wind blows, this stuff can get into people's lungs. So that's why the tailings have to be stabilized and capped."

"Any dust storm anywhere can give people silicosis, a disease caused by rock particles getting into the lungs," Delgado said. "The EPA requires the rock and clay treatment of the tailings, so we're doing it, but not because of radioactivity. People have exaggerated the whole radiation thing. They need to understand what the effects are and how much exposure you have to have to get some kind of effect."

He told us that the state environmental department monitors the site and that workers must wear dosimeters. "They're monitored, and we all get regular physicals. That's been the policy here for a long time. We've got a lot of old-timers who've been here for decades. The majority of our workers have stayed in good health. Very few of us here now are young. But we're doing fine."

At the end of our tour we returned to the administration building. "Next week a demolition team will take down the buildings," Terry Fletcher told us. "They have guys who operate a huge shearer. It weighs two hundred forty thousand pounds and is seventy feet high. It's like a big excavator with a scissors to cut metal. When we're done, the land will look as natural as the surroundings and then it will be sold as a ranch. By 2006 all the tailings piles in New Mexico will have been reclaimed and the only work left will be remediation of groundwater. That might take a decade or more. Every mill will be gone and you won't even know where it stood. It will be the closing of an era."

He spoke about exposure to radon and what it meant to miners. He himself had worked in the mines as a teenager. "The industry didn't do a good job in the 1950s educating people in the mines about the dangers. From 1972 on, everyone was required to wear a dosimeter, but miners didn't want to, because if they received a certain amount of exposure over time, they'd be relocated. They wanted to stay in their jobs. Nobody quit because of health hazards. I'm still here. There's definitely a segment of people who got sick. But every one of those guys would come back to work on the Colorado Plateau if they could."

By the end of the afternoon, I understood that the uranium industry was at present practically nonexistent in New Mexico and in other states, that in the early days people were unwittingly exposed to radiation or had disregarded warnings, and some who mined in the early days now were suffering from lung cancer and had received federal compensation for that terrible misfortune. Epidemiologists and the National Academy of Sciences agree that, since 1972, safety regulations and education have made uranium mining no more dangerous than any other kind. The commercial tailings and mine sites are being covered and restored to a natural appearance, and the Navajo Nation has taken care of its hazardous sites as well.

Exactly how much uranium is available worldwide remains uncertain, because nations conceal the size of their stockpiles of minerals and yellowcake. Most experts believe there's enough to run a thousand or more reactors for hundreds of years, and, with reprocessing of spent fuel, hundreds of years more. John M. Deutch and Ernest J. Moniz, authors of the MIT–Harvard report, *The Future of Nuclear Power,* assert that "there appears to be ample uranium at reasonable cost to sustain the tripling in global nuclear power generation that we envisage with a once-through fuel cycle [that is, virgin uranium goes through the reactor once and is not reprocessed] for the entire lifetime of the nuclear fleet (about forty to fifty years for each plant)." In the sector of the nuclear world that keeps track of uranium supplies—the World Nuclear Association, the World Energy Council, the IAEA—I found general agreement that uranium is so ubiquitous that supply depends on the will to find it and the price paid for it, and that we indeed had enough resources to run many plants for thousands of years.

"Most future uranium mining will use in-situ leaching, with far less surface environmental impact," Rip said. "Future impacts on the aquifer are unknown, but leaching won't require anything close to the fossil-fuel combustion that conventional mining and refining equipment do and it won't leave tailings. All mining is a dirty business. Uranium mining is probably cleaner today than most other kinds."

I looked over my shoulder to the west. The brown pall above the mesas and mountain ranges seemed worse. New Mexico's oil extraction industry as well as power plants on or near Navajo land were burning millions of tons of soft brown coal extracted in part by Native Americans from a gigantic open-pit mine. From this massive quantity of matter, a tiny fraction of energy stored in ancient organisms was being converted into heat to make steam to turn turbines to generate electricity for nine hundred thousand people in the Southwest, and the remains of the burned fossils and their component gases were being dispersed into the sky, the soil, the water, and our bodies.

Though the Diné have voted to end uranium extraction on tribal land, they are being courted by an energy corporation to allow construction of another huge new coal-fired plant that might bring jobs and some prosperity to the Navajo Nation, which has voted against the establishment of gambling casinos—they have enriched many other tribes. From what I was beginning to understand, the new plant would only add to existing risks to the health of the Diné. Apart from atmospheric pollution from the stacks, fly ash would harbor uranium and other toxic heavy metals like arsenic and lead. The risks could be worse than those from the early days of uranium mining, and certainly would affect many more people. The smokestacks of the new plant would pump out more unregulated carbon dioxide and other greenhouse gases.

"A shift is now going on in the environmental community about nuclear power," Terry Fletcher observed as we prepared to leave the uranium mill, "and some opponents are beginning to see that it's really a clean source of energy."

In 2004, Hugh Montefiore, a retired Anglican bishop in the U.K. who had been a board chairman and was a trustee of Friends of the Earth, an international environmental organization that had begun when a group splintered off from the Sierra Club, wrote: "I have been a committed environmentalist for many years. It is because of this commitment and the graveness of the consequences of global warming for the planet that I have now come to the conclusion that the solution is to make more use of nuclear energy." He was promptly forced to resign. Using analytic studies that have been roundly debunked but continue to be circulated by the environmental movement, Montefiore's former colleagues issued a paper claiming that nuclear power was scarcely cleaner than coal combustion because high-grade uranium ore would one day be exhausted, fossil fuels are burned when uranium is mined and processed, and to mine lower-grade ore would require even more fossil-fuel combustion. Therefore, Friends of the Earth argued, nuclear power would eventually be contributing so much carbon to the environment that its emissions would match those of coal.

This assertion has been dismissed by economists, geologists, and other scientists. Dr. Mohamed ElBaradei, director general of the IAEA, has said, "Nuclear power emits virtually no greenhouse gases. The complete nuclear power chain, from uranium mining to waste disposal, and including reactor and facility construction, emits only two to six grams of carbon per kilowatt-hour. This is about . . . two orders of magnitude below coal, oil, and even natural gas." Studies of carbon dioxide emissions from the nuclear fuel cycle indicate that they are from 0.5 percent to 4 percent of the emissions from the equivalent generating capacity of coal-fired plants. And, unlike coal mining, uranium mining is rarely associated with the release of methane, the most heat-trapping of greenhouse gases.

Often when I read about the role of fossil-fuel combustion in global climate change or our dependence on fossil fuels, or the human lives and billions of dollars spent keeping a large military presence in the Middle East to protect the flow of oil in our direction, I remember a poster on display in the lobby of the now-vanished uranium mill in the middle of Ambrosia Lake. I was to see that poster again and again on the nuclear tour:

> One fuel pellet contains the same amount of energy as
> * 149 gallons of oil,
> * 157 gallons of regular gasoline,
> * 17,000 cubic feet of natural gas, or
> * 1,780 pounds of coal.

What is the significance of this in everyday life? To make the amount of electricity needed to keep a single 100-watt lightbulb on for a year, you'd have to burn 876 pounds of coal, about 350 pounds of natural gas, 508 pounds of oil, or 0.0007 pounds of uranium enriched to 4 percent. Each of us annually consumes, on average, the same amount of electricity as fourteen 100-watt bulbs left on all the time for a year.

To convey the energy density of uranium, two energy specialists, Peter Huber and Mark Mills, write:

> A bundle of enriched-uranium fuel rods that could fit into a two-bedroom apartment in Hell's Kitchen would power the city for a year: furnaces, espresso machines, subways, streetlights, stock tickers, Times Square, everything—even our cars and taxis, if we could conveniently plug them into the grid.

On our way out of Ambrosia Lake, we came to a crossroads where a faded billboard stood. It must have been meant for the truckers hauling uranium ore to the mill or transporting away drums of yellowcake to be turned into enriched uranium.

LIMITED VISION DEMANDS UNLIMITED CAUTION.

Part 2 THE INVISIBLE STORM

Nothing in life is to be feared, it is only to be understood. Now is the time to understand more, so that we may fear less.

—MARIE CURIE

4 MOTHER NATURE AND FENCEPOST MAN

ON OUR WAY BACK TO ALBUQUERQUE, Rip, Marcia, and I drove through an all-encompassing incandescent orange sunset with streaks of crimson, a spectacle that would have been in bad taste if represented on canvas or velvet but in nature seemed elegant. We pulled in at a truck stop on Interstate 40. The jukebox was playing country-and-western music. One section of booths was reserved for truck drivers, another for smokers. Almost every male not sporting a trucker's cap was wearing a cowboy hat and boots. Just like characters in a truck-stop scene in a movie, thoroughly lipsticked, sassy, motherly waitresses rushed around delivering menus and big glasses of iced tea. At a corner table, alone, a bearded hiker hunched over a book, his mint-condition backpack on a chair for company.

"OK, the people we met at the mill seemed healthy," I said after we sat down. "But what about Marie Curie? She died of radiation poisoning finally, and her furniture is still radioactive."

"Madame Curie had no way of knowing the risks she was taking when she isolated radium and polonium from pitchblende," Rip said.

I mentioned that, when I was growing up, there would be a public announcement that a cloud of fallout from a Soviet or Chinese bomb test might drift over dairy herds. We were warned to avoid drinking milk for a few days. The Atomic Energy Commission kept revising the amount of radioactivity it considered tolerable to humans. Later, after I moved to New York, I began to hear that *no* level of radiation was acceptable; it was all bad. But how much had I already soaked up and how much more could I take? Had radiation caused the cancer my mother had? What about other family members, neighbors, friends, and parents of friends who had developed malignancies? Lew Critchfield's father had died of cancer; he'd been at Trinity when the first atomic bomb exploded and had worked for years in Los Alamos.

"Are we humans playing with something that we're ignorant about?" I asked.

"We now know a lot about radiation," Rip replied. "It's been closely studied for over a century. We know more about radiation hazards than about biological and chemical ones. We know more about radiation than about any other carcinogen. We know that as a carcinogen it's a weak one. And we now know that all the time, people everywhere are absorbing radioactive materials and radiation like mad."

"Where's all this radiation coming from?"

"Mother Nature, mostly," Rip replied. "We and everything else in our environment are made from the dust of stars that have exploded and blasted all the elements around the universe, and some of those elements are radioactive."

After we ordered, he went on. "Life evolved in conditions of higher radioactivity than we're exposed to today. It's lower now because of decay. The fossil record shows that life thrived in the presence of the higher levels of U-235 and other uranium and thorium isotopes and daughters. In other words, the world was much more radioactive billions of years ago. It's very, very important to understand that radioactive contamination is different from radionuclides and from radiation. Each of these three things is different from the other two. Most people confuse them. Radioactive contamination is the uncontrolled release of radioactive material. Its unstable atoms—radionuclides—shoot out particles or rays that are moving very fast, and that's what we call radiation. Say you run across a combat zone. The flying bullets are the exposure and the bullets that strike you are the dose. The exposure is the amount of radiation coming out of the radionuclides. Dose is the amount of radiation a person absorbs. The closer you are to the source, the bigger your exposure, and also the bigger the risk of a greater dose. Contamination is caused by radionuclides that wind up in the wrong place. The radionuclides with very long half-lives do not pose a high external exposure risk. A radioactive atom containing the same number of protons and neutrons always gives off the same signature—the same type of radiation with the same energy. Because of this rule, it's easy to detect and identify radioactive materials."

We are all temporary collections of atoms bound together by an electrical charge. Because our bodies are mostly empty space, radiation usually passes through us uneventfully. But when it hits an atom or an electron, it deposits its energy. If it's moving fast enough, and the deposition is large enough, the atom that it hits will be ionized. In the second or two it took the waitress to slap down our plates, our cells were receiving thousands of energetic depositions from nature.

"Damage from ionizing radiation and particles is just another way of saying that a chemical reaction has occurred," Rip said. "An electron has been knocked out of the outer shell around an atom, and the particle or radiation that did that has lost some of its energy. When a particle or a ray, from either a natural or a man-made source, slows down enough, it becomes part of the natural environment and is no longer dangerous in any way. Certain forms of radiation cause more destruction to human tissues than others. It all depends on the type of particles or rays that are emitted. Some lose their energy rapidly when they pass through matter. Remember, for a radionuclide, the shorter the half-life, the faster the energy is released. The shorter the half-life, the more interactions in material—a rock or a person—the radionuclide has, and those interactions may not necessarily cause damage in a person. Radionuclides emit particles and rays and have half-lives that can last for a few seconds or for thousands, millions, and even billions of years."

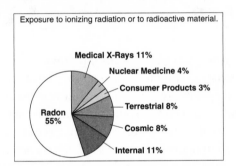

All charged particles, like electrons (beta particles), protons (fast-moving hydrogen atom nuclei), or alpha particles (fast-moving helium atom nuclei), lose energy as they penetrate matter. That's why the blanket of the atmosphere shields us from quite a lot of cosmic radiation and why you get exposed to more radiation when you fly in a jet or live at a high altitude than if you stay at sea level. Alpha particles come from decaying radioactive atoms and travel at different speeds, depending on which atoms produced them. A beta particle is always a fast-moving electron. Beta particles come from decaying radioactive atoms, too, and also travel at different speeds, depending on which atoms emitted them.

Gamma and X-rays are very-short-wavelength light. "White light that goes into your eye is made up of many different wavelengths," Rip said, drawing on a paper napkin a long horizontal line intersected by short perpendicular lines to indicate the spectrum.

"Each different wavelength in the visible part of the spectrum corresponds to a different color or hue that the eye sees. If you move along the wavelength scale to longer wavelengths, you move into radio waves, like

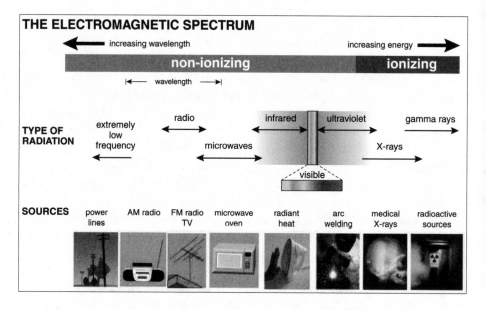

THE ELECTROMAGNETIC SPECTRUM

the ones that transmit to your car radio. If you go in the other direction to shorter wavelengths, you move through the ultraviolet light region to X-rays and, finally, to really short wavelengths called gamma radiation. Most gamma radiation comes from decaying radioactive atoms. There are very concentrated X-ray sources that are from the decay of radionuclides and are used in many industries. And X-rays are generated in medical or dental facilities by machines. When they're turned off, they're not radioactive and they don't give off radiation."

Rip now turned to flying particles. As far as subatomic particles are concerned, an alpha particle is massive—a bare helium nucleus made up of two neutrons and two protons. When it passes through a material it leaves a wake of ionizations along its short flight path. As a result, an alpha particle doesn't penetrate a great distance into any material. "If you start out with a hundred packets of energy and a unit of distance," Rip said, "an alpha will make many times more ionizations, spending its energy before it gets very far. It's like someone who takes all his money out of the bank and blows it at the first bar he enters. That's why I could hold a piece of plutonium metal—it gives off alpha radiation—on my palm on a piece of paper without any risk. The reason alpha particles are so easily stopped is also the reason why a radioactive material emitting alpha particles is very dangerous if it gets inside the body, especially in the lungs: *all* of the energy of the alpha particle is used up as it collides with the body tissues. The actual damage is usually done by the creation of ionized atoms, free radicals that chemically break the bonds of any organic molecule they hit. An alpha does a lot of damage over a very short distance."

Polonium-210 is an alpha-emitter that was used in the deliberate poisoning of a Russian former KGB agent in 2006. Another alpha-emitter, americium-241, is used in home smoke detectors, and for that reason Ralph Nader, the consumer activist, once tried to get them banned. But the alpha particles are stopped by the housing that contains the isotope. Many thousands of lives have been saved by smoke detectors, and nobody has ever been killed by alpha particles from inside a smoke detector.

Next Rip described beta radiation. "The electron in beta radiation is much smaller than the alpha particle and it hits far fewer molecules over a given distance of travel. So a beta can penetrate farther before giving up all its energy—in effect, a beta could spend a little money in several bars before he was broke. Most beta particles can be stopped by a piece of aluminum metal one-quarter inch thick." Beta-emitters such as carbon-14 and iodine-131 can be dangerous if they get inside the body in any quantity, but they're also dangerous outside the body, because, unlike alpha particles, they can travel relatively long distances—a meter or more—through the air. The potassium-40 in salt substitute is 89 percent a beta-emitter and the rest is gamma, but it has a half-life of over a billion years and so is only weakly radioactive.

In comparison with alpha and beta particles, X-rays and gamma radiation waves have no charge and therefore have greater penetrating power and can travel a longer distance through most materials and pass through almost any amount of air before stopping. You need several inches of lead or other dense materials to stop X-rays and most gamma rays. Gamma rays interact less with matter than alphas and betas so have the fewest ionizations per unit of track length, but they can penetrate farther than alpha or beta particles. Each ray that has a different energy will penetrate to a different depth. A few feet of concrete or several feet of water will stop both kinds completely. Because they can penetrate long distances in tissue, X-rays are used to take pictures of the internal parts of the body. X-rays and gamma-radiation treatments kill tumors in cancer patients. All radiation (alpha, beta, and gamma) that has the same amount of energy will ionize the same number of molecules in whatever material it is absorbed in. Each type of radiation differs only in the distance it has to travel to ionize that number of molecules.

Because gamma radiation has uniform penetrating ability, it's used in food preservation—it kills parasites and bacteria (like the *E. coli* that three people who ate organic spinach in 2006 died of). In the United States, food irradiation for spices, white flour, meat, poultry, and produce has been approved in recent decades; the food must be labeled as irradiated. Stores are reluctant to carry these products because of public doubts about the process. Medical sterilization facilities, in operation for over thirty years in the United States (for example, to irradiate bone and other tissue for transplants), have never had an accident and are not radioactive, and neither are the tissues they've sterilized. The Centers for Disease Control states that food irradiation could prevent many important diseases transmitted through meat, poultry, produce, and other foods: "An overwhelming body of scientific evidence demonstrates that irradiation does not harm the nutritional value of food, nor does it make the food unsafe to eat. . . . Consumer confidence will depend on making food clean first, and then using irradiation or pasteurization to make it safe. Food irradiation is a logical next step to reducing the burden of food-borne disease in the United States." The gamma rays do not stay in the food after irradiation any more than the light rays stay in a room after you turn off a lamp.

After listening to Rip's explanation, I still was doubtful. "I hear all the time about high risks from radioactive contamination that will last for millions of years," I said.

"When you read in the newspaper or see something on the TV news about radioactive contamination," Rip said, "you notice that they almost never put it in context. They never provide a comparison. They say, 'The area around Chernobyl is giving off radiation.' People get scared. But that radiation may not be much above natural background level. Or it may be less than the natural background level somewhere else. In a given area there may be hot spots and the rest of the place is normal. All in all, the risk from that radiation happens to be smaller than the risks you face in traffic. If you don't have a context, if you don't know the range of normal exposure, you don't have any point of comparison. Journalists hardly ever report whether the Chernobyl area is more radioactive than, for instance, where we're sitting right now. It just so happens that on average we're getting more radiation from nature here than the people living around Chernobyl are getting on average."

Rip took a new paper napkin and drew a pie chart and wrote under it "360 millirem per year." He said that was the total average exposure Americans get from all sources of radiation. A millirem is a measure of the actual biological effects of radiation absorbed in human tissue.

Then he shaded in most of the circle to represent natural background radiation. According to the latest data from universally respected United Nations scientific committees and other international agencies responsible for the establishment of standards regarding radiation protection, the annual exposure of a human being anywhere on earth to natural background radiation averages out to about 240 millirem a year. In some places it's much higher, though. "In northeastern Washington State, people get an average annual dose from background radiation of 1,700 millirem. Here on the Zuñi Uplift, at this altitude, we're probably getting around 500 to 700 millirem."

That sounded like a lot.

"Well, it's more than double what you'd get on Long Island. The blanket of the atmosphere is thinner at our altitude, so more cosmic rays get through. If we were standing on top of Mount Taylor, the dose would be even greater than here, and if we were in a jet flying at thirty-six thousand feet we'd be getting still more."

"I've heard people say that the ones who got the most radiation because of Three Mile Island were the reporters who flew in to cover the story," Marcia said.

"The total *calculated* dose Pennsylvanians got from Three Mile Island was far less than the *measured* dose New Mexicans receive from nature every day," Rip said. "A chest X-ray gives you roughly a dose of about 10 millirem, and a round-trip flight from New York to California gives you a dose of about 3 millirem. People refuse to get chest X-rays or mammograms, but those same folks don't mind increasing their exposure to ionizing radiation by flying from New York to the Southwest and spending time at seven thousand feet or higher to ski."

"So a lot of ionizing radiation is striking atoms in our tissues right now?"

"Right. But remember, we're talking about a yearly dose. Here's a way to think about it that's chemical, not nuclear. Imagine that the natural exposure is the same as the sprinkling of salt the cook put on the hash browns on your plate. That's the amount most of us get, and are used to, and the body doesn't have a problem with that; it knows how to deal with it. If a few more grains of salt were added, it wouldn't make any difference. But if he doubled that salt, the potatoes wouldn't taste right anymore, and if he put a cup of salt on your potatoes and you ate them, you'd start getting sick. And that's also the way radiation works. So what we receive naturally is like a sprinkling of salt, and it's from nature that we get almost all of our radiation exposure."

A heavenly bombardment is constantly irradiating us with about 50 millirem per year. Cosmic radiation is penetrating, ionizing radiation that originates in space. These high-energy particles smash into the atoms in air molecules and, sometimes, in a nuclear reaction, kick out various unstable subatomic particles. Some are so fast and small that they rarely interact with matter, arrowing as they do through its interstices. But others knock electrons out of our atoms, leaving behind a trail of deposited energy in our tissues. Our DNA, accustomed to this assault since life began, can promptly repair the damage most of the time.

Most radiation exposure comes from rocks and soil as well as from building materials—bricks, marble, granite, and drywall—containing uranium. Grand Central Station is made from granite, and if you spent all your time there, you'd get a dose of nearly 600 millirem a year. The annual exposure to the public permitted by the Environmental Protection Agency (EPA) for nuclear facilities: 15 millirem.

I thought about the Sandias, their ancient granite crumbling and being washed down into the pink gravel-strewn arroyos where I'd once played in the relentless sunshine. And what about New York, where I'd spent most of my adult life? After taking trains through tunnels bored into Manhattan schist (basalt containing uranium), I'd walked to work past polished granite facades and, during one year, through Grand Central twice a day. And what about the water that flowed through upstate granite into city reservoirs? On average, Americans get 32 millirem a year from drinking water. And during my mother's terminal illness I began taking several plane trips a year to New Mexico and continued to do that for over a decade.

"On average, 300 or even 600 millirem add up to a very, very small sprinkling," Rip said. He then put his pencil on the small wedge of the chart. "Next comes man-made radiation. Most of that is from diagnostic and therapeutic nuclear medicine."

Radiation from medical sources has accounted for a yearly average of 40 millirem per person around the world and about 53 millirem in the United States, but the latest studies estimate that on average Americans are now receiving about as many millirem from medical imaging alone as they are from natural background radiation. The U.S. Food and Drug Administration regulates the permissible dose for an individual diagnostic or research procedure. A single organ dose for an adult must be kept under 3,000 millirem and an annual total dose must not exceed 5,000 millirem. For children, a single dose can come to no more than 300 millirem and the total annual dose, no more than 500. A set of dental X-rays can give you up to 39 millirem, and a gastrointestinal diagnostic series or a cardiac catheterization, 2,000 to 10,000. Patients who receive external beam radiation from X-rays are not radioactive afterward. Internal radiation therapy—injected or implanted by means of capsules inserted into the body—provides a dose that continues until the radionuclides decay or are excreted. Doses that target organs can be much higher than whole-body doses. Killing a cancerous tumor can require a beam of radiation that delivers 6 *million* millirem to the affected tissue and a whole-body equivalent dose that is much lower. Some radiotherapy patients receive 7 to 14 million millirem in fractionated doses. One of the most common radiation therapies is radioactive iodine treatment of the thyroid gland, which wraps around the throat, as a cure for certain diseases, including thyroid cancer. The patient receives 10 million millirem to the thyroid and about 20,000 millirem to the rest of the body. Until the radioisotope decays, patients who receive internal doses of radionuclides can expose others who are near them to mild radiation and set off alarms in places like New York subway stations and the Midtown Tunnel, where radiation detectors are set up to catch terrorists.

Decaying medical isotopes excreted by patients have contaminated all of our sewer systems. In fact, a nuclear plant outside Phoenix that uses treated sewage water in its cooling towers had to get its license amended because of radioiodine in the water that made it appear that effluent from the plant was contaminated.

I asked an old friend, Charles Steinberg, M.D., an internist and diagnostician at the New York Presbyterian Hospital–Weill Cornell Medical Center, what his job would be like without access to nuclear medicine. He usually talks rapidly and articulately, but he was so taken aback by the mere thought that he actually couldn't speak for several seconds. When he recovered, he said, "Practicing medicine as we do today without these tools would be unimaginable!"

Numerous life-prolonging diagnostic and therapeutic procedures rely on uranium. We have the technology in this country to produce medical isotopes of very high quality, but instead we import 90 percent of them, because, thanks to funding cuts and successful litigation by antinuclear groups, appropriate reactors have been shut down. We now export a hundred pounds of highly enriched uranium annually to close allies, who use it as reactor fuel or bombard it with neutrons to make isotopes that are then shipped back here. (Other target elements are also bombarded to make a variety of medical isotopes.) A physician in the nuclear medicine department at New York Hospital told me he blames this outsourcing on the tremendous amount of misinformation circulating about all things nuclear. "People quote so-called studies, but in reality they're hypotheses that somebody came across in a chat room or an online forum," he said. "People want the benefits of nuclear medicine, but they don't want the supposed risk of making the isotopes."

The permissible dose limit from medical and other artificial sources set by radiation protection agencies for the public in the United States is 100 millirem per year of continuous exposure and infrequent exposure of up to 500 millirem a year. Then there are exposures in the workplace, which made up a very thin slice of Rip's pie chart. According to the National Council on Radiation Protection and Measurements (NCRP), open-pit uranium miners (if any were still working today in the United States) would get an estimated annual dose of 21 millirem. An X-ray technician may receive around 500 millirem per year. The ones I have talked to wear dosimeters, and when I've asked them if a reading has ever shown up in the course of their work, the answer is always no. Present Nuclear Regulatory Commission standards permit nuclear workers to receive up to 5,000 millirem per year in addition to natural background radiation. Still, the average *actual* exposure of a nuclear power plant worker comes to only a fraction of the permissible dose—that is, between 180 and 240 millirem per year.

"People ask how to avoid excess radiation," Rip said, and as he began to speak about a ubiquitous man-made source he gestured toward the truck stop's smoking section, with its column of haze. "They could avoid tobacco."

Smoking releases radionuclides accumulated by the tobacco leaf from soil and phosphate fertilizers. Smoke of any kind picks up radionuclides and makes them available to lungs. Probably unknown to the originators of the smoke—and those close enough to inhale it—secondhand tobacco contains radium, radon gas, and other short-lived radium daughters like polonium-210 and lead-210 that are released when tobacco is burned. According to a report by the NCRP, the dose to the thin lining of the bronchial airways is 1,300 millirem per year for a pack-and-a-half-a-day habit, and according to a report by the National Academy of Sciences panel on the biological effects of ionizing radiation, the exposure is as high as 8,000 millirem a year. That's the equivalent of eight hundred chest X-rays. Two packs a day adds up to 16,000 to 20,000 millirem per year.

Dr. Raymond Guilmette, past president of the Health Physics Society and currently the team leader for internal dosimetry at Los Alamos National Laboratory, spent over twenty-three years as a radiation toxicologist at Albuquerque's Lovelace Respiratory Research Institute studying the effects of inhaled radioactive substances. He doubts that radioactive materials in cigarette smoke are a strong contributing factor to tobacco-

related lung cancer. "Both radiation and cigarette smoke are weak carcinogens, by amount," he told me, and pointed out that far greater
amounts of chemically toxic smoke products get deposited throughout
the conducting airways of the lungs than do radioactive products. "It's all
a matter of dose, and the dose of tobacco products is much greater than
the radiation dose."

"Then we have coal combustion," Rip continued. "One to 4 millirem
per year to the lungs." That's not just from the Four Corners power
plants that burn coal strip-mined from the uranium-rich Colorado
Plateau. Every year in the United States alone, coal-fired plants concentrate in their fly ash hundreds of tons of uranium and thorium and their
daughters. Every year enough uranium is concentrated by coal combustion to fuel all of our nuclear plants. "Coal-fired plants expel their
radioactive by-products into the environment and expose people to anywhere from one hundred to four hundred times more radiation than
nuclear plants do. Nuclear plants by law have to contain every speck of
their waste. Worldwide, a person on average gets less than 0.01 millirem
a year from nuclear power—the equivalent of eating one banana, which
contains a radioactive isotope, potassium-40.

"In the United States, if you lived within fifty miles of a nuclear plant,
you'd get an *estimated* trace exposure of 0.009 millirem per year. That
would be like half of a single grain of salt on your hash browns. In reality,
you wouldn't be likely to get any additional exposure because all the
material is shielded. Look, 0.009 millirem is a standard estimate that's
been set up and nobody is really sure about it. When you get into measurements of thousandths of a millirem, which is a very tiny fraction of
the 360 or so millirem that you're getting a year total, it's impossible to
tell how correct that figure is, because of the overwhelming amount of
exposure we receive from natural background and medical radiation.
You'd have to live next to a nuclear plant for over two thousand years to
get the same exposure as you get from a single X-ray. That estimate
about exposure to nuclear power in the States includes all man-made
nonmedical radionuclides."

He went on to describe a character who would pop up now and then
on the nuclear tour. "The EPA has a standard some of us like to call
Fencepost Man—a hypothetical individual that you assume lives all year
on the boundary of a nuclear site and grows all his food there and draws
all his water from its wells. The people running the site have to make
sure that it exposes Fencepost Man to no more than 15 millirem a year.
In reality nobody lives the life of a Fencepost Man in this country.
Nobody lives on the uranium tailings piles we saw and grows all his food

there, for instance. People do build houses next to mountains of coal fly ash in West Virginia, though, and they're getting a dose."

My thoughts drifted to the cancer deaths of people I'd known. "Do you think just living in Albuquerque, what with all the bombs on the base, and Los Alamos just up north, could increase someone's risk of cancer?"

"One, uranium enrichment and manufacturing of the pits—the fissile cores of nuclear weapons—and the bomb triggers didn't happen in New Mexico. Bombs in fortified storage don't release any radioactivity. The radiation emitted by reactors is stopped by concrete several feet thick. By law, every reactor in this country has to be enclosed by a containment vessel inside a containment building.

"Two, even if there had been a radioactive release, it would be very difficult to trace it as the cause of an individual cancer, since there could be so many factors, external and genetic. The body doesn't discriminate between natural and man-made radioactive sources. There's no signature in a tumor saying what started it. Could be environmental, could be genetic. So you look at what's happening in a large population—is a large group showing an increase in cancer? And are any of those cancers the kind that are radiologically caused? We know that uranium miners had excess lung cancer and can say with some confidence that in most of them it was the radon plus smoking. On average, a third to a quarter of all humans in any population get cancer. So you would look at whether there was an increase above that. The cancer rate in New Mexico happens to be much lower than the national average although natural background radiation is higher than in most states. The highest rates of cancer are usually around centers of heavy industry, especially chemical and petrochemical facilities. There's an area along the Gulf Coast called Cancer Alley for that reason. We don't have much heavy industry here, except where there's fossil-fuel extraction. For radiation to begin to damage cells enough to rapidly produce noticeable health effects, exposure would have to go *way* up from the figures we've been talking about. A person would have to receive *20,000* millirem within a few days. An exposure over a short time to 100,000 millirem will kill a person. But cases of exposure like that are very rare. You'd never experience them in daily life. We're talking about Hiroshima, Nagasaki, firefighters inside the reactor building at Chernobyl who got 80,000 to 1.6 million millirem within hours, a few medical accidents."

I asked Rip about newspaper accounts that cesium, an element made by nuclear fission, had contaminated the streams and plants around Los Alamos.

Rip's tone was patient. "I bet you would find some cesium right out-side here on the Zuñi Uplift, too, and in New York City. Some of that cesium came from fallout from atmospheric nuclear testing, and some came from the Chernobyl explosion. But that cesium is putting out only a tiny fraction of the radiation being emitted by the volcanic soil and rocks around Los Alamos. OK, so a detector can find minute traces with a cesium signature, but the cesium in the soil, though man-made, is less radioactive than the geological formation. This has been studied and studied, and there just is no demonstrable health risk. I suggest you look at information in peer-reviewed scientific journals—that's the best source of accurate information about these matters."

When the radioactive plume from the Chernobyl fire blew over Ore-gon, the EPA told people not to drink any rainwater because it might contain radionuclides from the reactor in a certain quantity per liter. But a liter of organic milk naturally contains at least twice that amount of radionuclides, and a liter of salad oil naturally contains about ten times the amount of radioactivity as the traces from the Chernobyl plume.

What else might be radioactive here?

Rip was musing aloud. "There's natural radon, radium, and uranium in these glasses of ice water. It probably came from a local well. There are showers available at this truck stop, so people could also pick up a small dose that way."

"Potassium-40 in the banana in the cream pie," Marcia said. "The potatoes and carrots we just ate contain potassium-40 and radium. I saw stewed apricots at the buffet table—they contain potassium-40."

"Yep," Rip replied. "My chicken-fried steak—any red meat—does, too. I didn't see any lima beans at the buffet, but those contain almost as much potassium-40 and radium as Brazil nuts." He glanced around at the other tables. "I don't see any shakers of No Salt, that salt substitute, but some cafés have it. Potassium chloride. It's the most radioactive thing you'll find in the supermarket—way above natural background radiation. If you work with anything nuclear, every year you're required to take a course in radiation safety. The instructor uses a radioactive source for teaching purposes. But post-9/11 safety regulations forbid uncontrolled movement of radioactive materials. So instead of bringing some yellow-cake or something like we used to have in class, he went to the supermar-ket and bought salt substitute. Theoretically, by transporting it he was breaking the law. Anyway, it worked just fine and set off the detector like it was supposed to."

The list kept expanding. Americium-241 in the smoke detectors. Radon and radium from the natural gas that had cooked our dinner on the restaurant stoves and from the natural gas furnace that warmed the room on this chilly afternoon. Household natural gas exposes a person on average to 9 millirem per year. Wallboard and gypsum—building materials—contain uranium, thorium, and potassium-40. Around here there were plenty of gypsum and cement plants. We were being exposed to radon from the concrete floor and the walls, their cinder blocks being made from fly ash from coal-fired plants. Thorium from the Brazil nuts in packages of mixed salted nuts by the cash register, tritium—a radioactive isotope of hydrogen—in the glow-in-the-dark dials on wristwatches worn by some of the patrons, tritium in the luminous exit signs over the doors . . .

I glanced over at the hiker. If the book he was reading was, say, an old edition of *Walden*, the binding glue, which used to be made from animal bones and hooves, was radioactive.

"And let's not forget the mildly radioactive thorium, radium, and polonium isotopes circulating in the blood that are giving off eleven or so millirem a year," Rip said as we got up to leave. "And the potassium-40 in our bodies—it has a half-life of 1.26 billion years and emits gamma rays."

Right now, our own internal irradiation, from isotopes that are part of the makeup of our living tissues, is giving us a dose of about one-third of what we receive from rocks and soil. Right now, in the upper atmosphere, as you read this, cosmic rays are colliding with nitrogen atoms and producing a radioactive isotope, carbon-14. It makes its way through the air and into plants we eat and into our lungs the second we draw our first breath. We keep cycling carbon-14 through our bodies, and some stays there in equilibrium until we stop breathing for good. The amount remaining and its rate of decay allow archaeologists to estimate from organic remains the approximate year of death.

In the parking lot, we inhaled exhaust from the burning of diesel by the trucks idling there or whizzing past on the interstate. Every time fossil fuels or any other organic materials, like wood, are burned, they release radionuclides. Plenty must have been issuing from the tailpipes of a huge, idling flatbed truck with a drill rig on the back. It was either going to or from the oil patch in the San Juan Basin. "That guy may have a radioactive logging source," Rip said. "It's used to measure the properties and types of rock and soil in a drill hole to learn whether there's oil or gas or water where a well is being dug. Also, with drilling, the pipe that turns the bit collects radionuclides down in the ground. On the surface of a drilling site you find radioactive water."

As we rode off into the starry night, I had a lot to consider. Learning that radiation was such an intimate and pervasive part of all of life, all of nature, I felt reassured, as if the world was indeed suited to us and we to the world—that indeed, as the poet Rainer Maria Rilke wrote around the time Marie Curie was doing her work, we have been "set down in life as in the element to which we best correspond." If exposure from nuclear power generation is extremely tiny compared to all the other sources shedding particles and rays that collide with our atoms, if the human body has been successfully coping with natural background radiation, some of it quite high, for hundreds of millennia, if today our second biggest exposure is from medical sources, then isn't it more reasonable to worry about health risks from coal-fired electricity generation rather than from the other reliable, relatively large-scale supplier of electricity, nuclear power?

We exist in a storm of energy, of particles so fine and waves so subtle that we can't perceive them. They rain down from the sky and well up from rocks and dirt, from the desert floor and from the remotest of canyon walls. Radioactive minerals wash into the rivers and oceans and concentrate in the seabed; they turn up in yogurt, wheat germ, granola, acorn squash, oranges, and soy protein. To avoid radiation, you would have to leave your own body and you would have to leave the universe.

5 UNDARK

RIP EVENTUALLY BEGAN TO PLAN our next trip—a tour of experimental reactors. In the meantime, I set about learning more about the health effects of radiation. He provided me with literature, I looked up other documents, and I talked to specialists with many years of experience. If I was going to continue the nuclear tour, I had to know more about the exact risks radiation could entail.

Nature does all it can to harm us while at the same time doing all it can to heal us. Her ionizing radiation can change cells by shattering the electron bonds that hold molecules together; it can damage a DNA molecule by knocking electrons out of it or out of some other molecule that interacts with the DNA. The direst result is either cell destruction or a mutation whose consequences lie in the future. Having evolved in a bath of radiation and toxins, our cells come equipped with several defenses. They differ from cell to cell, tissue to tissue, and system to system, and their effectiveness depends on the type and dose of radiation, the magnitude and duration, the part of the body exposed, and the sex, age, and physical condition of the recipient.

Years ago, when I'd acquired the conviction that all radiation was terribly dangerous, I made no distinction between high-level and low-level exposure. Most radiation biologists, radiologists, and biophysicists do not consider the exposures we receive as we go about our daily lives harmful. Health effects appear at a dose level that's much higher.

The largest group in the world that has received radiation greater than the normal level consists of people who have been treated with nuclear medicine. Millions have benefited from the early detection of ailments thanks to X-rays and computerized tomography (CT) scans and from radiation treatments. But what was the worst that high-level radiation unleashed by war, accident, or ignorance could do to a large population? At what level did a higher dose become lethal or cause serious illness? A great deal of that information has been gathered from tragedies borne by a few relatively sizable cohorts: radium-dial painters in the 1920s; residents of Hiroshima and Nagasaki in August 1945; uranium miners of the

Colorado Plateau in the 1950s and 1960s; and workers and firefighters at the Chernobyl reactor when it burned in 1986. Were the consequences so grave that it was pointless to consider nuclear fission as a significant way to reduce greenhouse gas emissions? Was the worst scenario regarding radiation more catastrophic than damage to public health and to the environment from our largest baseload resource, fossil fuels?

The first radiation casualty was probably Marie Curie. She coined the word *radioactivity* from the Latin verb *radiare*, which means "to emit rays or glow," and comes from the noun *radius*, which means "spoke of a wheel" or "ray of light." At the time, Marie Curie was a young chemist from Poland working on her doctoral thesis in physics in Paris. In 1897, experimenting with uranium rays, she determined that they behaved consistently no matter what the physical or chemical state of the uranium was, and so concluded that the emanations had to be at the level of the atomic structure. She also found that the more uranium a compound contained, the more intense its radioactivity. Her discoveries revolutionized physics and won her a Nobel Prize in Physics (she was later to receive one in chemistry as well).

She began testing minerals for their radioactive properties and, with her husband, Pierre Curie, a fellow chemist, originated chemical methods of isolating elements and discovered two new radioactive ones, radium and polonium, for which they shared a Nobel Prize in Physics. While teasing flecks of radium out of tons of pitchblende in a lab set up in a poorly ventilated shed in a Paris courtyard, she found that radioactivity burned the skin. No one knew the degree of harm this indicated. Over time, substantial exposure to beta particles and gamma rays damaged her fingers, and she started to rub them constantly. She also developed other symptoms caused by external exposure to radiation that radium and its daughters emitted, accidentally inhaled radon gas that compromised her lung cells, and inadvertently ingested enough radionuclides to affect her bones, bone marrow, and, probably, her kidneys.

The discovery that X-rays could be used for medical purposes inspired her to bring this new technology to the battlefield during World War I. Organizing a fleet of vans outfitted with X-ray machines, she headed for the front. The equipment, very primitive in comparison to today's, was unshielded, and she unknowingly received considerable exposure from these as yet poorly understood rays that magically permitted physicians to locate and extract bullets and more successfully treat wounds and fractures. Her body burden of radioactivity must have been even further increased when she trapped radon in small glass tubes that were implanted in the wounded soldiers' bodies to heal them. She devoted much of the rest of her career to medical uses of radioactivity, and for this she became internationally revered. After Pierre observed that radiation damaged tissue, physicians began to expose tumors to radioactive materials and to treat many other diseases with them as well.

While experimenting with the helpful aspects of radioactivity at higher levels than natural background radiation, the Curies suffered from its assaults. Pierre's fingers became covered with radiation-induced sores and he was racked with pain until his death in a street accident. Plagued by fatigue and cataracts, Marie died in 1934, at the age of sixty-six, from blood abnormalities mistakenly diagnosed as pernicious anemia. Their daughter Irène, a scientist who continued her parents' research, died in her fifties of leukemia, now known to be linked to exposure to high-level radiation.

Picture, in the late 1910s and early 1920s, big, high-ceilinged rooms around the country full of young women at worktables applying paint made luminous by radium to dials for watches and clocks. Companies were manufacturing many radioactive items, like radium corsets and radium water, to promote health, and a magical liquid called Undark that you could use to repaint your own watch or clock dial. Maybe you would do that once or twice in your life. But the women dial-painters, some of them disabled and unfit for other jobs, kept at it all day long, year after year, doing delicate work that required a steady hand and a camel's-hair brush as finely pointed as possible. They did what painters have always done—they used their lips or tongues to draw the brush hairs together. Sometimes, for fun, the women would apply the glow-in-the-dark stuff to their lips.

A New Jersey dentist whose patients were employed at a watch factory noticed that some of the women had developed ailments of the mouth. He concluded that the paint was the culprit. Physicians from universities began to investigate and soon established a link between the radium in the paint and the illnesses—mainly bone and nasal cancers. The factory managers explained the risk to their employees, and they took precautions about the paint. In 1929 investigators from the Department of Labor surveyed thirty-one plants involved in dial painting or in the manufacture of substances containing radium and found thirty-two fatalities related to radiation. Nineteen other people had been poisoned by radioactive materials that made them ill. Thirty-three of the fifty-one cases were dial-painters. All of them had ingested radium.

The whole group, approximately two thousand people employed over a period of sixteen years, wasn't the largest to be exposed to radium from a source greater than natural background levels. For instance, also in the early part of the twentieth century, the public consumed four hundred thousand bottles of a radium-laced mineral water, with some health enthusiasts drinking several bottles a day; the founder of the company that produced it perished from radiation-related disorders. However, the dial-painters were the first radiation casualties in the world to be studied in depth, because they could be easily tracked, thanks to detailed employment records. Moved by the plight of the "radium girls," researchers continued to follow their cases for the next seven decades.

Epidemiologists examining all relevant data from cloths used to wipe brushes, from various human samples, and from exhumations in order to make estimates about the amounts of radium different victims had ingested finally arrived at an important discovery. Calculations proved that at least half the time the greater the amount of radium ingested, or the greater the woman's chronic proximity to the source, the higher the risk of cancer. This led to a better understanding of how to establish radiation protection standards.

Ultimately, some women carried radium in their tissues for forty or fifty years. During World War II many continued painting luminous dials for use in military aircraft. Federal radiation protection standards that arose from data collected in the field were firmly enforced and no new cases appeared. These regulations have been revised over the years as methodology has improved and knowledge, based in part on findings about the chronic exposure of the dial-painters, has increased. In 2006, the last surviving "radium girl" celebrated her one-hundredth birthday.

In August 1945, the United States, seeking to end the war with Japan, detonated atomic bombs over Hiroshima and Nagasaki. The exact number of people in those cities when they were attacked remains unknown, so it's difficult to ascertain how many were killed by the blast and how many from the flash of gamma rays and neutrons. The explosion would have instantaneously destroyed those closest to the hypocenter, and large numbers of people would have also died in fires and building collapses. The total number of deaths probably came to between 150,000 and 222,000. According to a national census in Japan in 1950, about 280,000 survivors stated that they had been exposed to radiation. For sixty years, 120,000 of them have been studied in joint projects of Japan and the United States, first by the Atomic Bomb Casualty Commission (ABCC) and for the last thirty years by the Radiation Effects Research Foundation (RERF).

To learn more, I turned to Evan Douple, a genial, bearded man who holds a doctorate in radiation biophysics and specializes in research in radiation carcinogenesis and in the molecular mechanisms of radiation-induced damage to DNA, cells and tissues, and, in radiotherapy, tumors. For twenty years he was a professor of medicine, pharmacology, and toxicology at the Dartmouth Medical School and director of its radiobiology laboratories and biomedical research program in the Norris Cotton Cancer Center.

"I was trying to understand radiation damage and improve the use of medical radiation to treat cancer," he told me. Then in 1992, he was recruited by the National Academy of Sciences (NAS) to join the National Research Council, in Washington, D.C., where he became director of the Board on Radiation Effects Research in 1997. One of the board's duties is to advise federal agencies, the public, and Congress on scientific issues and another is to oversee studies at the National Academies on the effects on humans of nonionizing radiations (such as radio-frequency electromagnetic radiation and extremely low-frequency electric and magnetic fields) and ionizing radiations.

The fifty thousand survivors who were closest to the atomic explosions and therefore received the largest doses of radiation have suffered the highest rate of deaths from leukemia attributable to radiation, Douple told me. Excess cancer deaths—that is, a greater number than in a comparable, unexposed group—continue to occur among these people. Those who were farther away received a lower dose and have had fewer cancer deaths. "Some were far away from the hypocenter and got doses of radiation so very low that for all practical purposes they're a control group," he said.

He paused. His voice softened. "The bombings were terrible events that we hope are never, never repeated," he said.

He went on to explain that a sound body of information had been accumulated, thanks to the fact that the survivors were from families that had resided in Hiroshima and Nagasaki for generations. The Japanese tend to live where they were born and the government traditionally has kept detailed records on all households and their occupants. "That helped make a database for the study that was very strong."

He continued. "Most people who were close enough to the bomb fireball to receive a lethal radiation dose from the prompt radiations probably were killed by the blast pressures and heat. Since the Americans deliberately exploded the atomic bombs 1,650 feet above the cities to maximize the destructive power of the blast, this minimized radioactive contamination of the ground." The fireballs were thrust higher by updrafts. About 10 percent of the few kilograms of nuclear material in each bomb turned from matter into energy and the remaining 90 percent shot up even higher with the ascending fireball. Most of the vaporized uranium, used in the Hiroshima bomb, and plutonium, used in the Nagasaki bomb, dispersed downwind along with other radioactive products. For these reasons, even the earliest measurements in the two cities did not detect high levels of radioactivity, and today levels are so minuscule that they're difficult to distinguish from trace amounts left by fallout from weapons tests. In effect, the cities were not contaminated with high levels of radioactive elements and the major radiation doses came from gamma rays, electrons, and neutrons emitted at the moment of the explosion.

Often missing in studies of exposed populations is precise information about the dosimetry—the measurement of the amount of radiation deposited in the body. "The atomic bomb has been very thoroughly analyzed in regard to the radiation it delivered and that has made possible a relatively accurate estimate of the dose," Douple said. "Survivors remember quite well where they were when the explosion happened, and because of that RERF scientists could calculate the doses survivors received. Most other exposed populations that have been studied also have dosimetry, but not as thoroughly reconstructed. But let me add the downsides. One: this is a Japanese population that was exposed while suffering from the conditions of wartime, with its particular characteristics, and some people claim that certain results wouldn't necessarily apply to other populations, so that the information from the results of the study might not be universally applicable. Two: in Japan the exposure was instantaneous, whereas other exposed populations around the world— such as radiation workers and those exposed to fallout—have usually experienced protracted, or fractionated, exposures, or lower doses delivered over an extended time."

He spoke of the many misconceptions people have about what happened in Hiroshima and Nagasaki. "I've always felt strongly that there needs to be a better effort to communicate technically complex scientific issues with the public. We scientists don't always make that extra effort. I've spent many years trying to inform people about the facts of radiation risk to humans. Most people think that the population received widespread, incredibly high—unrealistically so—doses. But most survivors did not get the highest doses—those who did were likely killed by the heat and pressure shock of the blast. The farther people were from the hypocenter, the lower the dose. Five thousand feet away from the hypocenter, the dose was low enough to permit survival. People tend to assume that all the survivors of the bombs died soon after the attacks. Of the original cohort selected for study more than fifty years ago, more than 40 percent are still alive today."

In 1945, those survivors were infants, young children, or teenagers. Now they're entering their cancer-prone years. "The next twenty years will be very crucial to understanding health effects in survivors exposed as children compared to those exposed as adults," he said. "As they go through this latter stage of their lives, measuring the incidences of cancer is so important. Fortunately, Congress has understood this importance and has decided to keep contributing to the RERF research."

Douple had been undermining a number of my long-held beliefs, and now he told me an astonishing fact. "The numbers of cancers in the survivor population are much lower than most people would probably have expected and low compared to the nonexposed general public. This suggests that radiation is a relatively weak carcinogen, not a strong one." He went on to say that in a population of about 80,000 atomic bomb survivors studied who were living in the city, there have been 13,454 observed solid cancers. "This is an excess of only 853 compared to the number expected in control populations, an attributable risk of about 6 percent increase."

Douple debunked the idea that radiation from atomic bombs has caused an abundance of genetic mutations. "There have been tales of three-headed cattle born near the atomic test site in Nevada, that kind of thing. A lot of those possibilities have been clearly ruled out by science. Among Japanese children born to one or both parents who are survivors, there has been no observable increase in defects or abnormalities appearing in the children in the first generation to suggest the appearance of visible results of mutations."

As a believer in plant mutations around Three Mile Island, I had to ask, "Where did this notion of mutations come from?"

"Mutations have been clearly demonstrated in plants and animals after high doses of radiation and the results of mouse experiments suggested that there might be mutations appearing in the offspring of the survivors," Douple replied. The Human Genome Project, which became a logical extension of the U.S. Department of Energy's interest in determining how human DNA reacted to radiation, coupled with some additional investigations of mouse reproduction and genetics, enabled scientists to pinpoint some differences that distinguish man from mouse. Genetic effects may occur, but they might be so small that they're undetectable.

An intensive investigation facilitated by the NAS, with funding from the U.S. DOE and Japan, of the biological effects of ionizing radiation on nearly seventy thousand offspring of atomic bomb survivors is continuing to look for potential increases in cancer or genetic effects.

The death toll from Allied bombing during World War II far exceeds the number of deaths, and the pain, suffering, and ruin caused by the atomic bombs in Japan, but the horror these strange new devices unleashed with their huge thermal and compression blasts left deep scars. Douple spoke of a group of Japanese survivors who received a high-enough thermal exposure to produce gruesome damage to skin. The atomic bomb differs from conventional bombs chiefly in its grim efficiency—a single weapon can level a city. He pointed out that the effects of an atomic bomb aren't necessarily nastier than those of a big conventional bombardment, except in its potential to increase cancer rates by a relatively small percentage over the many years in the lifetime of the people exposed.

"Why do you think people are so fearful of radiation?"

"You can't detect it with the human senses—can't feel, see, smell, taste it," he replied. "Radiation has a mysterious aura that you don't encounter regarding most other threats. You can see anthrax powder. Well, you can see a suspicious powder and then test it to find out if it's anthrax. The same is true about most other nasty agents. And there's another problem: to really understand radiation and its effects in people, you need some education in very basic elementary physics and a little math, but even in good secondary schools today, people avoid physics and get their science credit by taking advanced biology."

Jon D. Miller, director of the Center for Biomedical Communications at Northwestern University, is a political scientist who conducts research about public knowledge of science. He has found that one in five Americans thinks that the sun revolves around the earth and that only one in ten knows what radiation is.

Douple, concerned about wrong notions and skewed risk perceptions that abound in public discourse, tries to encourage medical students to communicate better with lay people. At an annual lecture he delivers at the Dartmouth Medical School, he begins by describing a woman who is worried about the radiation that she'll receive at her mammogram appointment. In her nervousness, before leaving home she eats a peanut butter sandwich, drinks several cups of coffee, and smokes two cigarettes, and as she makes the five-mile drive to the hospital, doesn't use a seat belt. "She's already accepted five risks greater than the mammography exam and she's accepted carcinogens—from the cigarettes and from the aflatoxin, which is a liver carcinogen, in peanuts—that might be far more likely to cause cancer than radiation. Coffee contains caffeine, a known mutagen. If ever we are going to put people's minds at ease and get ourselves to the middle of this century with enough energy—that is, with a return to the concept of nuclear power as an important component to reduce pollutants and global warming while providing a solution to our energy problems—then the public has to realize that there are things we should elect to do for good reason and we should all try to understand the benefit-to-risk ratios for particular choices and for comparing alternative choices."

Surprised to learn how relatively normal the health of the majority of atomic survivors was, I asked Douple what the biggest surprises of his career had been.

"When I came to Washington, D.C., I was shocked to learn that the newspaper editor who writes the headline is not the person who writes the story. Headline-writers do not even have to think for one second. They try to sensationalize, spin, and choose words that will catch the attention of the audience. And with the level of digesting information in this country, some people don't get beyond the headlines or the twenty-second 'story' in the TV news. That was a shocker. The second surprise was how Congress can make policy based on so little science. Congress will require agencies to take a stand or a direction when the actual information that the members have received to base that position on is often nonscientific and very weak."

Around the globe, the estimates based on atomic-bomb survivor data have been the basis of radiation protection standards for radiation workers and the general public. The bomb-survivor data is also used to calculate risks associated with exposures of populations and individuals to accidental radiation releases. Other studies have reinforced the RERF's findings. The National Council on Radiation Protection and Measurements (NCRP) and the International Commission on Radiological Protection promulgate recommendations for these science-based standards, updating them periodically. That's the science. Enter politics.

"The Congress passed legislation for the Radiation Exposure Compensation Act (RECA)," Douple said. "It contains a list of cancers, and if you get one of them, and you happen to have lived in certain counties in Utah, or some other areas in proximity to the Nevada Test Site, you are eligible to be compensated by the federal government. It doesn't matter whether you're a cigarette smoker. If you have one of several cancers, you qualify for compensation. RECA is a relatively old act. Politicians keep amending it and adding more cancers. But fallout does not recognize county and political boundaries, and we know that fallout from the Nevada Test Site came down over the entire United States, particularly in areas of high rainfall. While almost every cancer has been shown to be elevated in populations if they're exposed to high radiation doses, the probabilities of inducing cancers at low doses are quite low, especially compared to the background incidence of cancer that is presumably due to causes other than radiation exposure. With respect to radiation induction, leukemia tends to be a more sensitive cancer. But it's rare to start with, and the absolute risk for leukemia following radiation exposure is still quite low. There is also a latency period, or time delay, associated with radiation-induced cancers; cancers of the blood—leukemias—might appear five to ten years after exposure, but most solid cancers can take twenty to forty years to develop."

The very worst that can happen to a nuclear power plant occurred in 1986, when the most poorly designed and risk-prone of reactors had a core meltdown. The Chernobyl facility in Ukraine, then part of the Union of Soviet Socialist Republics, overheated and failed because of wrong decisions made by poorly trained operators. The very thought of this accident sends a shiver through most people.

I tracked the ghastly predictions as the event unfolded and worried that my daughter, in college in Massachusetts, might somehow be exposed to the plume. I assumed that in eastern Europe tens of thousands had died. Shortly after the accident, headlines in the *New York Post* announced that the bodies of fifteen thousand victims had been bulldozed into a pit. The Associated Press claimed that Ukraine said over three million people had been exposed to radiation, seventy thousand in that country alone had been "disabled" by it, and that over a thousand square miles had become "the most poisoned land on earth." In 1995 the Ukrainian Ministry of Health attributed 125,000 deaths to Chernobyl. But were any of these figures correct?

"Chernobyl has got to be the worst man-made accident in the twentieth century," I told Rip.

"Chernobyl was serious, but it wasn't the worst. Remember Bhopal?"

He was referring to the release of a toxic gas in 1984 from a Union Carbide pesticide plant in India that had killed at least 3,849 people, left an estimated 100,000 permanently disabled, and created lingering concern about contamination.

"After Bhopal nobody talked about shutting down all the chemical plants," he said. "If you want to talk about accidents in relation to electricity generation, dam failures in the United States alone have killed far more people outright than Chernobyl did. Chernobyl had a graphite-moderated reactor, the worst design for a power plant. It had been set up to make bomb-grade plutonium as well as electricity." The Soviet design had a built-in instability at low power that doesn't exist in Western reactors. During a poorly conceived experimental test of the electrical system, the reactor crew lowered the power after disabling the automatic shutdown equipment. The supply of water that kept the fuel cooled and stable failed, and an excursion—a sudden rapid rise in power caused by a runaway chain reaction—released enough heat to rupture fuel elements, causing a steam explosion that lifted the cover plate off the reactor and sent fission products flying. A second explosion expelled pieces of hot graphite and fuel elements that landed on the roofs of the plant and set them on fire. Outside air rushing into the hole and through channels in the graphite caused it to ignite as well.

"Here's the most important thing: *there was no containment building,*" Rip continued. "The reactor was basically a pile of graphite blocks in a Quonset hut. People need to know that. Every reactor in the United States has to have a very sturdy containment building. The containment building and the reactor containment vessel kept the meltdown at Three Mile Island from putting crap into the environment. You also need to know that a Chernobyl-type accident could not happen here. Chernobyl did not have to happen. It's an example of the worst design, worst engineering, worst shielding, worst safety management practices, and worst weather conditions—the wind blew every which way. The guy running the place was a political appointee who didn't know shit and had been responsible for a reactor accident at another plant. You had people making decisions who were so damned ignorant of the risks needing monitoring in a power plant that they didn't know that if you suddenly lost coolant in a reactor of that design you'd wind up with an uncontrolled chain reaction."

When a reactor core splits uranium atoms, it makes fission fragments, such as cesium and other radionuclides. When the Chernobyl reactor core melted, 13 to 30 percent of its 190 metric tons of nuclear fuel and fission products evaporated. And then the explosions shot them into the atmosphere, where the plume of radionuclides spread out at an altitude of 3,300 to 36,000 feet. Over 77,000 square miles were contaminated to varying degrees, and over 70 percent of that area was in the three most affected areas, Ukraine, Belarus, and Russia. Most of the radionuclides had short half-lives and have now decayed away. However, some remain.

"The radioactive contamination was much worse than in Hiroshima and Nagasaki," Rip said. "Chernobyl spewed a hell of a lot more radioactive materials over the landscape than the A-bombs in Japan did. Even so, the damage to health and the environment was a lot less than anyone thought it would be."

Boris Scherbina, the Soviet official handling the crisis, flew in from Moscow. His first move was to refuse to evacuate some forty-eight thousand civilians at risk, because he considered the panic that would ensue more worrisome than any radiation danger. Grigori Medvedev, who had been one of the directors of Soviet nuclear power plant construction, describes in his book *The Truth About Chernobyl* an astonishing moment in which the bureaucrat, who alone was empowered to decide on the fates of so many and to classify the accident as a nuclear disaster, arrived at the scene. "He still did not realize that the air around him—in the street and inside the room—was saturated with radioactivity, emitting gamma and beta rays that penetrated whoever happened to be in their way—ordinary mortals, Scherbina, or the devil himself." Eventually Scherbina did understand the enormity of the crisis and began ordering workers and dignitaries alike to start filling sandbags that helicopter pilots then dropped on the crater where the reactor had once been. All of these courageous workers received large and sometimes fatal doses of radiation. Finally, over a period of three days, evacuations took place. The reactor burned for ten days.

Fred A. Mettler Jr., who has a master's in public health in addition to his medical degree, spends almost all of his time saving lives, one way or another. In addition to his day job as a radiologist at the Veterans Administration regional hospital in Albuquerque, he's the U.S. representative on the United Nations Scientific Committee on the Effects of Atomic Radiation (UNSCEAR) and holds the title of Academician of the Russian Academy of Medical Sciences. He also serves on the Centers for Disease Control (CDC) committee on guidelines for terrorist incidents involving radioactive materials and has held high posts on the NCRP and with the Department of Homeland Security Working Group on Radiological Dispersal Devices, and he composed a key report on radiation protection in the event of terrorism involving radioactive substances. In addition to turning out hundreds of articles for scientific

journals and writing seventeen books on nuclear medicine, medical management of radiation accidents, and the medical effects of ionizing radiation and radiology, Mettler has also won a number of awards and a listing
in *The Best Doctors in America* in the categories of nuclear medicine and
radiation accidents and injuries.

Mettler, a cheerful, tall, fit-looking man in a dark blue pullover and
chinos, met me at the Department of Radiology at the University of New
Mexico Medical Center. He had just returned from meetings at the UN
International Atomic Energy Agency in Vienna.

We entered a cubicle that had a computer and a wall of shelves
crammed with mementos, tomes, bound journals, and a copy of the latest
edition of his own book, *Medical Effects of Ionizing Radiation,* co-authored
with Robert D. Moseley Jr.

One shelf held the Mettler museum of everyday radioactive sources
and included red FiestaWare dishes and old Coleman lantern mantles
that resembled silk socks for dolls. Enclosing the gas flame, a mantle
fluoresces a brilliant white light. This glow once came from a thorium-
rich solution; the mantles now contain the nonradioactive yttrium oxide
instead, but imports from Asia can still wake up a radiation detector.
There was a vial of potassium chloride, the mildly radioactive salt substitute, and a piece of trinitite, the sand fused by heat from the first atomic
bomb, detonated a few months after Mettler was born. An object resembling a small tambourine held together with masking tape was connected to a little box with a digital readout. He turned it on and pointed
it at a flame-colored saucer, and numbers appeared on the display
screen.

"At Chernobyl, I swapped the radiation sensor I was carrying in my
pocket for this clunky device that a worker had made," he said after offering me a breakfast Pepsi and having one himself. "He was happy with the
trade."

The son of a professor of neuroanatomy, Mettler studied mathematics
at Columbia and as an undergraduate landed an internship with the
Atomic Energy Commission in New York during the era of atmospheric
nuclear testing. He worked at a lab in Manhattan that studied fallout
pumped in from the roof. "Fallout could come from anywhere in the
Northern Hemisphere," he said. "We also checked milk samples."

Eventually he entered medical school and became a radiologist, went
on to get a master's degree in industrial health from the Harvard School
of Public Health, and also took courses in nuclear engineering. After
moving to Albuquerque to teach at the University of New Mexico,
Mettler did research in cancer diagnosis and treatment and eventually

chaired the radiology department of the medical school. Now, in addition to all his other duties, he goes wherever his expertise in radiation effects and radiological accidents is needed.

In 1977 he joined UNSCEAR. Originally founded by the UN General Assembly in 1955 to investigate the effects of fallout from nuclear weapons tests, the committee now assesses problematic sources of ionizing radiation and what it may do to humans. Twenty-one nations contribute scientists and special consultants who, coming from so many different countries, are deemed to be less susceptible to local biases and political partisanship. UNSCEAR is almost universally regarded as the most objective source of public information about radiation and health. When a UN member state suffers a radiological mishap, a joint team from the IAEA and the World Health Organization (WHO) arrives to determine the severity. Mettler has led a number of these teams. The researchers determine trace elements and radionuclides in the general diet and in specific foods, as well as in biological specimens such as hair, kidney, and liver, and in inorganic materials in the environment—coal ashes, rain, lakes, rivers, oceans, sediments, and air. The investigative team then reports to the UN and to the general public. An agency spokesman, Dr. Neil Wald, Professor of Radiation Health, Environmental and Occupational Health, and Human Genetics and Radiology at the University of Pittsburgh Graduate School of Public Health, has said, "It is important that public misperceptions be reduced as much as possible in this area, because unwarranted perception and fear of harm can itself produce avoidable health problems, as well as erroneous societal benefit-versus-risk judgments."

One of the shelves in Fred Mettler's cubicle held a glass box that contained a cloth badge that said "IAEA Chernobyl Medical Team," a medallion with a bas-relief of a helmeted man fighting flames, and a plaque expressing the USSR's appreciation for Mettler's services.

At meetings of UNSCEAR, the U.S. delegates sat next to the USSR delegation. During Mettler's early days on the job—he was then assistant to Robert Moseley, the U.S. representative to UNSCEAR—he attended one conclave devoted to a compendium of sources and effects of radiation. The Soviets presented for inclusion a document containing a claim about the interaction of electromagnetic fields and radiation and their influence on menstrual cycles. When people were invited to comment, nobody said a word. Then Mettler questioned the conclusions and the underlying biology and asked where the rest of the data was. "This was the first time anyone had criticized the study—I don't think anyone had thought about it seriously," Mettler told me. Finally Leonid Andreyevich Ilyin, the head of the Soviet delegation, spoke up: "Comrade Mettler is right. Take it out."

In 1985 Mettler became the American representative to UNSCEAR. The following year, the Chernobyl accident caused the deaths within a few weeks, Mettler told me, of thirty workers and radiation injuries to hundreds of others. The Soviet government evacuated about 116,000 people from the area and later ordered the permanent relocation of about 220,000 from Belarus, Ukraine, and the Russian Federation.

"The Soviets just sat on the whole Chernobyl thing for a few years," Mettler told me. "But by 1988 they'd had so much difficulty that they called a meeting in Kiev, the nearest city to Chernobyl. There was a big controversy among Soviet scientists about what was or what was not happening in the region as a result of Chernobyl. Then in '89, the Soviet Ministry of Health brought in the World Health Organization, which had only one doctor in its radiation branch. WHO got together a team of four and went into the area, spent two weeks, and decided that the physical effects of Chernobyl were no big deal, that there weren't going to be tens of thousands of deaths or acute cases in the next few months. Those guys got tarred and feathered and thrown out by Ukrainian and Belarusian officials who were not pleased that Chernobyl wasn't being declared a major catastrophe of monumental proportions. Then the Soviet Ministry of Atomic Power and Industry asked the International Atomic Energy Agency for help. Not to be outdone by WHO, IAEA spent a year there and sent four hundred people in five teams to investigate environmental and health aspects, potential ways to fix the problems, history, and dosimetry."

After casting around the Soviet Union for someone to head the IAEA teams, the Soviets agreed that even though Mettler was an American, he had a reputation for demanding good science, and they chose him. There was no budget. "Our deal was that we'd get all the free tickets we wanted on Aeroflot to anywhere in the world and that in the USSR they'd drive us wherever we asked and they'd feed us. It took a year to put the whole thing together. I assembled all sorts of additional teams. There was one just to look at cancer data and the cancer registry, and to really study the documents—the death certificates. Other teams were to examine all other kinds of health effects and find who had original data on studies."

For the medical team, Mettler needed a pediatrician, an endocrinologist specializing in thyroid cancer, and a hematologist. He phoned medical societies in several countries to find specialists who wanted a free Aeroflot ticket to the "most contaminated place on earth" and persuaded medical companies to supply free testing equipment and other essentials.

Mettler wanted the study to be as unbiased and as objective as possible. He insisted upon complete access and the ability to collect data from any area he chose—no place could be off-limits. Before leaving for Chernobyl, he pored over a huge, detailed map of the Soviet Union along with a contamination map. The map Mettler now spread out before me had yellow and red contours that indicated the extent of the contamination produced by Chernobyl. The hottest areas appeared in red, the less contaminated ones in yellow: the graphic resembled two irregularly sketched daisies with red centers and yellow petals. The center of the bigger daisy represented the location of the reactor explosion. The smaller daisy, detached from the first one, lay to the north and east of the bigger one. Usually in the region, the wind blows out of the northeast, but on the day of the reactor fire it blew out of the southwest. Then it shifted four different times. It also rained.

"When you put the two maps together, you see how inhomogeneous the event was," he said. "You could be within thirty kilometers of Chernobyl and get no contamination, or you could be three hundred kilometers away and get it. It's not what you would predict. So we know if we have a radiological event in the United States we can't hope for a neat plume."

He had randomly selected villages inside the contamination zone so that he could compare them with the closest villages in an uncontaminated area. "Look here," he said, tapping a dot on the map. "I chose Polesskoe, a village right smack in the middle of the contamination zone and within ten kilometers of a village that was out of the zone. I chose other villages, like Veprin, in the zone, and like Khodosy, outside of it. The reason for doing the comparison like this was so we could determine that the things wrong in the zone weren't due to, say, an iodine deficiency. In that region, iodine deficiency has always been a problem, and it causes thyroid disorders. For purposes of the study, it helped that wind and rain had made the distribution of radioactive material so inhomogeneous."

The Chernobyl nuclear power plant was situated on the Pripyat River, near the now-abandoned town of Pripyat, 110 miles from Kiev in a flat, mainly agricultural region dotted with lakes, marshes, and villages of wooden or stone houses. In 1990, the second phase of the investigation began with Mettler and his group traveling from village to village on a bus.

Upon arrival, the investigators would first hold a meeting in the town hall. Their task was to study real, living versions of the hypothetical Fencepost Man so often used in estimates of radiation exposure. The people in Mettler's study had not only been present when the reactor had burned, they'd also continued to consume local food and water.

"First we asked the people what their concerns were." Mettler shook his head. "They came up with all kinds of things we hadn't thought of! They didn't give a damn about themselves, they didn't mind living in a contaminated place, but they were worried about the future health of their children. We had a physician whose parents were survivors of Hiroshima, and the villagers were happy to see that he was normal."

Lead shielding had been put on the reactor, along with sand and boron to absorb radionuclides, and the villagers wondered whether the fire had dispersed enough lead to poison their children. Mettler, unprepared for that question, contacted specialists at the U.S. CDC in Atlanta.

"They told me to test the blood of two-year-olds, because they're lead magnets. They're always crawling around in the dirt, and they eat it. All over the area, lead levels were higher than here in Albuquerque, probably because they still used leaded gasoline. It turned out that lead poisoning was nothing to be concerned about."

To further ensure objectivity, Mettler wanted to study a cross section of the population and a cross section of age groups. No one in these hamlets had any identification except a name and a birth date. "I'd go to a village and look at the birth registers and ask for the first ten two-year-olds born in this year, the first ten twenty-five-year-olds born in that year, and so on. But everyone wanted to be tested. The local doctor would want his wife and family tested, for instance. It was a good thing I brought extra supplies."

Collection of samples and information was quality-controlled; everything was put into numerical codes. The villages were coded by number so that data analysts would not know their identities, and each villager was assigned a number set up in such a way that it would be evident if anyone made mistakes entering the numbers. The data went into the researchers' laptops immediately, and for safekeeping Mettler sent copies to Japan and the United States.

The team went over questionnaires with the villagers—whether they had headaches, how often they ate, what their education level was. "Some of the questions had nothing to do with radiation effects, but the answers would tell us if people were kidding," he said. "We asked, 'Do your hands get numb?' All the ladies said, yes, both hands got numb. When we inquired further, they would say that when they dug turnips in the winter their hands got numb. We looked at diet, the number of meals a week they ate. We wanted to know how much milk they drank each day because radioactivity in grass eaten by cows can wind up in their milk."

The thyroid uses iodine the body ingests in food to make thyroid hormone. Salt is iodized in the United States as a way to prevent goiter caused by iodine deficiency. But radioiodine, created in reactors, can cause thyroid cancer. The Soviets claimed that they had given potassium iodide to everyone to block the thyroid's uptake of radioiodine, but as the accident unfolded this became a political matter, with some bureaucrats afraid that distribution of pills would produce a panic. If potassium iodide is taken in advance of exposure, it prevents uptake of radioiodine; after the fact the remedy is far less effective. In countries that did pass out the pills, like Poland, the rate of thyroid cancer did not increase. "This being the USSR," Mettler said, "it wasn't politic to ask whether the people actually *got* the potassium iodide, so we asked, 'Did you take potassium iodide?' We asked, 'Do you drink hard liquor?' No one wanted to say they did—the only people who circled 'a lot' were drunk. One man asked me, 'Is it true that vodka gets cesium from the reactor out of your body?' "

A belief had spread that ingestion of alcohol would cure radiation exposure. Mettler found that people drank more heavily in the contaminated villages. All the villagers wanted to move—not because of the contamination, though. "They wanted to go somewhere else to get better jobs or houses, or to avoid future accidents like Chernobyl. This was true whether or not they lived in the contaminated zones."

He arranged for independent analyses of the data by the two most eminent radiation-epidemiology organizations in the world, the RERF in Hiroshima and the Imperial Cancer Research Fund in London. These two institutions, working with the coded data and numbers they had been provided, came to the same conclusions, and each contributed to the result section of the major report, a hefty IAEA publication titled "The International Chernobyl Project: Assessment of Radiological Consequences and Evaluation of Protective Measures." It was presented in 1992 by the IAEA at its headquarters in Vienna.

The findings did not correspond with what most people had come to believe about the reactor accident. The severest casualties occurred among plant workers, two of whom died from scalding. On the first day of the accident, about 1,000 emergency workers, mostly firefighters, received the highest doses of radiation. In the first three months, before the hottest radioactive materials would have decayed, 134 people out of that group suffered acute radiation sickness and were given treatments to mitigate its effects. Within a few weeks, twenty-eight perished. Soldiers, miners, and construction workers were drafted to clean up the plant and the thirty-kilometer zone surrounding it. They also built settlements, dams, and water filtration systems. Certified as "liquidators," this group of about 226,000 individuals received relatively low exposure—an average total-body dose of 1,000 millirem, less than they would have received from nature if they'd moved to northeastern Washington State for one year. Approximately 374,000 other liquidators, who worked outside the thirty-kilometer zone, received exposures that were considerably lower—about 100 millirem. Studies of the liquidators have failed to find any direct correlation between this increased exposure and a rise in cancer or death rates, and a recent study of radioactivity in their teeth suggests the dose they received was lower than had been estimated.

Relying on findings about the Japanese atomic bomb survivors, researchers have conservatively estimated that, over a lifetime of ninety-five years, deaths attributable to leukemia or solid cancers caused by the exposure would come to about 2 percent. Leukemia manifests in the first ten years, and no excess cases were detected. Solid cancers take longer to manifest. The health of the liquidators continues to be monitored.

Average exposures to evacuees and to other citizens of Ukraine and neighboring countries ranged from between 1,000 millirem and 7,000 millirem. Mettler pointed out the dose rate from natural radiation ranges between 100 millirem and 2,000 millirem per year in most countries, and up to 15,000 millirem in some inhabited regions. It bears repeating that in regions high in natural background radiation, no studies have found a radiation-induced increase of cancers and hereditary diseases.

AVERAGE IONIZING RADIATION DOSES CAUSED BY THE CHERNOBYL DISASTER	MILLIREM PER YEAR
Chernobyl (1992)	490
Pripyat (1992)	2500
FROM NATURAL SOURCES (SOIL, ROCKS)	MILLIREM PER YEAR
Average in Poland	240
Grand Central Railway Station in New York City	540
Kerala, India	900
A region in Norway	1,000
A region in Sweden	3,500
Guarapari, Brazil	3,700
Tamil Nadu, India	5,300
A house in Ramsar, Iran, built over 100 years ago	8,900–13,200

Source: UNSCEAR, Jovanovich, Sohrabi.
Data from 1993, converted to millirem

Scientific studies and conclaves have continued to uphold the findings of the first IAEA teams. In 1996 in Vienna, the United Nations, through its Department of Humanitarian Affairs, was joined by UNSCEAR, UNESCO, WHO, and IAEA, and also by the European Commission and the Organization for Economic Cooperation and Development (OECD), in holding an international conference called "One Decade After Chernobyl." These agencies and some other international bodies agreed with the earlier conclusions about the consequences of the accident.

In 2000, UNSCEAR issued an updated report of the potential long-term health consequences of radiation exposure from the accident; it confirmed previous findings, gave particular attention to rates of cancer, and came to several conclusions:

> There is no scientific evidence of increases in overall cancer incidence or mortality or in non-malignant disorders that could be related to radiation exposure. The risk of leukemia, one of the main concerns owing to its short latency time, does not appear to be elevated, not even among the recovery operation workers. Although those most highly exposed individuals are at an increased risk of radiation-associated effects, the great majority of the population is not likely to experience serious health consequences from radiation from the Chernobyl accident.

UNSCEAR also found that "the accident had a large negative psychological impact on thousands of people." Fear, born of ignorance of real risk coupled with anxiety about imagined harm, fostered epidemics of psychosomatic illnesses and elective abortions—perhaps as many as 200,000—because of dread of genetic mutants. Long-standing anxiety about government directives was exacerbated by the actions of Soviet officials during the first two years after the incident. Better management of the emergency, including adequate dissemination of facts, probably could have prevented much of this psychic damage, which also plagues war veterans and people displaced by natural disasters like major hurricanes and floods.

"Most of the world community agreed that the conclusions were what had been expected," Mettler said. "The Russians were in agreement. But at first, Ukraine and Belarus went nuts and claimed that all sorts of other things were happening. When we presented the findings in Vienna in 1992, the Ukrainians said that rates of TB, measles, and other diseases had gone up. I asked then and there for their data, and added, 'And, oh, by the way, include your data from the years preceding Chernobyl.' I got some data and had it translated and showed it to the audience. Certain diseases were down, others were up. So that data didn't support the Ukrainians' blanket statements. There was a great deal of discussion of thyroid cancer and leukemia. The claim that cancer rates in general went up didn't cut it for us, for a number of reasons. As I said, those separate groups, one in Hiroshima and one in London, independently analyzed the data and made the conclusions. The conclusions have been borne out and continue to be borne out. There's been no change."

In 2005, at a large conference in Vienna held by IAEA and WHO, Mettler presented the latest assessment by a team he had led in analyzing the health effects for the Chernobyl Forum, which was composed of eight UN agencies, the World Bank, and representatives from Ukraine, Russia, and Belarus. More than one hundred scientists reviewed all the data that focused on the most important potential radiation effects.

"The Chernobyl Forum findings are nothing particularly new," he told me. "The death toll stands at fifty directly attributable to radiation exposure sustained by reactor staff and emergency workers. Of the two thousand cases of thyroid cancer, nine children have died. There still have been no increases in leukemia or in birth defects. A slight rise in the cancer rate among liquidators does not differ statistically from that of the general population. The report makes the assumption, based on the most conservative hypothetical estimates, that over time about four thousand fatalities in excess of the norm may occur from latent cancers due to radiation from the accident—if," Mettler added, "people live long enough for them to develop." He was referring to the very difficult conditions that prevail in these countries and are likely to shorten life span.

The Chernobyl Forum noted that significant psychological reactions continue to afflict anywhere from one hundred thousand to two hundred thousand people, and that the large compensations paid out to about seven million Eastern Europeans have affected regional growth and created a culture of dependency. "People have come to think of themselves as helpless, weak, and lacking control over their future," the Forum report concluded.

"People have developed a paralyzing fatalism because they think they are at much higher risk than they are, so that leads to things like drug and alcohol use, and unprotected sex and unemployment," Mettler told *The New York Times*. After the collapse of the Soviet Union, shortly after the Chernobyl accident, life expectancy in the region plunged. Mettler thinks that the impacts of the accident may be difficult to detect against the background noise.

Prior to his presentation of the new report, I'd asked him where he thought the stories about huge numbers of deaths because of Chernobyl had originated. "Now we have the UN agencies and the governments of the affected countries finally in agreement about the health effects and how to deal with them," he said. "But in the 1980s people made claims based on inaccurate information."

According to a paper included in a 1996 report issued by the European Commission, IAEA, and WHO, predictions about deaths were uncertain, but their uncertainty was not accepted "by the decision-makers or by the affected populations. The citation of these projections without any understanding of the assumptions or uncertainties, the time frame, the size of the population concerned, and the number of normally occurring fatal cancers was misleading. Any estimate of the total number of fatal cancers resulting from the accident will be at best crude and will depend on the assumptions made. Such forecasts of numbers should not be accorded the status of scientific objectivity. . . . The largest part of the radiation dose from the accident has already been received." The authors, leading scientists from several countries, noted that the "general deterioration in public health in the countries of the former Soviet Union since 1987" had been wrongly blamed on the Chernobyl accident; they debunked claims that "tens of thousands of people 'have already died,' implying that they were victims of the Chernobyl accident. However, the total death rate in 1990–1992 among 'liquidators' . . . did not exceed that for the corresponding age group in the Russian Federation as a whole."

Mettler plucked from a high shelf UNSCEAR's *Sources and Effects of Ionizing Radiation* and found a graph of cancer rates in the area in the six years prior to the accident. "You see that the rate of all cancers was steadily going up and that the slope stayed the same after Chernobyl in 1986. It hasn't changed. Except for leukemia and thyroid cancers, which appear relatively quickly, most cancers from radiation occur twenty years after the exposure. Officials from Belarus and Ukraine were saying that Chernobyl caused the increase. We replied, 'Yes, cancer rates were going up, especially in cities. Your reporting system is no good.' In villages, there are no autopsies and there's less medical care than in cities. Cancer rates are higher in Poland than around Chernobyl. Is it just more common there? Or is there some protective effect in the Ukraine? Worldwide, about thirty percent of the population gets cancer. The reported rates in the Ukraine are fifteen percent. But how do you know what people have died of? Suppose a man dies of lung cancer. The Ukrainian villager doing the report could write down 'pneumonia' for lung cancer."

According to Mettler, statistical predictions of leukemia based on doses did not pan out. Even with thyroid cancer, there was no clear, documented evidence of an increase during the time the IAEA study took place. "And the Soviets," he added, "used to include thyroid cancer data in the category of 'all other,' and that category included prostate cancer, which isn't caused by radiation. But we couldn't exclude an increase in thyroid cancer. In a village of ten thousand we'd expect twenty-four thyroid cancers, mainly in children. Now, when you screen children more for thyroid cancer in a normal population, you'll find more. But it's clear that the magnitude of the number of cases—about two thousand—was not simply due to screening. We got a lot of crap for projecting thyroid cancers and leukemia when, as some critics said, 'You didn't find any increase!' So we got shot at from both sides. I figured we must be doing something right."

He added, "Russia and Belarus are now off the kick of 'radiation causes everything.' With time we'll see. I told my team, 'Talk to me in twenty years and we'll see who's right.' "

The original 1992 IAEA report actually stated that there had been no *demonstrated* increase in thyroid cancer. "But we projected one, and of leukemia as well," Mettler said. "The increase in cancer will probably be too small to see. Most solid cancers due to radiation will begin to appear in the next ten years." The predicted increase, based on the data from studies of the atomic bomb survivors and on the as yet unproven hypothesis that the health effects at high-dose levels extrapolate downward in a linear fashion to those from low-dose exposures, would be five thousand cases—one-half of one percent of the whole cancer burden of the exposed population. "These cancers will be impossible to find, especially with the life span in Eastern Europe going down due to alcohol and tobacco consumption and suicide. You might see half a percentage point of change. But with the way people smoke there, we may never be able to tell."

Mettler mentioned that in Ukraine there's a law now that any death in the Chernobyl area has to be attributed to the accident. "An official said to me, 'How else are we going to get aid?' Unfortunately, there's been very little compensation for anyone who suffered from Chernobyl. There are some who pretend that everyone is dying of radiation and others who really do believe that. The same thing would happen in the United States. There's supposed to be special compensation for villagers. If you are declared a Chernobyl person, you can ride public transportation cheaply, or for free. I know people who have 'Chernobyl disabilities' who were nowhere near the place. It's similar to the problems of compensation for American uranium miners."

"What surprises did you encounter?"

"We all wore dosimeters in Chernobyl, so we know how much exposure we got. We were in some areas that were giving off 10 millirem per hour. In a village if you watched people carefully as they were out walking around you'd notice that they would all cross the street at a certain point. They would avoid certain areas. They had maps that showed where the hot spots were. Kids had them memorized and would run around them when they were chasing a ball. They were avoiding doses. The villagers are getting less of the dose than if it were averaged and you assumed everyone wandered around normally."

When I told Rip how the villagers avoided the hot spots, he observed, "They're uncontrolled Fencepost people, keeping away from the places where they might receive a higher exposure." The hottest spots are decaying the most rapidly.

In the evacuated zone wildlife is abundant. "I talked last week to a radioecologist who said that there are big herds of deer and that other animals are practically falling out of the trees there are so many," Mettler said. "The background radiation is very low in the whole region. There's no granite, which would increase the level. Chernobyl is a giant, flat swamp." No studies of plant or animal mutations were made there prior to the accident, so it's unclear whether particular characteristics observed today might be the result of mutations caused by ionizing radiation.

Mettler showed me another graph. "In the zone around Chernobyl with the highest contamination, the average dose of radiation over a lifetime of seventy years from both background radiation and from the accident totals about 30,000 millirem. In many of the evacuated areas around Chernobyl, the current dose is lower than it is naturally in parts of France, Britain, and Spain, and much lower than in Finland—there the average lifetime dose from natural background radiation is 50,000 millirem."

Although tens of thousands of people were relocated, just over one million remain in significantly contaminated areas. "The explosion released cesium, and it gives off gamma rays. There are different amounts of cesium in different places. How much radiation will people get from cesium on the ground compared to cesium in the food chain or resuspended in the air if dust is kicked up? The answer is: not much. Ten-year follow-up studies of the people living in contaminated areas do not indicate physical problems."

An emotionally charged American-made documentary, *Chernobyl Heart*, which won an Academy Award in 2004 and which is recommended for screening in schools by the National Education Association, showed infants with heart deformities that the narrator attributed to radiation exposure. However, this claim remains totally unsupported by the scientific studies conducted by international agencies; in fact, birth defects are lower in the contaminated zones, and overall birth defects in the region have dropped. In any case, Mettler noted, spontaneous congenital defects occur everywhere, and no congenital heart defects of any kind have ever been found in association with radiation exposure.

The Chernobyl Forum limited its calculations about potential deaths to the populations living in the most exposed areas. On the twentieth anniversary of Chernobyl, two major new reports were announced. One, commissioned in Europe by the Green Party and written by two British radiation experts, denounced the findings of the consensus of more than one hundred scientists in the international scientific organizations that make up the Chernobyl Forum. The Green Party report claims that vastly greater numbers of people were harmed by low-dose radiation than the figures put forth by the IAEA, UNSCEAR, WHO, and five other investigative agencies. Extrapolating from doses that are extremely small, and about a thousand times lower than natural background radiation, the Green Party report predicted that sixty thousand deaths throughout Europe would occur as a result of the accident. Using this same method, it would also be possible to show that people living in areas with higher natural background radiation, like the Colorado Plateau, would also be at high risk. However, this is not the case. Furthermore, the people who received the highest doses from Chernobyl have not manifested cancers at the rate that had been projected by most experts in the early days of the crisis.

Mettler considers it highly problematic to try to extrapolate total deaths worldwide attributable to Chernobyl by using global exposure data from the estimates in the earlier studies. He points out that the 2005 Chernobyl Forum relied on the best scientific data it had on *actual* cases of exposed people rather than hypothetical ones. He told me that there was a major difference between studying a person who received a particular, demonstrable dose, on the one hand, and making large estimates about how many people around the world could in theory perish because of the Chernobyl plume, on the other hand.

Rip explained to me that there is an assumption underlying the way potential radiation exposure of a population is calculated and thus, by extension, an assumption in the prediction regarding theoretical cancers that might be initiated. In this assumption, a very small exposure multiplied by a very large population can produce large numbers of hypothetical cancers. "Suppose I've got a bathroom scale and I weigh myself and then I eat a peanut and get on the scale and expect to see the added weight of the peanut. Background radiation is my weight, and I add a peanut more, and that additional half a gram or so is hidden by background."

Most radiation protection experts say that many uncertainties arise when making any calculations about the effects of low-dose radiation, because it's difficult to distinguish particular exposures from the much greater natural background level. The other major 2006 report, this one on Chernobyl's health effects on Europe, came from the International Agency for Research on Cancer (IARC), which had established an international working group to evaluate the likelihood of increased cancers throughout Europe from the low-dose fallout from Chernobyl. IARC relied on several tools: updated estimates of radiation dose, a comprehensive examination of trends in cancer incidence and mortality, and state-of-the-art risk prediction models developed from studies of the atomic bomb survivors and other populations exposed to radiation in other settings. The report concluded that, hypothetically and over several decades, a total of sixteen thousand cases of thyroid cancer and twenty-five thousand cases of other kinds of cancer could be expected and that about sixteen thousand deaths from these cancers may occur in Ukraine, Belarus, and the most contaminated territories of the Russian Federation.

"While these figures all reflect human suffering and death, they nevertheless represent only a very small fraction of the total number of cancers seen since the accident and expected in the future in Europe," said Dr. Elisabeth Cardis, a radiation specialist with IARC. "Indeed, our analysis of the trends in cancer incidence and mortality does not demonstrate any increase that can be clearly attributed to the Chernobyl accident." Cardis considers the apparent increase in thyroid cancer to be the sole exception, noting that over a decade earlier excess cases had been found in the most contaminated regions. The IARC says its report contains "the best estimates to date of the impact of the Chernobyl accident on cancer in Europe. To put it in perspective, tobacco smoking will cause several thousand times more cancers in the same population."

Our interview about to conclude, I asked Dr. Mettler what people could do to protect themselves from excess radiation.

"What people should be worrying about is cigarettes. Obviously if you could prevent smoking, you would get a lot bigger bang for your buck than chasing millirem around."

If he were forced to choose, which would he live next to—a nuclear plant or a coal plant?

"The nuclear plant," he replied. "Even if you have scrubbers at a coal-fired plant, you're still getting a bigger dose of radiation than you would from a nuclear plant. They're really tightly regulated. The coal plants are not."

And if he could appear on TV before millions, what message would he want to give?

"I would tell them that radiation has risks related to dose level, and that many of the things we use radiation for have undetectable risk. Some have higher risks. People have to make decisions about benefits versus risks. Is the benefit of nuclear medicine greater than its risks? Are the benefits of nuclear power greater than its risks?"

Mettler appears to think that they are, and so does Ukraine. That nation has continued to rely on its nuclear plants for 50 percent of its electricity, has become a party to international nuclear and radiation safety standards, and plans to expand nuclear generating capacity.

In addition to the radium-dial painters, the citizens of Hiroshima and Nagasaki, and the people around Chernobyl, I'd heard about some other groups who may have been subjected to unusual doses of radiation.

For years stories have circulated about "cancer clusters" around nuclear facilities. In 1991 the National Cancer Institute reported on an epidemiological survey it had done of 107 U.S. counties in which people lived near sixty-two different facilities—commercial nuclear power plants, DOE research and weapons plants, and a fuel reprocessing plant. Some of those counties had lower rates of cancer than the control counties and others higher rates; no evidence of an increased risk of death from cancer could be linked to living near a nuclear facility.

But what about exposures to New Mexicans, given the nuclear weapons complex? I called on Dr. Charles Key at the Cancer Research and Treatment Center at the University of New Mexico Medical Center. After his service for many years as medical director of the New Mexico Tumor Registry, he is a full member of the Cancer Epidemiology and Prevention Program at UNM as well as a professor of pathology.

In the 1970s, Dr. Key helped prepare a network of cancer registries set up by the National Cancer Institute, called the Surveillance, Epidemiology, and End Results (SEER) program, to monitor cases of the disease throughout the nation. New Mexico was among the first participants. SEER collects and publishes information on incidence, deaths, and survival rates documented by pathology laboratories, hospitals, and physicians.

"Much of what cancer registries do is calculate cancer rates—trends over time," Key told me. He had a gentle demeanor and chose his words with care. "For example, incidence of thyroid cancer has increased. The theme, these days, is that cancer is an environmental disease. The popular scapegoats for cancer increase are air and water pollution and radiation. People neglect to take into account lifestyle: exercise, nutrition, and other factors. Ionizing radiation can increase the number of thyroid cancers. On the other hand, if you were going to pick a cancer, thyroid is better than most—it has a high survival rate."

An assumption about the thymus gland—now known to be an important part of the immune system—as the cause of breathing problems in newborns led doctors in the 1930s and 1940s to treat infants with X-rays. The thymus is radiosensitive, and X-rays shrink it. As these infants grew up, they had a higher risk of thyroid nodules, including cancerous ones. People with skin conditions were sometimes given therapeutic radiation as well. "Every other study—of atomic bomb survivors and people exposed to fallout from nuclear weapons tests—has associated ionizing radiation with thyroid and other cancers," Key said. "Radiation to cure one cancer can cause another." A large study conducted by the National Cancer Institute reported in 1997, "The overall average thyroid dose to the approximately 160 million people in the country during the 1950s is estimated to have been about 2 rad [about 2,000 millirem]. To put this amount of exposure into perspective, routine medical use of X-rays fifty to sixty years ago exposed children to anywhere from 5 to several hundred rad."

Natural background radiation in New Mexico exposes people to 0.3 rad per year (about 300 millirem per year).

Dr. Key outlined how, over a long period, the SEER program has been collecting data in New Mexico and how SEER epidemiologists have done comparison studies among groups possessing different genetics, lifestyles, and environmental exposures. The respective populations of different groups show different cancer patterns. He confirmed that New Mexico continues to have one of the very lowest cancer rates in the nation.

I mentioned fears that had been aired for years in the local press regarding a nuclear waste repository called the Waste Isolation Pilot Plant in southern New Mexico. Dr. Key said it was hard to imagine a safer or more isolated place to store radioactive materials. "After seeing the bubble tents in Los Alamos containing stacks of barrels of radioactive waste covered by tarps, I think that stuff is better off deep underground at WIPP. Although the fact that at Los Alamos the waste is inside a compound with a high fence around it makes it an unserious threat to me."

He said that he would have no problem with living next to a nuclear plant. "Other countries are more dependent on nuclear power than we are, and they're managing the materials. You can't assume nuclear power is a hazard and coal is not. Coal trains, for example, catch on fire once in a while and there are other fossil-fuel transportation accidents, and coal-fired plants put radionuclides into the atmosphere. The fact that no new nuclear plants have been built in decades means that in the United States we're falling further and further behind in the implementation of this power source that, in the long run, is less hazardous than what we already deal with all the time with coal. Nuclear power could provide electricity for transportation, for instance. I'd like to see electric cars and trucks instead of these big gasoline trucks driving on our streets."

Some New Mexicans, particularly watchdog groups with their gazes trained on Los Alamos National Laboratory, maintain that secret doings at the lab have radioactively contaminated the land in the area and caused "cancer clusters." Key had participated in the Los Alamos Cancer Rate Study that had been commissioned subsequent to a spate of newspaper articles about this threat.

Los Alamos, hidden in the Jémez Mountains in a once-secret location where scientists gathered during World War II to work on the Manhattan Project, may be one of the most unusual cities in the nation. Many of those people—some of them European refugees—and their descendants have continued to live there along with other scientists, engineers, and technicians in what has become something of a cultural and intellectual outpost. "Ah, the Los Alamos cancer clusters," Key said. "Everything about Los Alamos is off the scale—the SAT scores of the students, the performance of their swim team. They're not high achievers because of exposure to radiation."

The population there is aging. The rate of thyroid cancer in Los Alamos detected in the late 1980s and early 1990s increased. "Instead of two to four cases a year, we were seeing five to eight. But if you find people who are over the age of forty in the mall and line them up and examine each of their thyroids microscopically, about five percent to ten percent of them will turn out to have occult [hidden; not normally detectible] thyroid cancer. If you use ultrasound and take biopsies of their thyroid nodules, you'll find more cancer. It's sitting there, like prostate cancer in maybe half the men in their eighties. Many people outlive their thyroid or prostate cancer. Lung, cervical, and stomach cancer are lower in Los Alamos than in the rest of New Mexico and the United States."

The cancer-cluster rumors started in 1991, at a public hearing held by the DOE in Los Alamos, when a local artist mentioned a number of brain tumors that had occurred in one neighborhood, which he thought might be due to radiation from the laboratory. The *Albuquerque Journal* then compiled a list of more than ninety cases of brain cancer. "Some people had moved elsewhere after being born in Los Alamos forty years ago," Key said. "One elderly woman had come there recently to live with her children. Others were cases of metastasis—cancers from the lung or breast that had spread to the brain. Some on the newspaper's list didn't even have brain tumors; they had leukemia or lymphoma. But the newspaper took the number of cases and claimed that there was an astronomically huge number of excess brain tumors. The journalist was counting things over an extended period of years and including cases of people who were not residents at the time of diagnosis but then making comparisons with people who were.

"You can't take a set of data collected under one set of rules and compare it to a set of data collected under another set of rules. In public meetings about the cancer-cluster issue, people would be surprised when I'd cite how radiation does increase cancer risk. But then I would ask them to apply the same criteria to levels of exposure. These are hard to document and hard to monitor. People had the assumption that those measurements had been kept secret."

Multiple studies were conducted at a cost of millions of dollars. Dr. Laurie Wiggs, who is an epidemiologist at Los Alamos National Laboratory and who participated in an examination of the cancer clusters, told me that the New Mexico Health Department had done a study of cancer in Los Alamos in conjunction with the New Mexico Tumor Registry. Increased numbers of brain and thyroid cancer cases were found to have occurred during a period of a few years. She attributed the depiction of these results in the media as "cancer clusters" to the Texas sharpshooter method. "The sharpshooter fires at the barn and then goes over and draws the bull's-eye around the bullet holes that are closest together," she said. "Sometimes there are real clusters, but it can be difficult to identify the reasons for these clusters. Statistically speaking, if you have a small population and a limited number of expected cases, small increases can make things appear out of line."

Yet another body, an investigative working group, finally determined that there were no environmental carcinogens in Los Alamos and that there was no cancer epidemic. As a result, community fears were allayed.

I'd heard that a tiny trace of plutonium could kill you. On August 16, 1994, *The New York Times* published an article about the seizure of ten ounces of plutonium at the Munich airport. "A tiny speck of the fine powder can cause lung cancer in anyone who inhales it, and a small amount in the water supply of a large city like Munich could kill hundreds of thousands of people," the article claimed. This information has been replicated by other news media and antinuclear-power groups.

Six scientists at the Lawrence Livermore National Laboratory wrote in a paper that the claim that inhalation of plutonium "can cause cancer in anyone" is misleading and the assertion that ingesting plutonium from the water supply could kill vast numbers of people is false. After a discussion of scientific research on the health effects of plutonium, the authors concluded:

> The claims of dire health consequences from the introduction of plutonium into the air or into a municipal water supply are greatly exaggerated. The combination of rapid and almost complete sedimentation, dilution in large volumes of water, and minimal uptake of plutonium from the GI tract would all act to preclude serious health consequences to the public from the latter scenario. And although the dispersal of plutonium in air (as the result of a fire or explosion, for example) would cause immense concern and cleanup problems, it would not result in widespread deaths or dire health consequences, as terrorists might hope. Dissipation due to wind and air turbulence would rapidly dilute any respirable aerosol. Only people within a few meters of the source could receive a prompt lethal dose. Delayed effects in the form of fatal cancers outside this region would probably not appear in affected individuals until years later. For a vast majority of the population of any city, the increase in cancer risk arising from exposure to plutonium aerosol would be a fraction of that arising from other, more common health hazards.

The Health Physics Society (HPS), whose members spend their time in the field measuring radiation health effects, has stated that the radiological hazards presented by plutonium are equivalent to those from the far more common elements used in commercial applications—radium and thorium. The HPS notes that many people have inhaled plutonium without any discernible impact to the lungs and that in any case public exposure to plutonium is extremely unlikely.

Humans first created plutonium during the Manhattan Project. Los Alamos National Laboratory remains one of the few places on the planet where, as a metal, it is fabricated into weapons components. Only people who are carefully screened and found able to do the job safely are permitted to work with plutonium, and they have to use a glove box or manipulators in an isolated, protected location. In the early days, a number of lab employees handling the metal accidentally received significant doses when shielding was breached. The worst exposures in the history of the lab befell twenty-six men who worked at Los Alamos going back to the Manhattan Project. A study cohort was formed in 1951 and the men have been examined every five years since then. A few who accumulated a significant body burden of plutonium became members of the UPPU Club ("You pee Pu," Pu being the chemical symbol for plutonium), because traces of plutonium in urine can indicate the amount remaining in the body—it tends to migrate to lung, liver, and bone tissue—and these employees were required to provide periodic urine samples and to undergo frequent examinations. The average exposure was over 1,000 millirem. Researchers report from time to time on the health and mortality of these men. Their diseases and physical changes as they age are consistent with those of a male population of their age group, as is the number of cancers and the death rate. One person developed osteosarcoma—bone cancer—perhaps related to the plutonium exposure he'd received. A study done fifty years after the accidents, when the average age of the exposed group was seventy-two, found nineteen of the twenty-six UPPU club members in reasonably good health; most of them have lived well into their eighties. They produced normal children and grandchildren. The findings of the study differ "from some popular misperceptions that large health risks occur with any exposure to plutonium."

"People make extrapolations about risk," Key said, "and if the risk had been as high as those extrapolations had it, all of those exposed would have died." He politely expressed disappointment at most of the reporting in the popular press about radiation. "I would treat radiation with a great deal of respect, but I think you need to be realistic. Compared to tobacco, gasoline, drunk drivers, or being a couch potato, radiation is of very little risk to most of the public. Twenty-five years ago or so there were books published that gave a balanced look at the question. But since then there have been a lot of popular ones that rely more on emotion than science. Over the years, the standards for radiation protection have changed. The level of allowable exposures has been lowered. People point to this and say that they were misled. I rather think that over time, you learn more. Perspective changes. You become more cautious. The technology has become more refined. Some people are very eager to assign sinister motives. But I don't think there are any."

6 INTO THE STRANGE CITY

SUPPORT FOR NUCLEAR ENERGY in the scientific community has generally been consistent. In 1980, a year after the Three Mile Island incident, social scientists at Smith College and Columbia University surveyed 741 scientists selected at random from *American Men and Women of Science* about their position on nuclear power. Whether or not industry or government gave them funding, and whether or not government, industry, or academia paid their salaries, 89 percent of all the scientists asked were in favor of nuclear power. If their specialty was energy, approval was 95 percent, and, among radiation and nuclear scientists, 100 percent. The scientists were also queried about whether they would accept the construction of a nuclear plant near where they lived. Sixty-nine percent assented, and among those in the energy field 80 percent did, and 98 percent of those in the nuclear world.

In 2002, a survey of 865 American members of the American Association for the Advancement of Science (AAAS) and 1,332 members of AAAS in the then fifteen states of the European Union found that respondents considered the benefits of nuclear power to outweigh the risks. The information sources on nuclear matters that the scientists found most trustworthy were international or regional agencies, followed in order of veracity by national government agencies and environmental groups. The public media, utility companies, and political parties were ranked the least reliable. I was to find throughout the nuclear world that scientists and engineers were similarly skeptical of these latter three sources. Since most of us get our information from the public media, that might explain why I possessed assumptions about risk that were different from those of the scientists I encountered.

Do scientists, in particular those involved in the study of ionizing radiation, know something that the rest of us do not? How did the decision come about that some levels and some kinds of radiation were not risky whereas others were lethal? The biologists, chemists, geologists, physicists, radiologists, engineers, and technicians I met never mentioned being frightened of radiation. In general, they seemed as cautious about risk as anyone. They used seat belts and did not smoke and wanted a safe, clean environment for their families and themselves. None of the people I met would enter a setting that they considered life-threatening unless there was an emergency. I wanted to find out what they knew about the working life of uranium that kept them tranquil and what worries they did have. I also wanted to learn how they thought about risk, what their greatest concerns were, and how they arrived at standards for safety.

Dr. Leo Gómez has published papers on everything from effects of radiation on humans to Russia's radioactive contamination of northern oceans. After working at Oak Ridge National Laboratory and Los Alamos National Laboratory in the fields of public health and cancer research and treatment in the 1970s, he was hired by Sandia National Laboratories. Until his retirement in 2000, he participated in several major programs on nuclear waste disposal and environmental and health risk analysis. He's served on review panels for global climate change programs for the DOE and as editor of the multinational journal *Radioactive Waste Management and Environmental Restoration*. He now consults on various research projects and performs numerous peer reviews of studies done by the DOE and other agencies, among them the National Institutes of Health's plan for triage of a population affected by radiological terrorism. For years he's lectured at elementary schools about atomic energy and radiation risks, and he's a founder of the Society for the Advancement of Chicanos and Native Americans in Science. He finds time to travel occasionally to Japan with his wife Mary Burnett de Gómez, who teaches the Japanese art of flower arranging and keeps a shop full of Japanese objets d'art in Albuquerque. For several years he's been on a mission that may result in a far greater understanding of the science of radiation protection.

I joined Gómez one day for blue corn enchiladas at an outdoor café in the South Valley, near where he'd grown up. He has a neatly trimmed gray beard, black hair, dark, almond-shaped eyes, and a merry spirit combined with an iconoclastic point of view.

I asked him about the arguments of the antinuclear-power groups and the impatience of scientists in the face of such charges.

"Any time that a person gets 'religion' about a project, reason and balance take a backseat to accomplishing the 'religious' goals," Gómez replied. " 'Religion' is available to both the left and the right, with equal fervor. And to pronuclear energy people and antis. People with 'religion' accept anything told to them by their heroes and don't feel the need to verify any of the claims."

He went on to speak about prejudice, and how belief systems tend to avoid fact and lead to bias, whether we're scientists or ordinary folks. "In the early days of nuclear energy, a politician proclaimed that the mind of a nuclear scientist was to the mind of an average person as a mountain was to a molehill." In Gómez's opinion, this attitude gave license to considerable arrogance on the part of the Atomic Energy Commission and its successors, the DOE and the Nuclear Regulatory Commission (NRC). "People fell into the trap of believing that they were too dumb to understand how nuclear power and radiation worked," he said.

Gómez once made a presentation at Sandia National Laboratories to scientists, engineers, and mathematicians about public responses to the prospect of the Waste Isolation Pilot Plant, which had yet to open in southeastern New Mexico. He and others on a team Rip had assembled had determined through extensive data collection, computer modeling, and risk analysis that the repository had far less than a one in a million chance of failure—something closer to one in a billion. Nevertheless most media coverage emphasized the potential dangers, and New Mexicans responded by worrying. The technical community couldn't understand why odds that they considered so supremely favorable were scaring people.

Gómez then asked the group what they would think if Sandia provided free, around-the-clock dental care to all of its employees. Everyone in the meeting approved of that. "But what if the dentists all had HIV and there was a one in a million chance that if you went to have your teeth worked on you'd get the disease?" he asked. "As an employee, you would have the option of using their free services or driving an hour to Santa Fe to pay to see a dentist. What would you do?" Almost everyone responded that they would go to Santa Fe and pay for their own dentistry before taking a one in a million chance of getting an illness that at that time was untreatable.

"They were experts on radiation but not on AIDS, which scared them to an unreasonable degree," he said.

As a small boy Gómez became interested in science when he saw a science fiction film, *Them!* "It was a horror movie about giant ants that had mutated at the Trinity atomic test site. There were anthills in my yard, and I'd feed different concoctions to the ants to see if they grew. They never did." He read voraciously at the public library, and from the age of fourteen worked up to three jobs at a time and participated in sports.

A high school teacher introduced him to scientists at Sandia Labs and encouraged him to attend college. No one in his family had ever done that. He worked his way through the University of New Mexico, dominated entirely by Anglos at that time, and in 1963 got a degree in biology. The biology department chair told Gómez not to bother applying to grad school. "He said, 'You're not graduate material.' Prejudice at its worst is psychological genocide. It affects how you view yourself."

Gómez went on to obtain a fellowship from the Public Health Service and to acquire a master's degree in radiation biology and health physics and a Ph.D. in radiation biology at Colorado State University, a leading school in radiation studies. Oak Ridge National Laboratory hired him to find better ways to detect low levels of transuranic elements in contaminated workers.

"I determined a worker's whole body count," Gómez said. "He'd stand in a kind of closet and get scanned. I'd compare that with the readings from his personal dosimeter, which would tell me his total radiation dose from natural background and man-made radiation. A dose is the amount of radiation someone is exposed to. An absorbed dose is how much energy the radiation actually deposits in the body." Sometimes workers would have to put on all their radiation protection gear and enter hot cells—shielded cubicles containing high-level radioactive materials that were normally handled remotely by manipulators. The men were only permitted to remain inside for a minute or so, and could only do that a limited number of times.

"They were 'good ol' boys' from east Tennessee," Gómez went on. "I'd buy them coffee and get to know them, and they'd open up to me. They got bonuses for working in radioactive areas. After talking to them, I knew I could never calculate their whole body doses from their personal dosimeters. They were supposed to wear them, and at a certain point, after they'd received a certain dose, as shown on the dosimeter, they'd never be permitted in hot cells again and would be transferred to another, lower-paying job. So they'd leave their film badges outside the cubicle in order to get in more runs. They'd been going into the hot cells for many years. They were mostly smokers, and two packs a day adds up to 16,000 to 20,000 millirem per year to the lining of the lungs. And as far as I could tell they were in good health, even at high exposure rates for a very short time."

Gómez has been working in recent years to establish a laboratory in a location with very low natural background radiation, where scientists from around the world could isolate and observe the effects of low-dose radiation on cells, tissues, and small mammals. This might sound like a byway of science, but actually the potential impact of results from such a laboratory could be enormous, as I was later to discover. Many in the radiation protection field told me that at present there is no experimental, scientific basis for our present low-dose standards and that much more research is needed.

Questions about the risk from low-dose radiation lie at the center of the controversy about nuclear power and the disposal of nuclear waste. What should we be concerned about? On the one hand, there's the intense bombardment of high-level radiation suffered by the Curie family and some other experimenters during the early days of radiation research, by a portion of the radium-dial painters, by residents of Hiroshima and Nagasaki in 1945, by the employees and firefighters at Chernobyl on the first day of the accident in 1986, and by some patients treated aggressively with nuclear medicine. On the other hand, there's the low-dose radiation we all receive from nature and from some artificial sources. Although radiation has been studied for over a hundred years, and much is known about damage to health from high doses, mystery surrounds effects caused by low doses—that is, those below 10,000 millirem per year.

A recent survey of 1,737 scientists from the DOE's national laboratories and the Union of Concerned Scientists showed that most scientists think that below a certain threshold, as yet to be determined by research, radiation does not harm the body, thanks to innate defense mechanisms that are triggered by exposure and kill damaged cells. The majority of the scientists polled said they question the validity of present low-dose standards, which are based on conservative estimates extrapolated in a linear fashion from data about atomic bomb survivors.

In the survey, only 36 percent of the scientists in the field subscribed to the assumption that the health effects from high-level doses may also be occurring at the same rate—or perhaps at an even higher rate—in the low-dose realm, into which no one has yet been able to peer because of the veil of natural background radiation. Referred to as the linear non-threshold hypothesis, or LNT, this assumption states that radiation does damage in a linear fashion from high dose to the lowest possible dose and that there is no threshold below which radiation is harmless. However, only 18 percent of those polled thought that the LNT should be used as the basis for regulation.

Although the majority of radiation protection experts appear to subscribe to the threshold hypothesis, they believe that since what happens
in the low-dose realm is unknown, the worst should be assumed when
setting safety standards. So until verifiable data can clear up the unknowns, LNT will underlie our regulations about radiation protection,
including the way we build and manage nuclear plants and nuclear waste
facilities and the way we clean up contamination. Currently, the Environmental Protection Agency (EPA) requires the annual exposure to a
member of the public from a nuclear facility to be limited to 15 millirem.
But even people in the LNT camp I met are dismayed at the way the
LNT hypothesis can be misapplied by special interest groups and the
media, exaggerating risks about what is in reality a weak carcinogen.

Scientists of each persuasion agree that exposures above 200,000 millirem per year can overwhelm the body's defense and repair systems. At
exposures of 100,000 millirem, some health effects become apparent. But
the range of responses to lower doses is very difficult to sort out.

The entire issue is complex. Regulations, policies, and predictions
about health impacts from radiation are not necessarily based on hard
data but rather on assumptions—on extrapolations, conservative estimates, and even useful fictions that are understood by the experts employ-
ing them for statistical purposes but can be misunderstood or misused by
others who wish to make a case for one side or the other. The result has
been something of a lack of agreement among agencies such as the EPA
and the NRC about precise low-dose radiation exposure standards.

Governmental panels of scientists periodically issue their estimates
about radiation, and these numbers influence policymakers. Radiation
biologists, biophysicists, health physicists, and others involved with radiation protection are divided into roughly three groups when it comes to
a definition of the risks from ionizing radiation. The first group, embrac-
ing a model relying on certain experimental evidence that there is a
threshold dose below which ionizing radiation is unlikely to produce
harm, assumes that only above this threshold does the probability of cancer increase in a linear fashion as the dose increases.

As you read this, your body's repair systems are mending your DNA
and other molecules whose bonds nature's rays and particles are constantly shattering. Sanguine about this repair response, the researchers
who support the threshold hypothesis note that only above a certain
level, as yet to be accurately determined but perhaps around 100,000 millirem, does radiation begin to cause irreparable damage. These scientists
cite thriving populations around the world who are naturally exposed to

relatively high levels of radiation that exceed our radiation protection standards. No peer-reviewed studies have found ailments traceable to excess natural background radiation exposure. Not only does nature bathe Finns with a dose three times higher than a person would receive in the exclusion zone surrounding Chernobyl, but also in parts of Brazil and India a variety of radioactive minerals emit up to four hundred times the average natural background radiation of the United States. Southwestern France and parts of China also have higher levels. Ramsar, Iran, high in natural radium, outstrips all the other locations. People in some areas there receive as much as 26,000 millirem a year and they, like the residents of other areas who are getting a high natural dose, do not appear to suffer any ailments as a result. Rather, they appear to be healthier and longer-lived than control groups who do not live on such formations.

The threshold hypothesis camp also notes that the long-term studies of the radium-dial painters indicate that those who received doses above a certain limit developed bone cancer whereas the women whose exposure had been less did not. Threshold proponents say that applying LNT is analogous to saying that if you put your hand in water heated to 212 degrees Fahrenheit you'll get a very bad burn and that if you put your hand in water 36 degrees Fahrenheit you'll also get "burned," but less so. "Worse yet," said one health physicist I met, "LNT is used to 'prove' that if a million people put their hands in 36-degree-Fahrenheit water, at least five hundred will get third-degree burns."

Within the threshold-dose camp, there's a group of proponents of radiation hormesis, the concept that high levels of exposure to ionizing radiation may be dangerous, but low levels are essential. Hormesis advocates mention experiments with mice exposed to low-dose radiation that inexplicably outlived the mouse control group mostly shielded from natural radiation, and refer to experiments with bacteria that, when shielded from radiation by lead bricks, appeared to undergo changes. Stating that the DNA-repair mechanism can be stimulated by radiation, the hormesis supporters argue that conservative risk estimates based on LNT may be in error because they have failed to take into consideration the potentially beneficial effects of low-dose radiation, which in certain situations can reduce radiation damage.

A few years ago, the DOE undertook an investigation of the effects of low-dose radiation that revealed interesting phenomena, thanks to innovations in technology and biological research techniques. Preliminary evidence indicates that low levels of radiation exposure may not be as harmful as the LNT hypothesis has assumed, and that low doses of radiation indeed stimulate DNA repair mechanisms. An administered small "tickle dose" of radiation might under certain conditions provide some protection against a subsequent second dose. This effect has been demonstrated in certain cell systems and may result from the tickle doses making DNA repair enzymes more plentiful or more efficient in their response when the second, larger dose is received. The timing between the two doses is extremely critical, and the second dose has to be relatively large for the effect to be seen.

"Whether this works in humans under realistic exposure conditions is still unknown," Evan Douple had told me. He keeps to the middle ground. "But it's very unrealistic to adopt the position that the adaptive response is a reason to relax radiation protection standards."

Recent low-dose studies have also uncovered cellular intracommunication. "If you hit one cell with radiation, it can cause effects and changes in cells that were not exposed to the radiation, such as cells five cells away," Douple said. "We're still learning how that phenomenon occurs and what it may mean. When you get down to a low dose, it equals levels that humans have been living with for millions of years."

The second group of radiation experts, whose position is influential in antinuclear power circles, believes that there is no safe level of exposure and that risk for cancer remains high even at low doses. This worst-case assumption results in extreme scenarios such as the one in the report on predicted deaths from Chernobyl that was issued by the Green Party. The authors of this report used LNT to make their estimates.

Between the hormesis-advocating faction and the strict LNT faction that assumes harm from even very low doses are researchers in a third group who put their faith in a version of the linear nonthreshold model that assumes that fewer cancers occur as a result of exposure at lower levels and that in rough proportion to the increase in the radiation dose the number of cancers increases. This model underlies most decisions made about radiation protection standards set by scientists in this country and internationally and also informs the calculations about predicted deaths from Chernobyl by the Chernobyl Forum and other international research groups. All but a few people I spoke with in the field think that this model is unrealistically pessimistic. But they don't want to be quoted. (Similarly, earth scientists who seem much more worried about global warming than their public statements would suggest say in private that as scientists they have to maintain a conservative stance until more data reduces uncertainties in the models.)

As is usually the case on the frontier of any scientific discipline, more data will be needed to come to a resolution. Gómez says that until then, the argument will remain a "religious" one.

How are decisions made about regulating radiation exposure? Who decides what is "low dose" and "high dose"?

In 1972, a controversy arose about how much additional exposure an expansion of nuclear power would give the public. In response, the National Research Council of the National Academy of Sciences established an advisory panel to study the biological effects of ionizing radiation. Those scientists reported that existing evaluations of radiation health effects were mainly based on extrapolations ranging from the imprecise to the invalid. Since then, the National Research Council has periodically convened Biological Effects of Ionizing Radiation (BEIR) panels to review the latest information and provide updated estimates.

Evan Douple served as the study director for BEIR VI, which reported in 1999, and was the director of the Board on Radiation Effects Research that oversaw BEIR VII, which reported in 2005 and 2006. "Epidemiologists need giant populations in order to measure what is happening under such low exposures," he said. "BEIR VII faced the challenge of examining all of the world's epidemiology data to estimate the current risks to people exposed to radiation and then used the new biology data to see what the radiation risk estimates might be at doses below which epidemiology cannot provide statistically significant estimates."

The BEIR VII committee was asked to develop the best risk estimates; it was not asked to recommend whether there should be changes in the permissible dose levels for the general public, for those receiving medical radiation, or for nuclear workers. That's a policy decision that must be made by governmental and regulatory bodies. "We know more about radiation than most other physical, chemical, or biological agents that cause harm," Douple told me. "Some agents like polio virus have been examined thoroughly, but for the most part, we don't know as much about most toxic agents as we do about radiation, especially regarding long-term effects and dose effects."

He observed that when encountering studies of populations with exposure to low doses, an analyst could be misled by a phenomenon called the healthy worker effect. "When you go looking for disease, if you don't set up your experiment properly you can get that effect, which gives an indication of something like hormesis [a beneficial effect]. But it could also mean that the control population is not well matched. For example, look at airline crews. By the way, they're the true radiation workers today—they receive measurable radiation doses as a result of their work. Most people who work in the nuclear fields come back with film badges that aren't recording significant exposures above background levels. But aircrews do get radiation due to the time they spend at high altitudes, particularly on the longer flight routes. The pilots tend to be very healthy—they have to be able to pass medical exams on a regular basis. Pilots won't have the cancer incidences that groups like, say, welders might." (Welding is associated with excess risk of lung cancer.)

Besides the scientists who strongly believe that even a tiny dose can harm you, there is that other outspoken camp arguing that radiation standards are far too restrictive, estimates of risk are far too pessimistic, and low-dose radiation has hormetic effects. "The hormesis group searches the literature and pulls out only the experimental results that support their point of view," Double said. "But when you look at the experiments they're using to build their case, or when you see what experimental results they've discounted and elected to not use, you start to have doubts. The results in one study could be a minority group of data compared to a vast body of evidence with different results. Or the experiments could be flawed, especially if they rely, for example, on animal work done early on, before mouse populations were better controlled. You can end up building a theory and supporting your evidence in a way that is truly not sound science."

It turns out that in early experiments, mouse populations were erratic in regard to the internal flora and fauna they harbored in their tissues, so results might not be consistent. Today strict quality-control regulations govern how lab animals are raised and experimented upon. Lab mice now are required to have defined flora and fauna. "The data is now much more reliable," Double said. "In radiation research, as in any kind of research, a good scientist tries to record all effects and to report whatever he or she finds. Sometimes it's harder to prove negative results that don't show effects, since most journals will favor accepting manuscripts reporting statistically significant effects rather than the absence of effects."

The 2005 conclusions by the BEIR VII panel evoked strong responses. Activist groups declared that since any exposure to radiation was potentially damaging, nuclear power ought to be abolished. One spokesman for this faction is Arjun Makhijani, president of the Institute for Energy and Environmental Research and consultant to antinuclear-power organizations. He announced that the new BEIR report indicated that the risks of radiation exposure called for stricter regulations. On the other side, Paul Genoa of the Nuclear Energy Institute asserted that the report only confirmed that, as the industry has long known, the risks were very small and that current regulations were quite protective of public health and safety.

The BEIR VII panel, using nearly fifteen years of the latest findings regarding the Japanese atomic bomb survivors along with other epidemiological and biological data, altered the analyses of the previous panel in some small ways, slightly increasing the estimates of low-dose potential cancer deaths in a given population following a specified exposure level. For example, if a hundred thousand people were exposed to a radiation dose of 10,000 millirem, BEIR VII risk estimates would predict that in the lifetime of those persons an excess of one thousand radiation-induced cancer deaths might be expected—but that in that same population the death rate would be far below the background cancer rate: approximately forty thousand persons would also die from cancers owing to the various factors that produce this disease in a typical population. (Worldwide, the biggest killers, apart from malnutrition, are influenza and malaria; in poor countries people tend not to live long enough to develop cancer.) The panel concluded that the greatest source of man-made exposure today still comes from medical radiation and it stressed that there's no clear evidence of radiological harm to people from higher-dose procedures such as CT scans. Nevertheless, if a person were to be exposed to multiple whole-body CT scans, the total dose received might approach the equivalent of a thousand chest X-rays, and his or her risk of developing cancer from the exposure would be expected to rise significantly though it would still not approach the level of the background cancer rate.

From the standpoint of radiation hazards, the average citizen is likelier to suffer more from unregulated or misguided medical diagnostics and treatment than from any other source, BEIR VII concluded. Although medical radiation accounts for 95 percent of the exposure we get from man-made sources, less than 5 percent of medical exposure is regulated, and the use of that 5 percent of medical exposure is regulated only in the broadest way. It's not subject to dose limits or dose constraints. There are two billion X-ray exams a year. CT scanning of the whole body of healthy people has become popular at clinics for the "worried well."

Fred Mettler, who has begun an international campaign to educate operators of diagnostic medical equipment in order to reduce radiation exposure to patients, told me, "The buttons on those machines are being pushed by millions of people. Digital radiography has replaced films and has the potential of reducing doses of radiation by half. But in fact doses have gone up because the technician will take multiple images. Without films for the radiologist to look at, there's no record of overexposure to X-rays, no idea of how many times that button was pressed. More people are having positive emission tomography [PET] scans. There are now more chest CT scans. All in all, there are more CT scans per patient than ever before. Worldwide, there are billions of patients getting medical radiation—and more of it each year."

A large and vocal coalition believes that the government has been keeping the truth about radiation from the public and that it's far more dangerous than is generally understood. These activists have been known to demand that nuclear sites be cleaned up so thoroughly that their radiation level is below natural background. Radiation protection experts reply that to do this would require the cleanup workers, and in some cases the general public, to be subjected to far greater exposures than they now are: any debatable benefit would not be worth the excessive risk and cost.

Radiation protection standards based on hard data rather than on the present pessimistic—and inconsistent—estimates regarding low-dose radiation could potentially save billions of dollars now being spent for cleanup and shielding that may turn out to be unnecessary. I began to think that it would be better to take some of that money and apply it instead to protection of the public from coal-fired plant emissions, waste that is inarguably harming hundreds of thousands of people today.

"Even though there are tons of data suggesting that there is a practical 'threshold' dose, below which radiation damage is either zero, or is repaired, or is handled in some other way, old perspectives die hard," Gómez said, referring to how LNT continues to dictate standards of radiation protection in this country. "There is no evidence of human cancers from exposures below 10,000 millirem."

Gómez and many of his colleagues believe that the paucity of information regarding the effects of low doses and low-dose rates and the lack of understanding of the mechanisms that cause radiation damage to cells could be remedied by experiments in a facility as impervious as possible to background radiation but still accessible to humans. The low-dose studies done by DOE were conducted against a background of natural radiation higher than the ultra-low-dose lab Gómez envisions. Raymond Guilmette, the past president of the Health Physics Society who specializes in internal dosimetry at Los Alamos, experimented years ago with cultured human lung cells surrounded by lead shielding. He originated the idea of an ultra-low-dose facility in the United States and has been cheering Gómez on in his quest. Guilmette and other researchers around the world on every side of the low-dose controversy are eager to get to work in such a setting to determine which theories are correct.

Others in the radiation protection field counter that for practical applications of radiation protection research is needed regarding the risks of background doses plus small doses of additional irradiation; these scientists believe such studies can be done in conventional surface laboratories. Gómez's reply is that public perception will not change until very clear information definitively shows what sort of risk low-dose radiation really poses and that to arrive at that information requires an ultra-low-dose lab.

For now, radiation protection standards will remain conservative. "Things are pretty linear for cancer down to five rads," Fred Mettler told me. "We know that from Hiroshima and Nagasaki and from breast cancer data. The argument really becomes what happens below that level. It looks like things happen in linear effect at low doses. As the dose increases, you get more mutations and breaks in DNA, but not necessarily cancer. The National Council on Radiation Protection and Measurements (NCRP) recently reported on this and said that you can't exclude the linear hypothesis. We assume it for protection purposes. I don't see that changing. There's a function of science, and then there's protection."

As I tried to distinguish between the consequences of high-dose radiation and low-dose and to acquire some perspective about the range of risk, I wanted to find out what the consequences would be if terrorists set off a radiological dispersal device, or dirty bomb.

Fred Mettler, who was an appointee to a Department of Homeland Security group addressing that very question, told me that radiation in a terrorist attack, in the form of nuclear materials dispersed by conventional explosives, should be easier to handle than chemical or biological agents.

"With things like anthrax, we're far less prepared. With dirty bombs, though, the best thing to do is to prevent exposure in the first place. Between 1943 and 2000, there were a hundred and fifty times when people got nailed for smuggling or trading in illicit nuclear products. But we don't catch people one hundred percent of the time. And there's a lot of loose stuff out there, particularly in the former Soviet Union." He talked about industrial radiological sources, in particular a device used for checking welds in pipelines. "There are a couple of million of these floating around the country. The source is radioactive enough to send gamma rays through one inch of steel. Now, that's radioactive."

Mettler clicked on his computer screen and a photo appeared: lower Manhattan on September 11, 2001, with smoke blowing toward Brooklyn. "What do you do if that smoke had contained radioactive material? At least you have a meter to detect any radiation, you have trained people, and you have a history of what happens. In any scenario, ninety percent of the dose will be external. Not much will be taken inside the body. Even with the guys at Chernobyl breathing crap in the reactor, most of their internal contamination dose didn't make a blip on the screen. Same with the people living in the contaminated villages around the reactor. Chernobyl was basically a big dirty bomb, and therefore the best model for the cesium dose a dirty bomb would cause. The findings of the Chernobyl Forum bear upon how we must respond in the event of a major radiation release.

"There, cesium is on the ground and the dose it's delivering is over a period of thirty years," Mettler said, referring to Chernobyl. Cesium's harm could come from ingestion, but the amount ingested would be extremely small. "Maybe one percent of that is the ingested dose over seventy years."

Mettler described his concerns about a dirty bomb: "One—dispersal. Two—panic."

In 1987, in Goiânia, Brazil, scavengers pried open a canister from a discarded radiation therapy machine that contained powdered cesium-137, a hard-gamma emitter. It glowed in the dark. People played with it, rubbed it on their bodies, ate sandwiches with the powder on their hands, and shared the blue, luminous substance with others.

"It caused four deaths and contaminated two hundred and fifty other people who were then treated with a drug that can reduce the internal dose of cesium," Mettler said. "How many people showed up to be checked by doctors? One hundred and ten thousand! In any terrorist situation, how do you deal with that many people who come to your hospital? In Brazil, about sixty-seven square kilometers were contaminated. The cleanup involved the disposal of four thousand cubic meters of waste. And this was a small accident. Remember, 9/11 produced many, many times that amount of waste—about one cubic kilometer (1,000,000,000 cubic meters). What if that 9/11 waste had contained radioactive materials? Strongly radioactive sources are not that hard to find. Industrial sterilization facilities have radioactive sources that are as hot as hell. They emit millions of rads in about five minutes."

Mettler told me that most doctors don't know what a radiation injury looks like. "They need to learn about that," he said. "A guy came into the emergency room here at the medical center with burned fingers. Most doctors would have given him a prescription for a cream and sent him home. But the woman who looked at him just happened to know something, and she called me and said, 'We have a radiation burn.' He was working for a private company on some equipment, and a capacitor had failed, and he was reaching inside the machine, trying to fix it, and he got a dose. Doctors need to know all these symptoms in case somebody comes into the ER with suspicious burns—this needs to be reported so that person can be investigated. What was he doing with a source? Where did he get it?"

Mettler began pulling out documentation to prove his point. "Here's a picture of a guy who stole iridium-92 in a third world country from an industrial radiography facility. . . . Here's a picture of the hole that one of those pipeline weld devices containing a radioactive source made in the leg of a Peruvian who stole it. . . . Here's a guy, a Russian who was a drug addict with a prison record. He was caught trying to sell an iridium-192 battery."

Not only does Mettler want all doctors to be able to recognize radiation exposure, he wants them to learn how to sort out dosed people arriving at hospitals. "We want to teach doctors about potassium iodide." Though potassium iodide blocks the uptake by the thyroid of radioiodine, it can't protect the gland—and other organs—against other radionuclides. "Doctors need to learn when to give potassium iodide, why, and to whom. It's a big problem. We have the data from Chernobyl, the thyroid cancer incidents. The younger you are, the higher the risk from exposure to radioiodine. There's been no increase in thyroid cancer there in adults. You want to give potassium iodide to pregnant ladies and to kids. But then there's the political stuff of everybody wanting it. There's a question of dose limits. No one would pay attention to them if there were a disaster." People over forty are advised by the Centers for Disease Control to take potassium iodide only in the event of major radioiodine contamination, since risk of allergic reaction is greater among older people and risk of thyroid cancer is lower.

In the event of such a radiological disaster, how do you determine where the radiation is and what you do about people who may have been exposed?

"The first responders—police, firemen, other safety personnel—have to carry a pocket sensor that starts to beep around radiation. There are also radioprotective drugs for the first responders. They're given now to people having radiation therapy. When first responders and hospital people approach a patient, they have to assume that there's not only radiation, there's also smallpox and anthrax. We at Homeland Security and the Centers for Disease Control have to use an all-hazards approach."

Mettler pointed out that the biggest problem in a radiological event would be communications. "It's the responsibility of the media to report things as they really are. But the media is often dependent on what the government says. The reporting for 9/11 was quite remarkable. It was reasonably accurate."

It's best to prevent postevent psychological effects in the first place rather than trying to fix them after the fact, and the best way to do that is to have in place detailed plans and to practice carrying them out in addition to establishing good lines of communication. Mettler referred me to the work of Steven M. Becker, a psychologist and professor of public health at the University of Alabama and director of its Disaster and Emergency Communication Research Unit who has studied responses to disasters. Dr. Becker told a 2004 meeting of the NCRP that surveys indicated public confusion about the advice to "shelter in place"—that is, to stay indoors in the event of a dirty bomb or similar event. And who was the best at communicating this and other information in a crisis? Becker's research indicated that the local TV weather reporter seemed to be the person the public most trusts for guidance, since he or she is unaffiliated politically, is well known, and understands science.

Mettler told me some of what Homeland Security is working on regarding counterterrorism. "We're trying to figure out how to mobilize medical teams, on-call rosters in case of emergency. And what about stockpiling drugs? How effective are they? How much do we need? When do we give them? What about sheltering and evacuation in case of a radiological incident? How effective is that? We're working on quick assessment of patients and on prioritizing the stockpile of drugs. Billions have been allocated for research and development. We need to focus on what will save lives. Radioprotection drugs, ways to figure out how much exposure someone has had. We're doing a document with the International Commission on Radiation Protection. With radiation stuff you have a certain set of parameters: where the nuclear power plants are, what the meteorology is, the probabilities of certain types of accidents. Normally you assume single events and you do the calculation. But terrorism has completely changed the equation in that you can't assume it's going to be a single event. The 9/11 attack was four events. Whatever planning you do, you can't allocate all resources to one place. The scenarios are different, too. We hadn't planned on suicide scenarios. Probability is out the window because terrorists are trying to upset probabilities. It's no longer a matter of just looking at radiation. You could get an event where anthrax, sarin gas, smallpox, and radiation were all mixed together. Whatever plan you devise has to deal with the most critical elements."

"What if we hear about a dirty bomb exploding?" I asked.

"OK. Say that something just happened and we don't know what it is. The best thing at that point is to shelter people—in their houses or in a building. You do not evacuate until you know where to go, which way the wind is blowing, whether it will rain. At home most people will have the most protection, they'll have food and be well sheltered. They will not have to worry about somebody infected with smallpox going through a camp of evacuees. The kids should go home. If I could say only one thing to the secretary of Homeland Security, it would be, tell everyone: Shelter, preferably at home, until we can figure out what happened."

Mettler read me a passage from a Homeland Security document stating in effect that dose reduction is best accomplished at home. He added a fact so vital that it should be emblazoned on bumper stickers and refrigerator magnets: "Doses can be reduced by a factor of a hundred or a thousand by staying indoors, removing clothes, and showering."

Rip had also told me that the best thing to do if a nuclear incident of any kind occurs is to remain indoors, or, if you are outside, to take shelter as soon as possible, because you will have more shielding. You should close doors and windows, and turn off any ventilation system that takes in outside air. The roof, walls, and windows will stop all alpha radiation, most beta, and, if distant enough from the center of the explosion, most gamma. If you've been outside, take a shower and wash your hair and wait for an all-clear announcement or orders to evacuate.

The scientists I talked to agreed that, based on simulations of dirty-bomb explosions in studies commissioned by the NRC and other agencies, the primary casualties would in most instances be caused by the conventional explosives, with only a few people—if any—being directly affected by radioactive contaminants. First responders have been drilled about what to do in the event of a dirty bomb. But the general public requires some training as well. This was done during World War II and during the cold war to prevent panic and deadly missteps. A coherent program of education should be developed. It could be presented in segments on the evening news or in spot announcements on the radio. The

George W. Bush administration and Congress ignored recommendations of the 9/11 Commission to allocate Homeland Security funding based on greatest probability of risk—the likeliest targets remain unprotected—and to rank the biggest risks and vulnerabilities in the country. During the five years following 9/11, the administration and Congress gave low priority to protecting the nation from weapons of mass destruction. In 2007 Congress finally voted to implement the commission's recommendations.

Terrorists could easily prey on a terrified and uninformed populace. The NRC, acknowledging that we really need to have some straightforward guidance, has issued a statement providing some perspective about the threat of a dirty bomb:

> At the levels created by most probable sources, not enough radiation would be present in a dirty bomb to kill people or cause severe illness. For example, most radioactive material employed in hospitals for diagnosis or treatment of cancer is sufficiently benign that about 100,000 patients a day are released with this material in their bodies.
>
> However, certain other radioactive materials, dispersed in the air, could contaminate up to several city blocks, creating fear and possibly panic and requiring potentially costly cleanup. Prompt, accurate, non-emotional public information might prevent the panic sought by terrorists.

I asked Jamie Gorelick, former Department of Defense general counsel under President Clinton, deputy attorney general of the United States under Janet Reno, and a member of the 9/11 Commission, for her thoughts. She now gives lectures on how we can better address large-scale disasters. "The 9/11 Commission found that preparedness and well-led consequence management capabilities are critical," she wrote me in an e-mail. "Both require a clearheaded understanding of the risks, stripped of hysterical rhetoric. We shouldn't be scaring people just to heighten concern. We should be educating, so that sensible plans can be put in place."

All the scientists I talked to, whatever they thought about the concept of the threshold dose versus LNT, were in general agreement that humans are less susceptible to radiation effects than had been previously assumed, that radiation is a weak carcinogen, and that once radiation hazards have been understood and safety procedures instituted, workers do not die or fall ill. I found the scientists who spend their days considering the best way to understand excess radiation exposure and to protect people from it to be careful, compassionate, and dedicated; they have been relieved to discover that the outcomes of radiological catastrophes have demonstrated much better survival rates than early calculations had predicted.

Rosalyn Yalow, a nuclear physicist who won a Nobel Prize for co-inventing the radioimmunoassay technique for analyzing blood and tissue chemistry, has said that there have been studies of populations living in areas with higher natural radiation, of radiation-exposed workers, of patients medically exposed, and of people accidentally exposed, and yet, "No reproducible evidence exists of harmful effects from increases in background radiation three to ten times the usual levels. There is no increase in leukemia or other cancers among American participants in nuclear testing, no increase in leukemia or thyroid cancer among medical patients receiving Iodine-131 for diagnosis or treatment of hyperthyroidism, and no increase in lung cancer among non-smokers exposed to increased radon in the home. The association of radiation with the atomic bomb and with excessive regulatory and health physics ALARA practices [As Low As Reasonably Achievable] has created a climate of fear about the dangers of radiation at any level. However, there is no evidence that radiation exposures at the levels equivalent to medical usage are harmful. The unjustified excessive concern with radiation at any level, however, precludes beneficial uses of radiation and radioactivity in medicine, science, and industry."

Just to review: we receive more radiation from nature and from nuclear medicine than from any other sources; it is extremely unlikely that most of us will ever receive high-dose radiation; the body does not distinguish between radiation from man-made sources and radiation from nature; and the consequences of low-dose radiation get lost among all the other natural and man-made assaults we live with. It's possible that some radiation is essential to health. Common sense tells me that the extra 200 to 300 millirem I receive during my stays in the Southwest are unlikely to harm me. Even conservative estimates based on LNT suggest that risk from low-dose radiation is tiny—even among nuclear workers. On a daily basis many of us subject ourselves to much more serious risks—like driving, smoking, or living downwind from a coal-fired plant or downstream from a chemical or petrochemical plant—than to any danger that might come from the generation of nuclear power. Despite the worst-case scenario that unfolded at Chernobyl—thanks to Soviet politics, inept management, a foolish experimental test, and an extraordinarily flawed plant design lacking reactor containment—the number of acute deaths the accident caused, and those conservatively estimated to occur over the next several decades, are expected to total just a tiny fraction of the 240,000 premature deaths that are caused decade after decade in the United States alone by coal-fired plants.

Part 3 **THE HIDDEN WORLD**

Out beyond ideas of wrongdoing and rightdoing there is a field. I'll meet you there.

—JALALUDDIN RUMI

7 RISK AND CONSEQUENCE

WE HAVE A POWERFUL LONGING for freedom from all risk. When it came to nuclear matters, I wanted certainty: a 100 percent guarantee of safety. Phrases like "highly improbable," when applied to my questions about whether a power plant might explode or whether nuclear waste would irradiate the biosphere, made me skeptical. Meanwhile, I was finding that most members of the nuclear community agreed that since the early days of the atomic age, radiation protection had vastly improved; that overconfident nuclear utilities, having made mistakes that they're still regretting, had lost public trust, and, despite having made many changes for the better, still had not fully regained it; and that nuclear power and its operators must continue to be stringently regulated and supervised. Because the nuclear tour was going to visit the hidden world of reactors, I wanted to know very clearly what the risks were.

On my next visit to Albuquerque, I learned from Marcia that in 2002 Rip had been asked to join a panel of independent scientists doing an internal peer review of the scientific findings of the licensing application for the Yucca Mountain Project. The final version of the application would be submitted to the Nuclear Regulatory Commission, which would decide whether this multibillion-dollar repository for nuclear waste was to be built in Nevada or whether the risks were too great.

I wanted to know if Yucca Mountain or any other nuclear facility could be made risk-free.

"The question really is how, when it comes to safety, health, and the environment, you can reduce uncertainties to a minimum," Rip said when I arrived at the FernAndersons. "To make a correct analysis of risk, you've got to have correct assumptions and as much accurate data as possible." He'd begun to apply this standard to his scrutiny of the claims made by the Department of Energy and its latest contractor, Bechtel SAIC, about Yucca Mountain.

In the 1970s, when the federal government began investigating appropriate geological formations for nuclear waste storage, Rip had led a preliminary data search on Yucca Mountain, located within the Nevada Test Site. He went on to head research into the feasibility of other waste-disposal solutions and ultimately to lead the team that got the world's first long-term, deep geologic repository for defense-related nuclear waste, the Waste Isolation Pilot Plant (WIPP), in southeastern New Mexico, open in 1999.

"Is Yucca Mountain going to work?" I now asked Rip. "I keep hearing that it won't."

He said he wouldn't know until he determined whether the science had been properly done and whether a risk analysis of the data would indicate that the repository could safely sequester nuclear waste for millennia, as Congress has mandated. WIPP had proven itself in that regard: the Environmental Protection Agency, agreeing with the analysis Rip and his team had done, certified that repository.

"But couldn't an unexpected event or chain of events make the nuclear waste there dangerous to the public?" I asked. "How can you be really sure it or a nuclear reactor or *anything* nuclear is safe?"

"Probabilistic risk assessment," Rip replied. He explained that all forms of energy generation pose hazards, some forms more than others. To enhance nuclear technology and improve human engineering in order to make the uranium fuel cycle as safe as is reasonably achievable, scientists and engineers have turned not only to better hardware and design but also to a method for reducing uncertainty.

He got out graph paper and drew a curve resembling the outline of the back of a turtle. It was the classic bell curve, which shows the normal distribution of data. Phenomena least likely to occur are represented by the extreme ends of the curve; those of highest probability by the curve's high point.

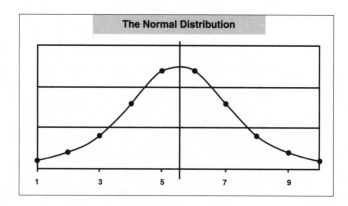

Probabilistic risk assessment (PRA) has its roots in ancient Egyptian and Chinese texts on gambling and in the work of the seventeenth-century philosopher and mathematician Blaise Pascal. In answer to a question from a gambler friend, Pascal devised a precise way to come up with the numerical odds of a future event. His method set in motion probability theory, which changed profoundly "the way we regard uncertainty, risk, decision-making, and an individual's and society's ability to influence the course of future events," as John F. Ross, a writer on risk, puts it. "Although this revolution has contributed to most of the marvels of the modern age, it has also introduced a host of new challenges and incumbent dangers as well as creating confusion, alienation and suspicion of modern science among much of the public."

In the early 1970s, in response to a request from Congress for a report on the probabilities and consequences of catastrophic reactor failure, the NRC sponsored a study directed by Norman Rasmussen, a professor of nuclear engineering at MIT. A panel of specialists relying on reactor performance data and on projections about the impact on public health of radiological releases made some predictions. These were based on the best information then available—at that time data from the real-world, day-to-day operation of power-plant reactors were limited—and on extremely pessimistic estimates. The loss of coolant to the hot core of a reactor resulting in overheating of nuclear fuel great enough to result in a meltdown—that is, reactor failure—had only occurred in controlled experiments with prototypes.

The Rasmussen study developed new concepts of risk analysis and performance assessment to predict potential failures of reactors in a probabilistic manner. Probabilistic risk assessment, called in most other countries "probabilistic safety assessment," centered on the analysis of a hypothetical chain of mishaps as well as a distribution of scenarios generated by interacting events that could lead to a core meltdown and a radioactive release. This method is known as a fault-tree, or event-tree, approach. What if the pumps that supplied the reactor core with coolant failed? What if the reactor's containment structures were completely breached and all the nuclear fuel dispersed in micron-sized particles? And what if this occurred during a storm that would disperse them over a heavily populated area? The report concluded that hypothetical risks to an individual were acceptably small compared with other risks that we already tolerate.

The Rasmussen study was published in 1975. In 1977, the NRC approved a peer review of the panel's findings that endorsed the PRA methodology as the best available tool. But the peer group warned that, because of incomplete data and those very conservative estimates, many uncertainties remained. In a 1991 study, the NRC applied PRA to specific nuclear plants in operation, in order to identify the likeliest potential risks, and concluded that, despite the study's use of conservatisms, American plants more than met the NRC's goals for safety.

"The conservative estimates not only lead to uncertainty but also lead to incorrect guidance from the analyses," Rip said, preparing me for a theme that would recur throughout our nuclear tour. "That is, the calculations suggest that something will happen that can't happen. For example, there's the fiction of Fencepost Man, who lives his entire life in a particular spot and gets all his food and water from that place, even if the site happens to be in the middle of a barren desert."

Some members of antinuclear advocacy groups have nevertheless selected information and conclusions from these old NRC reports, which preceded the problems at Three Mile Island and Chernobyl and the new information these accidents were to yield and which relied on assumptions that are mathematically tidy but do not occur in the real world (for example, complete and uniform dispersion of fuel in micron-sized particles). These theoretical events have an extremely low probability of occurrence but nevertheless have led some people—usually, antinuclear groups—to draw alarming conclusions.

Probability is a number expressing the likelihood that a particular event will occur, expressed as the ratio of the actual number of occurrences to the number of possible occurrences. "The traditional way of calculating such risks is deterministic," Rip said. "When you build a bridge, say, or set up an experiment or predict an outcome, you make crude estimates based on not very much data. Usually in big projects, things are calculated deterministically, not probabilistically." (That is, instead of a distribution of values, one value is used for each of the input variables you need, and it gives you one answer, one number. But if instead you choose the probabilistic approach, you take a sampling of the full range of each of the input variables so it gives you a full range of output variables.) He pointed to the bell curve. "And so it will tell you the probability of the occurrence of each of the answers you get. Probabilistic calculations are a huge number of deterministic ones summed up. But if you only use a deterministic calculation, and get one value, then what happens? Conflict. Arguments about my value being better than yours. You could never get people to agree on the value. But if you put aside deterministic calculation and instead used distribution of data sets and probabilistic calculation, and you covered all the values and all the combinations of values, the resulting curve would have all the information included in it that was in all of the sub-distributions.

"You start by looking at the subsystem. For a nuclear waste repository, for example, you'd look at the geology, hydrology, biology, and engineering—and then couple them into a total system." He drew a sketch of how subsystems are connected to form a total system.

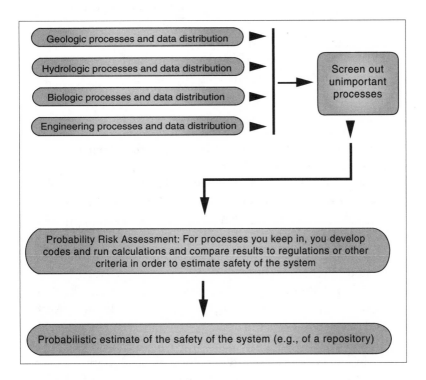

"When you do that, you have the whole picture. You can see what's important and what's unimportant. If anybody comes in, say from engineering, or geology, or microbiology, and argues, 'You didn't include my value,' well, you could show that you had used it and that it did or did not have a great effect. Probabilistic risk assessment stopped all that terrible arguing that somebody or other had used the wrong value and therefore your calculation was invalid. PRA is more precise. It first rules out subjectivity in estimating the likelihood and consequences of events."

After assigning a value to each factor that may affect the interaction of various events, scientists and mathematicians will run a huge number of computer simulations in which every event can be shown interacting with every other event. In this fashion they can learn what the probability of each interaction is. The result is a distribution of the risks along a bell curve. As dots in a pointillist painting add up to a field of poppies if you step back and regard the whole, the entire range of points of data reveals the big picture.

"You can see objectively what's happening at every point on the curve," Rip said. "The outcomes on the extreme ends of the curve are so unlikely—having a probability of occurrence of, say, less than 1 in 100 million—that you don't have to worry about them. PRA helps you guide the experiments you need to do, it eliminates the irrelevancies, the fluff, and at the end you've looked in detail at all the weaknesses and all the strengths and you've reduced the uncertainty, and you know where to spend time and funding most efficiently."

"This all sounds pretty abstract," I said. "What about PRA in every-day life?"

"People usually focus on consequence exclusively when they're in an unknown or new situation or one that's out of their control. If you're a passenger in an airplane, you don't control the outcome the way you do if you're driving a car. But the risk per mile traveled is actually much higher if you're in a car than if you're in a plane. We tend to equate or confuse risk with consequence. You can have a scenario with a high probability of risk and with very bad consequences. You can have a worst-case scenario with a low probability of it ever happening. PRA ranks these in the order of likelihood. It relies on common sense instead of guesswork."

"We all do PRA intuitively," Marcia interjected. "People might decide to work in a mine or a coal-fired plant despite the known health risks because the benefits and pay are good."

"And the people control the decision," Rip added. He explained that the risks in our everyday lives follow the same laws, both for events we understand and for ones that are unknown and scary. "For familiar events we unconsciously understand that risk is a multiplier of consequence and probability. We know that, for example, every time we drive a car we do not get into a wreck. A wreck is the consequence, but since we don't always crash, we know that the probability is small."

"But when the event is unknown and scary," Marcia said, "people automatically assume that the probability of it occurring is high and so they get riveted on the consequence."

"That's called a worst-case analysis," Rip said. "Troublemakers know we humans instinctively tend to do that, and they attempt to scare people into making bad decisions by always showing only worst-case consequences. For example, people tend to think that the worst event in the history of nuclear power, the explosion at Chernobyl, is the norm while ignoring the fact that worldwide, over twelve thousand cumulative reactor-years have passed in safety. To assess the risk for a sequence of events, the probabilities are multiplied together. Say you want to know the odds of missing an airplane flight you've booked. To calculate the risk, you could set up a sequence of actions and analyze them statistically to determine your odds."

> Event One: Go to the garage and start the car.
> Event Two: Drive on the highway.
> Event Three: Park the car in the airport garage.
> Event Four: Get on the plane.

"In Event One, assume that you own an old car and that once out of every 1,000 times you turn on the ignition it fails to start. So the probability of the car failing is 1 in 1,000. If that happens, you can't drive it to the airport so you'd miss the plane. In Event Two, the car starts, but if you got into an accident on the highway, you'd miss your plane. Assume the probability of having a wreck while driving to the airport is 1 in 10,000 trips. The risk of missing the plane from only Event 2 is then 1 in 10,000. In Event Three, the multistory airport garage collapses on you and your car, and you miss your flight. Suppose the chance of this garage collapse is 1 in 1,000,000. The risk of missing the plane from the garage collapse would be 1 in 1,000,000, and the risk of all three events occurring with this particular flight is 1 in 10 quadrillion (1000 × 10,000 × 1,000,000). In other words, that outcome is virtually impossible. In Event Four, the plane has a mechanical problem and the flight is canceled. Let's assume that the probability of major mechanical problems is 1 in 1,000.

"Again, if you took the worst-case view and only looked at consequence, you'd assume that you'd always miss your flight," Rip said. "The combined risk of all four events occurring on your trip to the airport would be 1 in 10,000,000,000,000,000. This is a vanishingly small number compared to a worst-case analysis of all four events, which would say that you'd never make your flight. Suppose you were planning to go to the airport and you heard a hurricane was coming. Then Events One through Four would not even occur, because the airport would be shut down. But when people make conservative estimates and link together worst-case scenarios, the result can be an overall picture of a situation that is impossible in the real world."

Conservative, rough estimates can therefore result in notions of extreme outcomes that may play to our worst fears and lead us to assume that consequences that are just about impossible in the real world are highly probable. PRA offers a realistic approach to evaluating potential risks, and recent refinements have enabled the NRC to hone its predictions. In 2006 the commission issued an official disclaimer about the now-outdated research of the 1977 and 1991 worst-case analyses:

> The U.S. Nuclear Regulatory Commission has devoted considerable research resources, both in the past and currently, to evaluating accidents and the possible public consequences of severe reactor accidents. The NRC's most recent studies have confirmed that early research into the topic led to extremely conservative consequence analyses that generate invalid results for attempting to quantify the possible effects of very unlikely severe accidents. In particular, these previous studies did not reflect current plant design, operation, accident management strategies or security enhancements. They often used unnecessarily conservative estimates or assumptions concerning possible damage to the reactor core, the possible radioactive contamination that could be released, and possible failures of the reactor vessel and containment buildings. These previous studies also failed to realistically model the effect of emergency preparedness. The NRC staff is currently pursuing a new, state-of-the-art assessment of possible severe accidents and their consequences.

"Looking at everything in a probabilistic mode requires most people to make a leap in their thinking," Rip continued. "We're wired to react to risks that are sudden and make a big bang. People who get all puckered about getting hit by a meteorite—the chance of that is something like 1 in 17 billion—might be unafraid to smoke, or to talk on a cell phone while driving without a seat belt, or to live downwind from a coal-fired plant, even though the risks from these things are much greater, the consequences could be fatal, and they're far likelier to happen."

Marcia said, "I learned from Rip that whenever you hear doomsday-type predictions made by troublemakers or con men who mislead people by using worst-case analysis, you should step back and ask what the probability—and therefore the risk—of the event is. He taught me that you don't just look at the consequences. If you can't make a reasonable estimate of the probability of an event, ask someone who's knowledgeable."

"At Sandia in the 1970s we all worked on probabilistic risk assessment concerning reactor safety or heard about it," Rip said. "As computer power grew, PRA advanced, and it became cutting-edge science."

Back in those days, he became the first to apply this way of assessing risk to something besides reactors when he headed a program to investigate the practicality of disposing of high-level nuclear waste thirty meters below the floor of the deep ocean. Every day he asked himself how this scheme could fail and he encouraged others to do the same. Serendipity intervened. One day Melvin G. Marietta, a slender Sandian with a pensive air and a somewhat pessimistic bent, appeared in Rip's office and said, "I'm an applied mathematician and I like to make mathematical models."

"Immediately we struck up a working relationship that goes on to this day," Rip told me. "We're doing the inside peer review of Yucca Mountain together."

Les Shephard, a vice president at Sandia National Laboratories and a former participant in the seabed program, once described to me the magnitude of the duo's accomplishment. "Mel Marietta is to me one of the world's leading experts in probabilistic risk analysis and performance assessment, and of course Rip acted as the mastermind. Rip saw the potential of Mel's modeling and how it could be applied to the long term. It's really big science. Rip is the father of probabilistic risk assessment as it's used throughout the world today in all kinds of applications." Shephard went on to say that Rip had recognized early on that the social and political factors affecting nuclear waste disposal had as much importance as the technical ones and had seen that the framework of probabilistic risk assessment could be used to engage the community, to understand its responses, and to build people's particular interests and anxieties into an overall assessment of risks.

"What led you to decide that the Waste Isolation Pilot Plant could safely contain nuclear waste?" I asked Rip.

"We were able to use PRA to determine which factors and combinations of factors would or would not pose an eventual risk. We had to identify all features in the environment—aquifers above the salt bed, sinkholes, things like that. We studied events that had occurred over millions of years, like floods and earthquakes. We analyzed all the possible processes, like rainfall and erosion. We collected data regarding these features, events, and processes and assigned numerical values to each aspect of each of them, made computer models of them, and used those models to determine whether something was likely or not likely to happen. A probability of nearly 1 means it will—practically speaking—always happen because it has already happened."

He gave the example of rain falling on the desert where the WIPP site is. "This has happened historically, so the likelihood of future rainfall was assigned a value of almost 1. A probability of zero means it will never happen. You could say the probability of rainfall continuing to occur during a period of ten thousand years is 0.9999999. Then there is only a 0.0000001 probability that you are not accurate in this prediction. Nothing is ever one hundred percent. The air might suddenly be sucked out of the room if all the nitrogen and oxygen atoms vibrated in exactly the right way, but the chance of that happening over the lifetime of the universe is extremely tiny.

"We had a jillion simulations going so that we could look at every possible scenario. For example, we had geological data about floods in southeastern New Mexico for the past hundred thousand years or so. Well, what if the temperature rose, glaciers melted, and rainfall increased so much that there were floods at the WIPP site—would they affect the repository deep underground? No, because there is a feature—a layer of impermeable, cementlike rock—capping the salt bed. So we could eliminate flooding as a risk."

WIPP met EPA's standards, won court cases, and got licensed. "It succeeded because we'd done over a thousand studies and could show with data, computer modeling, and risk analysis precisely what could go wrong and what could never happen. It was bulletproof. I don't know yet if the same will be true of Yucca Mountain. The internal peer review is a step toward finding out whether it's going to be safe."

Later we'd be visiting Yucca Mountain and WIPP. In the meantime, I wanted to know whether PRA could be called a mathematical system for making the most accurate predictions possible by using the most data possible.

"It's a mathematically driven *philosophy*," Rip replied. "You've asked why I think that WIPP is safe, or why I consider nuclear energy low risk as compared to fossil-fuel energy, or why I believe that a dirty bomb would be unlikely to expose a lot of people to harmful radiation, or why I know that a terrorist trying to crash a jet into a spent-fuel pool would fail to cause a disaster. Probabilistic risk assessment based on accurate data—and the more data, the better—shows what the vulnerabilities are, and the risks, and the consequences. There will always be uncertainty. That's the quantum nature of the universe."

Acceptance of uncertainty seemed to me a huge step out of my normal viewpoint, but one well worth contemplating as being more in harmony with the actual nature of human existence, rather than the imagined one in which we generally dwell. Pascal, in his *Pensées*, asked:

> For after all what is man in nature? A nothing in relation to infinity, all in
> relation to nothing, a central point between nothing and all and infinitely

far from understanding either. The ends of things and their beginnings are impregnably concealed from him in an impenetrable secret. He is equally incapable of seeing the nothingness out of which he was drawn and the infinite in which he is engulfed.

"In many cases, PRA identifies that uncertainty to a great degree," Rip continued, "and tells you how to minimize the risk in your design. Probabilistic risk assessment is about the future. You take a range of scenarios, you calculate how likely they are to occur by themselves or in combination with others, and if they do, what the consequences will be."

He pointed to the bell curve. "What is the worst case, and how many features, events, and processes would work together to produce it? And how likely is that worst case to occur? What's the likeliest, there in the middle part? You figure out what you really need to pay attention to and where to put your dollar. Probabilistic risk assessment changed my life. After I learned it I thought differently about everything."

The breakthrough PRA methodology Rip and Marietta elaborated and went on to employ in other major programs in the nuclear world, such as counterterrorism, continues to be refined and advanced and used elsewhere. Today, the EPA applies PRA to Superfund cleanup projects and other programs. NASA has set about to improve its PRA capabilities to enhance mission safety and performance. Engineering firms rely on PRA, and private industry has employed its software to study everything from whether contaminants affect an endangered dolphin species in Hong Kong Harbor through determining risks and consequences relating to food safety and airborne diseases to calculating the vulnerabilities of the Gulf Coast to major hurricanes.

The biggest PRA application of all may be the analysis of the complex interaction of systems leading to global climate disruption. For decades now, scientists from many disciplines have been collecting an immense amount of data regarding features, events, and processes all over the planet that signal anomalous changes caused by the increase in greenhouse gas emissions and the rise in temperature. Most predictions about how many degrees the temperature is going to climb and how many centimeters or meters the ocean is going to rise come from PRA. Computer modeling of all the information being gathered is creating a portrait of a change that thousands of scientists from many disciplines consider catastrophic.

Since a large number of these features, events, and processes are already occurring, there is a probability of nearly 1 that they will continue and interact. Over time, predictions about the consequences will grow more exact, but there will always be a measure of uncertainty. The word *uncertainty* can lead some politicians and the fossil-fuel industry to claim that we need to study the problem more before coming to any conclusions—despite an overwhelming body of data that supports anthropogenic greenhouse gases as a major cause of global warming.

Curiously, I discovered that the same environmental activists who implicitly believe in the models of global climate disruption that have been derived from probabilistic risk assessment nevertheless distrust that same methodology when it is applied to nuclear safety.

"The battle between scientists and environmentalists can be resolved by using and trusting a good probabilistic risk assessment, not by using extremes," Rip said. "Let's work together to arrive at the mean of the distribution." He was referring to the field in the middle of the bell curve. "That's where the real world is. That's where we can meet."

8 GOING TO EXTREMES

CLEARER ABOUT WHAT CONSTITUTED RISKS from radiation and about how those risks were assessed, and what the worst consequences were, I was ready to continue following the life cycle of uranium.

Uranium must go through a series of transformations to become usable fuel. Drums of yellowcake were once trucked from Ambrosia Lake and other mining districts to a plant in Oklahoma and chemically turned into uranium hexafluoride gas, which was then shipped to the Oak Ridge National Laboratory in Tennessee for enrichment at its gaseous diffusion plant there. I'd visited Oak Ridge as a tourist before it closed in 1985; now shut down, it's still undergoing major decontamination and environmental remediation. From an overlook on a rise, I was permitted to gaze down at the roofs of some four hundred vast structures covering 725 acres. The only other man-made structure comparable in scale that I've seen is Hoover Dam.

Begun during the Manhattan Project, the plant had executed one of the most difficult and painstaking tasks that humans have ever attempted. The uranium hexafluoride gas was pumped through a vast network of pipes and membranes—a maintenance nightmare—that separated atoms of the more fissionable uranium-235 from uranium-238 by taking advantage of an extremely slight difference in weight between the two isotopes. Rip compared the process to a race in which two groups of people run a marathon, and one group weighs slightly more than the other. Over a long distance, the heavier runners would probably be a little slower and the lighter ones would pull away and win the race. The other big American gaseous diffusion plant, in Paducah, Kentucky, is now the only producer of enriched uranium in the country. The operator, the recently privatized United States Enrichment Corporation (USEC), is presently constructing a state-of-the-art centrifuge plant.

Gas centrifuge technology is more compact than gaseous diffusion and requires only a fraction of the energy expenditure. To machine and set up the centrifuges properly is difficult, even with the best equipment and expertise possible. Uranium hexafluoride gas is fed into a rotor tank that spins inside a casing that has been evacuated of air. Centrifugal force causes the heavier U-238 atoms in the gas to concentrate on the outside wall, thus largely separating them from the lighter U-235 atoms. The enriched product is then bled off from the inner wall of the rotor tank into the next in a cascade of centrifuges that gradually increase the percentage of U-235 atoms in the mix. The process is tedious, and balancing and maintaining the cascade is a challenge. The end product, if enriched to between 3 percent and 5 percent of U-235, fuels nuclear reactors; if enriched above 80 percent it becomes highly enriched uranium (HEU), which can fuel the fission components of nuclear weapons.

I had assumed that any country with civilian nuclear power plants could make a nuclear weapon. But I learned that this route has never been taken by any of the eight nuclear nations, as those possessing atomic bombs are called. All of them built the bomb before establishing nuclear power plants to generate electricity.

"People like to say that nuclear power plants are the path to weapons," Rip said. "But that's not right. If you have a steel ingot, you can make a saucepan out of it or you can make a machine gun. To stop machine guns, should you ban steel? There are two types of reactors, a reactor designed to make energy and a reactor designed to breed plutonium. You can't make a bomb directly from either one. In order to do that, first you have to take the reactor's spent fuel out and reprocess it to separate the plutonium from the nonfissile U-238—it makes up most of the spent fuel—and the fission products. That requires a big, messy, complicated operation and special equipment for remote handling of the highly radioactive materials. Say you go to all that trouble. About one percent of low-enriched fuel turns into plutonium in a power reactor, but as I've told you, it's plutonium-240. The longer the fuel rods are in the reactor, the more contaminated they are with Pu-240. It's really hot, thermally and radioactively, and hard to handle."

Plutonium-240 builds up in reactor fuel over time. Low-enriched fuel remains in a power reactor about eighteen months. In bomb-production reactors, the rods are removed much sooner in order to obtain plutonium-239, which is weapons-grade. Using plutonium-240 from spent nuclear fuel, the U.S. national laboratories built and tested some bombs in the 1960s with mediocre results. Being so hot, such weapons have a shelf-life of only a few years and are therefore so impractical that no other country has bothered to go that route.

"Anyway," Rip continued, "you don't even need a reactor and a reprocessing plant if you can afford to build a uranium enrichment plant and use it to produce highly-enriched uranium. You can then make a uranium bomb. Configuring the trigger for a plutonium bomb and engineering a delivery system are very hard. Any terrorist hell-bent on making a bomb would go the uranium route, if he could get his hands on highly enriched uranium. But countries go the plutonium route."

The right to enrichment technology is guaranteed to signatories of the Nuclear Nonproliferation Treaty (NPT). A country that wanted to develop a bomb on the cheap might secretly attempt to reprocess power-plant fuel, but enrichment facilities are unconcealed, subject to international scrutiny, and a country caught diverting civilian nuclear fuel would be risking international sanction. This happened to North Korea.

The main secret of making an atomic bomb is that it can be done, and in 1945 that fact became known to the world. In the 1950s, the United States and the Soviet Union were racing to be the first to develop commercial nuclear power and to cash in on what was assumed to be a large, potentially global market for it. For this reason, and also to pull ahead in the ideological struggle between communism and capitalism, the American Atoms for Peace program gave universities around the world research reactors fueled with highly enriched uranium; the countries had to pledge to forgo military applications. To control this fissile material, the United States leased the reactor fuel and took it back for reprocessing when it was spent, but after 1964 the U.S. also started selling the fuel abroad, and in 1978 discontinued the reprocessing option.

The United States and Canada provided India with a heavy-water research reactor. Despite a pledge to use it for peaceful purposes, India eventually built its own reprocessing facility, covertly extracting enough plutonium-239 from spent fuel over the years to make a plutonium bomb, which was tested in 1974. Not a signatory to the NPT, India has since been excluded from international sharing of peaceful nuclear technology and materials—though not from safety information—and has resisted any outside interference in ongoing weapons production, relying on domestic resources; it has also created a successful civilian nuclear power program. A. Q. Khan, a Pakistani metallurgist who had worked in the Netherlands, stole Dutch centrifuge technology, returned home, and built an enrichment plant and began producing HEU in the 1980s. With help from China, Pakistan manufactured several bombs of tested Chinese design, and in 1998, immediately after India tested new weapons, Pakistan held its first nuclear tests in response. Like India, Pakistan has suffered a similar exclusion for failing to sign the NPT; its civilian power plant program, established after the weapons program, remains small. By 1993, North Korea, relying on Soviet technology, used its uranium mines, a 5-megawatt research reactor, and a reprocessing plant to make a quantity of weapons-grade plutonium sufficient for several bombs. The reactor, the only one in North Korea, was shut down in 1994 after extended negotiations with the United States and placed under the United Nations International Atomic Energy Agency seal. The United States learned in 2003 that North Korea had purchased a small stock of centrifuges from A. Q. Khan. (His penchant for exporting nuclear technology to poor countries is described by William Langewiesche in his 2007 book *The Atomic Bazaar.*) The U.S. accused North Korea of clandestinely enriching uranium, although no evidence has emerged that it has actually done so. The renewed conflict with the United States led North Korea to abrogate its previous agreement and to restart its sole reactor, a research reactor, which North Korea claimed was providing electricity to the public. But surveillance photos show that the site has neither a turbine building nor transmission lines.

In 2006 North Korea tested a plutonium nuclear device. Experts think it was either a dud or an advanced design of low yield. Israel, which has no nuclear power plants, refused to sign the NPT; it is known to possess a considerable arsenal of nuclear weapons.

Iraq began working on uranium enrichment after Israel destroyed Saddam Hussein's only nuclear reactor in a bombing raid in 1981. After the first Gulf War, under UN orders, the IAEA dismantled Iraq's nuclear weapons program, and it has been defunct since the early 1990s. Iran began constructing a centrifuge-enrichment operation in competition with Iraq, following the Iran-Iraq War. An Iranian anti-government organization eventually revealed the program to the world. Since Iran had signed the NPT and had agreed to protocols that permitted IAEA inspectors to look for additional facilities, the undeclared program was a treaty breach. Iran, invoking its right under the NPT to enrich uranium for peaceful use, has declined to halt its development, and refused an alternative offer of fuel for a peaceful nuclear-plant program to be supplied and reprocessed by Russia. The UN is applying diplomacy and imposing sanctions to curb both North Korea and Iran. The IAEA, whose mission is to encourage the peaceful use of nuclear energy and to prevent proliferation, has been awarded the Nobel Peace Prize and is highly respected for its achievements; its research and inspection teams are considered by the scientific community to be credible. The NPT previously restricted IAEA inspections to facilities that signatories had declared, but a new additional protocol, which most signatories have endorsed, allows surprise inspections to undeclared facilities. The UN Security Council will continue to pressure Iran and North Korea to accept a monitored fuel cycle and to abandon enrichment and reprocessing.

Harold F. McFarlane, who is Deputy Associate Laboratory Director for Nuclear Programs at the Idaho National Laboratory and the president of the American Nuclear Society, told me that the flow of knowledge about nuclear technology cannot be stopped; too many people around the world are working on it. Since that is the case, "it is curious and promising that only a few countries have chosen to develop nuclear weapons," as Richard Rhodes writes, noting that John M. Deutch, who later became director of the CIA, had estimated in 1992 "that some twenty to twenty-five nations had explored the acquisition of a nuclear-weapons capability and could begin building such weapons within a relatively short time—perhaps six months or less—but had decided not to do so. . . . Deutch's estimate implies that the fears of those who believe that nuclear-weapons acquisition is driven primarily by technology are unfounded. In short, it's not lack of technical know-how that stops people building bombs. What does make a country go nuclear? Historically, only a perception of a fundamental threat to a nation's survival."

Harold McFarlane told me that the Global Nuclear Energy Partnership (GNEP), an initiative of the DOE, has looked at the problem of rogue states and sub-national groups that might want to build bombs. "GNEP has come up with resolutions to some of the issues," he said. "You would not export sensitive technology—in particular, reprocessing and enrichment technology." The GNEP intends to set up an international consortium of nations with secure means of making and recycling nuclear fuel, which would provide it to those countries wanting carbon-free electricity and take the fuel back for reprocessing. This arrangement already exists informally to some extent, and the IAEA has proposed a similar idea. The GNEP would formalize and enhance it, researching and demonstrating technologies that would improve and secure the fuel cycle, enhance energy security and environmental safety, and promote nonproliferation.

I spoke to Jonathan Schell, who has written extensively on nuclear proliferation and whose forthcoming book is *The Seventh Decade: The New Shape of Nuclear Danger*. He argued that the right to enrich uranium emphatically enables proliferation. "If the enrichment process were internationalized, the problems would be solved, except for possible clandestine construction," he said. "The most important step would be to take the nuclear fuel cycle out of national hands and put it in international hands. Uranium enrichment and plutonium separation are the hardest technical tasks to perform on the way to the bomb. They're the Mount Everest standing in the way of a new state's acquiring fissile materials. But how do you prevent more countries from acquiring the fuel cycle?

Internationalization, which I would favor, is a nonstarter with the U.S. and other countries now doing enrichment. The problem is that the non-nuclear powers refuse to give up the right to enrichment as long as the nuclear nations continue to lay claim to it. This interferes with all serious nonproliferation efforts."

At present, the nations that reprocess spent fuel guard the small volume of plutonium that results. It is now in some cases being turned into reactor fuel, which further renders it useless for weapons. Mohamed ElBaradei, the director general of the IAEA, in tandem with the financier and philanthropist Warren Buffett, wants to establish a nuclear fuel bank as a more financially attractive and more secure choice for those countries now considering construction of their own enrichment and reprocessing facilities.

"ElBaradei says we must put the fuel cycle under international control, and everyone—including countries that already have nuclear power and the bomb—has to get fuel from the controlling body," Schell said. "But the nuclear nations don't want to give up their national fuel-cycle programs. And so there's a double standard: countries who want to enrich uranium or reprocess their spent fuel are told that they can't, but it's okay if France and Russia do. These are very big political obstacles. If nuclear technology were completely under international control, and all weapons banned, then the opportunities for proliferation would be radically diminished. Such an event would decrease one of the disadvantages of nuclear power. But as things stand now and for the foreseeable future, an expansion of nuclear power would increase the risks of proliferation."

To address this risk, new power-plant reactor and reprocessing technology is being designed to restrict opportunities for proliferation under cover of a civilian electricity program. Given that in the long run nuclear plants are likely to play a significant role in greenhouse-gas mitigation, with many more being built, some analysts have suggested a hub-and-spoke arrangement in which certain provider nations supply sealed reactor cores that can operate for a few decades and then take them back for recycling.

Since 1945, diplomacy and sanctions have prevented nuclear attacks, but it seems clear that any nation that is determined, beyond all common sense, to have the atomic bomb and can afford it will probably be able to get one if it is not stopped. The presence of civilian nuclear power plants seems not to be a deciding factor one way or another.

About half of all the uranium fuel in American reactors comes from retired nuclear weapons. As of April 2007, the USEC had transformed

300 metric tons of highly enriched uranium (about 90 percent U-235) into 8,774 metric tons of lowly enriched uranium (about 3 to 6 percent U-235) for power plant fuel, thus eliminating 12,000 nuclear warheads that had been purchased from the former Soviet Union as part of a program called Megatons to Megawatts. By 2013 a total of 500 metric tons, or the equivalent of 20,000 warheads, will be turned into low-enriched fuel with the energy equivalent of three billion tons of coal (thirty million coal cars). This not only reduces the inventory of nuclear weapons in the world, it also prevents the uranium from being cycled back into bombs, since once it passes through the reactor the uranium is no longer suitable for bomb use. The warheads once intended to blow up our cities are now providing about 10 percent of the electricity that lights them. The United States has agreed also to dismantle a portion of its warheads in the coming years, and they could supply a new source of fuel. In 2006 the NRC licensed an international consortium to begin enriching uranium for fuel at a centrifuge plant—to be built in southeastern New Mexico— that will operate far more efficiently and use less energy than the old designs.

Of the astounding effort to enrich uranium that began with the Manhattan Project, Richard Rhodes wrote in *The Making of the Atomic Bomb*, "No essence was ever expressed more expensively from the substance of the world with the possible exception of the human soul." For about thirty years after the inception of the Manhattan Project, nuclear scientists may have thought they'd bested nature by enriching uranium and using it in a sustained chain reaction on earth. But they were wrong.

One day I asked Rip the story behind the photograph of him in a wet suit diving headfirst into the transparent blue coolant water of a reactor pool. When he was studying nuclear chemistry and oceanography, he told me, he'd become a certified diver, and later became a qualified diver at Sandia. He descended into the pool containing the Sandia reactor to unbolt a metal plate so that some rods could be removed with manipulators. "It would have taken weeks to do that remotely," he said. The thought of anybody swimming in a reactor pool had shocked me more than his story of safely holding an egg of plutonium on his palm. But he assured me that prior to the dive, the chain reaction had been stopped, the core had cooled down, and he'd put radiation detectors on top of the fuel bundles and found there was no problem. "The fuel bundles have an inert section at one end that allows you to handle them. That provides some distance between the person and the fuel rod. I was probably five feet from the nearest uranium pellet and had a lot of water to shield me. So I knew I was OK."

"I don't care what you say," I replied. "There is nothing more against nature than reactors."

He replied offhandedly that long before humans appeared, natural nuclear reactors had existed on our planet.

This seemed unbelievable. He reminded me that uranium contains several isotopes and is mostly made up of the isotope U-238, with traces of U-235. He then told me that in 1972, a scientist at a uranium fuel plant in France made the alarming discovery that a particular batch of yellowcake from mines in Gabon, Africa, was quite deficient in uranium-235, the isotope essential to the chain reaction and therefore highly coveted. Only a tiny fraction of natural uranium is U-235. Since the U-235 isotope is so rare and takes so much effort to acquire, it's heavily guarded, and minutely and repeatedly measured by any agency or government that has gone to the enormous trouble and expense to extract it. French officials scrutinizing the absence of that much U-235 suspected theft by nuclear terrorists. Eventually geologists cleared up the mystery.

When the earth formed, 4.5 billion years ago, a great deal more U-235 was present, and it made up one-quarter of natural uranium rather than the 0.7 percent of today. Because U-235 has a half-life of 700 million years, there is less of it around today. U-238, which has a half-life of 4.5 billion years, now makes up almost all of the world's uranium. Around 2 billion years ago, when the natural proportion of U-235 had decayed to about 4 percent, a planetary crisis occurred: blue-green algae evolved, harnessing photosynthesis and in the process emitting a lethal gas that killed off most other organisms. Over time, as many new life-forms flourished, the air and water became rich in this gas—we call it oxygen. Richly oxygenated groundwater dissolved uranium out of rock. Algae growing in mats in a river delta in Oklo, Gabon, may have contained microorganisms with the remarkable ability to absorb the freed uranium ions from flowing water and concentrate them into deposits that became very abundant in all the isotopes of uranium, and especially uranium-235. This concentration of U-235 was enough, when moderated by water flowing through the deposits, to start a slow-neutron chain reaction, meaning that on a subatomic level in the natural reactors exactly the right conditions led to the expulsion of neutrons from atoms of U-235. These neutrons struck other nuclei, breaking them into pieces that in turn struck still other nuclei, freeing more and more neutrons and starting the chain reaction that became self-sustaining. This could only hap-

pen in the absence of contaminants—boron and cadmium, for example—
that absorb neutrons. The chain reaction was moderated by water, which
has the capacity to slow down neutrons enough to enable them to be captured by the nuclei of neighboring uranium atoms. The heat released in
the fissioning process turned groundwater to steam. When the water
boiled away, the reactor lost its moderator and the chain reaction
stopped. More groundwater then flowed through the deposit, the chain
reaction started again, and the process repeated itself—these natural
reactors cycled the way geysers do.

Over a period of 150 million years, sixteen reactors in Oklo produced
power at about the 100-kilowatt level, their self-regulation preventing
meltdowns and explosions. Finally, the fuel supply was exhausted.

A few samples of uranium dug out of the Colorado Plateau also have
been depleted of U-235, and some ascribe this to the presence of ancient
reactors as well. Rip was doubtful. "The concentration of uranium here-
abouts is sufficient to support natural reactors, but the rocks contain
boron and cadmium. Even in trace amounts they absorb enough neutrons flying out of the uranium atoms to prevent a chain reaction, which
is why boron and cadmium rods are used to shut down man-made reactors. The control rods are like the accelerator going up or down in a car.
At Oklo it just happened that there were no elements in the rocks that
had the capacity to bind the neutrons and poison the chain reaction." He
added, "All we're doing in man-made reactors is speeding up Mother
Nature."

"How can you make that claim?"

He reached for a paper napkin and got out a pen and sketched the
planetary model of a uranium atom, in which the electrons circle the
nucleus the way planets circle the sun. "It all gets down to neutrons."

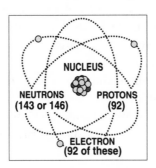

The protons and neutrons that make up the nucleus of any atom are
held together by the strong force, one of the fundamental forces of
nature that operates throughout the universe, along with electromagne-

tism, gravity, and the weak force, which has to do with radioactive decay. Most nuclei are stable. But the uranium-235 nucleus, relatively large and heavy, suffers from an instability caused by its protons, which, being positively charged, repel one another. A free neutron that's moving slowly among atoms and happens to be captured by a uranium nucleus can increase its tendency toward imbalance enough so that the strong force can no longer counter the force of repulsion among the protons. The nucleus ceases to cohere.

Fission occurs: the atom shatters violently, releasing fragments and energy. The strong force reasserts itself and two smaller atoms—and therefore two newly made elements—assemble themselves from the wreckage. Each being positively charged, they repel each other. As the fragments of the old nucleus, including a few free neutrons, rush apart, an extraordinary amount of energy is released as heat. If the speed of the neutrons can be slowed down by arranging for them to collide like billiard balls with the nuclei of atoms in some medium such as water or graphite, then they can be more easily captured by nearby U-235 nuclei, which in turn become unbalanced and wobbly and break apart and release still more fragments, neutrons, and energy. This process has its origins in the subatomic realm of quantum physics, where uncertainty rules.

The neutrons slow down if they collide with a nucleus that is much lighter than one in a uranium atom and will absorb some of the energy and cool off the neutrons—the hydrogen nuclei in water, for example. When a neutron has, say, one hundred energy units and has a direct and elastic collision with a hydrogen atom that is approximately the same size, the hydrogen atom flies away and now the neutron has fifty energy units and the hydrogen atom has fifty energy units. Then the neutron runs into another hydrogen atom and again loses half of its energy and now has only twenty-five units. With each collision, that neutron is being progressively slowed down. H_2O works well as a moderator because it has lots of hydrogen atoms. Hydrogen atoms have a single proton as nuclei, and protons are about the same size as neutrons, so the collision results in a large energy exchange and the neutrons quickly slow down. The energy released as the nuclei undergo fission heats up the water to a boil. Without water, the neutrons shoot away and disperse, ending the chain reaction.

The Oklo fossil reactors not only consumed U-235, they also bred more fuel for themselves. Neutrons liberated by the chain reaction can penetrate the nuclei of the more stable isotope uranium-238. As a result, it doesn't fission but does form a new element, neptunium-239, which decays in about four days to plutonium-239. After about twenty-four thousand years, half of the plutonium has decayed down to U-235. At Oklo, the plutonium, which remained close by, provided about a third of the fuel needed to keep the chain reaction going.

The natural reactors created twelve thousand pounds of additional uranium. The total released energy the Oklo reactors generated probably amounted to about 15,000 megawatt-years—the equivalent of what is generated today by a single large man-made nuclear reactor over a period of fifteen years.

About 1.7 billion years later, life had progressed in complexity from that of blue-green algal mats to that of the neuronal webs in the brains of those pioneers of nuclear enegy, Albert Einstein, Leo Szilard, and Enrico Fermi, none of whom lived long enough to hear of the discovery of the Oklo reactors but who may have been aware of a couple of papers positing the existence of such phenomena.

The extraordinary history of the discovery and application of nuclear energy is narrated in detail in Richard Rhodes's *The Making of the Atomic Bomb*. In the early twentieth century Einstein declared that matter was composed of energy. Szilard was a fan of the prophetic science fiction writer H. G. Wells, and in 1932 read his novel *The World Set Free*, which described, in Szilard's words, "the liberation of atomic energy on a large scale for industrial purposes" as well as the development of an atomic bomb, used in a 1956 war waged by England, France, and America against Germany and Austria, to destroy the major cities of the world. A year later on that famous stroll Szilard took through London, his vision of the nuclear chain reaction—mass being converted into energy on a nuclear level—set off a conceptual chain reaction that, fueled by experimental results, spread among physicists. In Italy, Enrico Fermi observed that neutron bombardment transformed the nucleus of almost every element and that particular substances had a moderating effect on neutrons. In France, Irène Joliot-Curie and Jean-Frédéric Joliot succeeded in creating artificial radiation and making isotopes by bombarding elements with alpha particles. In Denmark, Niels Bohr proposed a model of the atom that made it possible to understand theoretically how a uranium atom could split.

These various findings culminated in the discovery of nuclear fission, and, thanks to the labors of Lise Meitner, Otto Hahn, and Fritz Strassman, the uranium atom was first split in a laboratory in Germany in 1938.

Fermi, Szilard, and others saw that a chain reaction, if uncontrolled, would result in a huge explosion from a tiny amount of matter. Several of these scientists, fearing Hitler's genocidal policies and warmongering, left Europe for the United States and brought with them the worry that he would soon have at his disposal an atomic bomb. They had reason to be afraid, since these refugees had narrowly missed persecution and death because of Nazism, and they knew that colleagues who understood the principle of how to create such a weapon remained in Germany. The émigré scientists and their American counterparts went to work on what became the Manhattan Project. With astonishing speed they invented the very bomb they had feared—in order to put an end to Hitler's plan for a worldwide Nazi regime.

On December 2, 1942, in a squash court underneath the stadium at the University of Chicago, Fermi, Szilard, and other scientists, including a young American physicist named Walter Zinn, initiated the operation of the world's first man-made reactor—Chicago Pile-1. Four hundred tons of graphite blocks plugged with slugs of uranium were stacked in a pile about the size of a two-car garage. Cadmium rods could be moved into and out of channels in this pile to control the rate of the chain reaction. Fermi ordered the gradual removal of all the control rods except one, and when it was slowly withdrawn, the world's first engineered self-sustained nuclear reaction occurred. Moderated by the graphite and cooled by cold air thanks to the chilly weather, the neutrons, emanating from uranium slugs, obeyed all the laws of nature that had once brought into being the natural reactors in a primordial African delta.

Later, Chicago Pile-1 was dismantled and reassembled at Argonne National Laboratory near Chicago. During the next three years, the Manhattan Project secretly built and operated more reactors and enrichment plants to produce as rapidly as possible enough enriched uranium and plutonium to make three atomic bombs. By the time the first one was tested at Trinity Site in southern New Mexico in July 1945, Hitler had been vanquished, and President Truman, anticipating the deaths of hundreds of thousands of Allied troops in an invasion of Japan, as well as huge numbers of civilian casualties, made the decision to drop the two others on Japan in August.

In addition to atomic energy's military uses, some scientists also foresaw its enormous promise for doing good. Instead of deploying this extraordinary source of power to cause destruction, why not use it to help humanity? After the war, Zinn became the head of Argonne National Laboratory. Szilard refused to work there until assured it was not going to produce weapons. Argonne began research into peaceful uses of the atom in medicine and biology, physics, reactor analysis, applied mathematics, and nuclear engineering.

The scientists and the Atomic Energy Commission (AEC), which had strictly guarded the mystery of tamed neutrons, now decided to share it; they successfully communicated to politicians and the public the important advantages of nuclear power as a new means of generating electricity. To foster the construction of nuclear power plants by permitting formerly secret information to be used commercially, Congress passed the Atomic Energy Act in 1954. But could such plants safely and reliably supply power on a big scale? And why bother, when oil and coal were so cheap and abundant? The wartime reactors had been crude and hastily assembled. If atomic power was to succeed, utility companies needed a sound technology and a lot of assurance. The AEC gave Argonne the job. The lab found a big, isolated piece of land to test reactors in Idaho.

On an early March morning 1.8 billion years after the first natural reactors had started bubbling, about sixty years after the initiation of the first man-made reactor, and a few days after the United States began invading Iraq, Rip, Marcia, and I left Idaho Falls and headed toward what is perhaps the largest facility dedicated to studying the sustained chain reaction as well as being the birthplace of most reactor technology, called since 2005 Idaho National Laboratory (INL).

We wound up on a vacant two-lane, endlessly straight road that led across the sagebrush plain of the Arco Desert toward steep, jagged, snowcapped mountains: the Lost River Range, the Lemhi Range, and the Bitterroots. Nearby were fantastically contorted lava beds—Hell's Half Acre and Craters of the Moon.

"The government knew in '49 we'd be doing the worst things you can do to reactors, so we wanted a lot of land," our official escort, Harlin Summers, said as he drove. A tall, lanky, amiable fellow with a silver mustache, he wore a yoked white shirt with mother-of-pearl snaps, cowboy boots, and trousers sporting a large belt buckle that spelled out IDAHO. He'd grown up on a nearby farm and worked in infrastructure support at many facilities at the site we were approaching. Now retired, he still occasionally served as an experienced docent.

After an hour on the road, we saw a sign warning that only authorized vehicles were permitted. Cameras on poles monitored our progress. You can't just drop in. Rip and a couple of his colleagues here had arranged access, and it had taken time, because my background had to be checked. After 1983, when a terrorist truck bomb blew up a marine barracks in

Lebanon, protective measures at all national labs increased. Here, helicopter patrols began and guard posts were added. Today, because American tanks were rolling toward Baghdad, the Department of Homeland Security had ratcheted up the terrorist threat level from the usual yellow to orange. We learned only later that Harlin Summers's son was a member of the invasion forces.

"No need to worry about terrorists here," Summers now assured us. "Ours is the best SWAT team in the nation. Every year it's in contests with Navy SEALs and Delta Force teams and it always wins. We don't even notice an orange alert, because we're always at a much higher security level. Since 9/11 we've also been doing a lot of research for Homeland Security."

Acronyms and abbreviations, the sine qua non of every Department of Energy (DOE) facility, began flying at us. We'd entered INL, formerly the Idaho National Engineering and Environmental Laboratory (INEEL). It was also known in its infancy as the National Reactor Testing Station (NRTS), and then as the Idaho National Engineering Laboratory (INEL), and subsequent to our visit it joined with Argonne National Laboratory-West (ANL-W) to become Idaho National Laboratory (INL), which is how I will refer to it.

Bounded by high mountains on three sides, INL is 890 square miles of high desert steppe formed from layers of lava thousands of feet deep and covered in most places with dirt and sand. The land seemed untouched. Summers gestured toward some cinder cones on the horizon. "These were pilot buttes. Prospectors going to the gold rush and pioneers used them as markers when they were finding their way across the high desert. They were following a cutoff of the Oregon Trail. It's remained undisturbed since 1949. Best ruts around. Roads and buildings cover less than five percent of the site. The rest is in its natural state, and archaeologists and environmental people love it."

Rip's father's ranch, homesteaded by Rip's grandfather, was about twenty miles away, and on family trips along a road that skirted the site, then a naval ordnance facility, Rip would see gun barrels from warships being brought in by rail to be refurbished. Later, around 1949, he began to see distant buildings under construction at what had become the National Reactor Testing Station.

"Half the people I knew worked out there," Rip said. "Some were laborers, others had advanced degrees. Everyone got treated really well.

In the winter a lot of the farmers who needed money were taken out there and given jobs according to their background and skills."

"Were you curious about what was going on?"

"It was for me like Sandia Labs was for you—it was just there all your life, a no-never-mind. Secret and forbidden, sure, but you'd see your cousin and ask, 'What's going on there?' and he'd say, 'Well, today one of the reactors blew up.' That's how we knew they had reactors out there. After a while they had visiting days and I went. Later I found out that they had to do the worst things you can imagine to reactors—as you will soon find out—in order to figure out their weaknesses and come up with the safest technology. If it weren't for Hitler, reactors would have been used to make electricity instead of to enrich plutonium for bombs. Most of those Manhattan Project scientists started out wanting to use the chain reaction to do good for mankind, and after the war some of them went back to that, in Idaho. They designed reactors of different types and then tested them all the way to failure in order to eliminate flaws and come up with the best model for making electricity safely and efficiently."

As a child, I thought reactors were machines that turned atomic energy directly into electricity and included them with rocketry in my vision of a utopian future. But when I grew up, they remained associated in my brain with other dated 1950s objects like the hulking UNIVAC computer and tail fins on cars; I imagined reactors as huge, grimy edifices with chimneys belching deadly smoke. I'd never seen a nuclear plant and didn't know a reactor could melt down until *The China Syndrome* came along in 1979; the movie dramatized the theoretical possibility of a reactor core melting all the way down to bedrock.

Until I began to prepare for the visit to Idaho, I'd never considered how reactors were designed. It had never occurred to me that the operation of the first nuclear plants had been preceded by years of experimentation or that the technology might still be evolving. Certainly I was aware of tremendous transformations in my lifetime: my current laptop computer was far lighter and more powerful and sophisticated than the first one I bought in the early 1990s, and even that clunky model surpassed the computer that had guided the Apollo 11 mission to the moon and back. I now carried a cell phone that was the unrecognizable offspring of the black, bulky, dial-free, party-line apparatus at my grandmother's house.

When I had brought up with Rip the potential dangers to the public from radioactive doings at government facilities, he described how big those federal reservations were and how isolated their reactor sites were. But it was another matter to experience the immensity firsthand. We kept going for miles on end, occasionally passing a few cattle wandering among the sagebrush. The lab still permits local ranchers grazing privileges, although the days are long gone when experimental herds and their milk were monitored to determine how much exposure to radiation they might be getting from reactor releases.

"One reason the AEC chose this site is because radioactive materials have an affinity for sandy soil," Summers said, referring to the phenomenon of sorbing. "It holds them in place. The background radiation here after over fifty years of reactor testing and all kinds of other experimentation regarding nuclear materials is right at the national average." He mentioned a local factory that produces phosphate fertilizers, which have a high level of uranium and are mildly radioactive. But the phosphate plant's radionuclides go mostly unregulated, whereas INL has to sequester materials that are less radioactive than the rocks in the mountains around here. "And there's plenty of lava that any escaping radionuclides would have to migrate through before they'd get to the Snake River aquifer," he continued. "It's five hundred eighty-five feet down. Almost every trace of radioactive stuff would bind up with soils and decay before it got there. Any radionuclides that somehow made it that far would be extremely diluted anyway—there's enough water in the aquifer to cover the state of Idaho to a depth of four feet. That's another reason this site is ideal. You need a lot of water to cool a reactor."

Shimmering in the distance against the mountain ranges were widely separated islands of low buildings with an occasional two- or three-story silver dome, or a tower, or a big windowless structure rising up. When choosing this land, Argonne studied the bedrock, which had to support the enormous density and weight of a reactor building. Today all reactors in the United States must be anchored in bedrock. Regulations instituted during the Manhattan Project required each reactor to be surrounded by a controlled-access area and, beyond that, an additional, unpopulated zone whose expanse was determined by the size of the reactor and the nature of the environment. Similarly today, power-plant reactors must be surrounded by open land. At INL the cores of the reactors were sometimes situated in underground structures, so that soil and rock would provide shielding in addition to that of the reinforced concrete enclosures.

At last we came to an intersection with another narrow road that led to Midway—a town renamed by its citizens Atomic City in 1950, in the hope of attracting laboratory workers as residents. "In Atomic City, the highest recorded temperature is a hundred and four degrees in the summer, and the coldest forty-four below in the winter," Summers said. "That's without wind chill. People chose instead to live in Idaho Falls. We bus people in. They work rotating shifts. Until about 1992 we had almost fourteen thousand employees. Then came the budget cuts. We're now down to seventy-five hundred."

"Are any reactors being tested here now?" I asked.

"Negative," Summers replied. "After the tests were finished, most reactors were taken down. The last reactor went through initial start-up here in 1978. Of our fifty-two reactors, thirteen are still standing. Only three are operable. One makes medical isotopes. The other ten just sit there—we have no funding to dismantle them."

Eventually we arrived at a squat building with an unremarkable exterior. With evident pride, Summers pointed to a bronze plaque. "This building is a registered National Historic Landmark," he said. "It contains the first experimental breeder reactor in the free world."

Once Argonne had designed and built the first Idaho reactor, Zinn and his colleagues had to address a long list of challenges. Scientists at Hanford, a big nuclear defense facility in Washington State, had produced plutonium for bombs in low-temperature production reactors, but to make steam to turn turbines requires much higher temperatures. Susan M. Stacy notes in *Proving the Principle*, the official history of INL:

Making electricity had little in common with making bombs. Could a reactor be reliably controlled for long periods of time? What metals and materials could withstand the corrosive forces of heat and radiation for long periods of time? What form should uranium fuel take? What was the best way to carry heat away from the reactor? Could power plants be safe enough to operate near populated areas? Could uranium produce electricity cheaper than coal or natural gas? In sum, the science of nuclear reactors had to be developed from scratch.

Summers unlocked the door of a visitor center and escorted us past photos and exhibits into a room with an educational display. In a ritual of welcome to newcomers to the nuclear world that—like the safety posters, the many acronyms, and the drab, minimalistic governmental decor— evoked childhood memories, he passed the wand of a Geiger counter over the obligatory Coleman lantern mantles and orange-red Fiesta-Ware dishes as he described rays and particles.

I suddenly remembered that Rip, in addition to advising Marcia and me to wear pants, long-sleeved shirts, and sturdy shoes without open toes, had suggested we avoid polyester because the static charge attracts naturally occurring radionuclides from radon. Detectors could be set off at various facilities we were visiting; they don't distinguish between natural and man-made radionuclides. In any case, the dose would be low, of course. I'd made a risk-benefit decision and was in wrinkle-free microfiber pants, part of my uniform as we toured the nuclear world.

Summers now led us to a high-ceilinged room with concrete-block walls. There stood a black object whose size, bulk, painted metal skin, and contours reminded me of an old locomotive. Thirty years ago this earlier reactor prototype had been emptied of uranium fuel, thoroughly cleaned, and decontaminated.

Normally there would have been a reactor head—a sealed hatch at the top of the reactor—but it had been removed so that when we went up some steps we could peer inside. "We're looking down into the pressure vessel," Rip said. It was about the diameter of a fifty-five-gallon drum. "This is where the fuel was. It was packed close together and that made for a very hot reactor. The core was about the size of a football. It put out a heck of a lot of neutrons. They provided the heat the way coal or wood did in an old-time locomotive. The coolant circulated around the fuel rods to keep them from melting. It was an alloy of sodium and potassium that didn't slow the speed of the neutrons and that carried the heat away from the core to a heat exchanger just like you have in a car radiator. The pipes with the hot alloy were next to pipes with water. Heat was transferred over to the water pipes, cooling the liquid metal before it was circulated back to the core while boiling the water in the pipes."

At 1:23 p.m. on December 20, 1951, this prototype, officially called Experimental Breeder Reactor-I (EBR-I) and nicknamed ZIP, an acronym for "Zinn's Infernal Pile," made world history when Zinn and his team slowly brought the chain reaction to the level of criticality, and the fast neutron flux released enough heat to boil water.

"So reactors do not directly make electricity?" I asked. "They just . . . boil water? Like the Oklo reactors?"

"The thermal energy released from a lot of shattering U-235 nuclei heats up water and makes steam," Rip said.

"Steam? That's it?"

Moving electrons produce magnetism, and magnets moving past one another produce electricity by sending electrons down a wire. To spin a turbine shaft to move the magnets takes a lot of energy. Falling water or blowing wind can do it. Or steam can be forced down smaller and smaller pipes until it becomes a high-pressure jet that strikes a turbine's pinwheel blades, which turn the shaft that makes those generator magnets spin past stationary magnets, creating a current of electrons that flows along wires. Voilà: electricity.

EBR-I produced enough steam to run a 440-volt generator that put out enough electricity to light up four incandescent bulbs. In their glow, the team chalked their names on a wall, drank a bit of champagne, and went back to work.

Zinn tersely noted in a logbook, "Electricity flows from atomic energy. Rough estimate indicates 45 kilowatts."

Probably starting hundreds of thousands of years ago, early humans used sparks they struck from flint to ignite tinder. From then on, wood became the chief source of heat until about two centuries ago, when forests waned and coal took over. Steam from water boiled by burning wood or coal spun the recently invented turbine to drive locomotives and, eventually, run factories and power plants. Thus began the Industrial Revolution and the widespread manifestation for the first time of coal combustion—and its burgeoning output of greenhouse gases. But Zinn's experiment marked the first time in humankind's long journey that the nucleus of the atom rather than the outer shell of electrons was consistently supplying energy for human use. Rip, who as a teenager read about the breakthrough in the local paper and whose home had only been electrified about a decade earlier, thanks to a new hydroelectric dam, considers this moment equal in importance to the first time Michael Faraday successfully got his invention, the electric motor, to do physical work and to the moment Thomas Edison illuminated his incandescent bulb with electricity. All the same, until this windy spring day I'd never heard of EBR-I.

Next Zinn cranked the output of the reactor up to 100 kilowatts, enabling it to meet all the electrical needs of the whole facility for several years (today it relies on hydroelectric power). Zinn thought that a hot-enough temperature would cause the EBR-I to shut down automatically, preventing a runaway chain reaction. To prove this hypothesis required holding off the flow of coolant through the core and disabling the automatic safety mechanisms that would shut it down before it overheated. The test was risky and required precise timing. Unfortunately, an assistant in the control room misunderstood a command and it took two precious seconds for the operator to intervene and push the correct button. Within minutes, radioactivity in the control room set off alarms. Everyone hurried out of the building. No explosion occurred, and no one was harmed, but half the core had fused into a blob: the world's first meltdown. An analysis of the core traced the cause to the spacing of the fuel elements. Zinn used that information to make improvements. The AEC, after receiving Zinn's report on the meltdown, kept it a secret. But word got out, and an industry publication denounced the commission for covering up a matter that the public deserved to know about, prophetically warning that such a predilection for concealment would undermine public confidence in nuclear safety.

EBR-I received new fuel elements arrayed in a superior fashion and ran on them until 1962, when the uranium core was replaced by one of plutonium; it ran until 1964 and proved that if uranium were in short supply a substitute was possible.

Argonne had to address many other considerations as the laboratory developed the EBR-I and subsequent prototypes for civilian nuclear power. Every step had to be examined in terms of durability, efficiency, likelihood of failure, and the consequences of that failure. The fuel had to be tested in order to determine whether uranium oxide pellets were best or whether a liquid or gaseous form would work better. If pellets seemed the best choice, then what shape should the rods be that contained them and how should the rods be arranged for maximum efficiency? No one knew for certain what the life span of a fuel rod was or how an accumulation of fission products—the ashes, as it were—which tend to bind with nearby matter as they burst out of splitting atoms, would impact reactor components. The extreme temperatures and the fact that high-speed neutrons affect almost every element they encounter meant that all the materials to be used in reactors had to be tested before they could be included in any component.

Intense radiation of the kind that occurs in a reactor core does strange things to materials, warping, shrinking, expanding, or bending them. INL built the Materials Testing Reactor (MTR), which was used for neutron bombardment and irradiation of a variety of materials to determine how they would hold up in an environment that was unusual to say the least. The MTR could even duplicate the intense radiation of a nuclear explosion and so was used to evaluate components of equipment that would be used in atomic bomb tests. But the chief task of the MTR was to discover how reactor parts would perform under lower levels of radiation over longer time periods and how to make cores optimally efficient and safe. Years of research yielded the basic core designs and fuel element composition and configuration that are applied to most power reactor technology today: ceramic pellets of uranium enriched to about 4 percent U-235, clad in a shielding of zirconium alloy, and packed precisely into long, pencil-thin rods bundled together in fuel assemblies.

"Designs were mocked up here in Idaho," Summers said. "It was a place to practice before the real thing. Every creative in the world owes a debt of gratitude to the Materials Testing Reactor."

If reactors were ever going to provide electricity for the public, they had to be made as meltdown-proof as possible. Many creative experiments were conducted that pushed reactors into worst-case situations. What would happen if an accident prevented the reactor from being properly cooled and the speed of neutrons sufficiently moderated? The Argonne scientists theorized that the chain reaction would accelerate and the fuel would melt down. If the reactor couldn't be brought under control, nonatomic explosions might occur, flinging nasty fission products into the environment. But no one in the early 1950s knew for sure how such an event might play out in the real world. The scientists tried to anticipate all the human errors that could cause reactor problems and to forestall them by coming up with an inherently safe design.

Only at the Idaho lab did the government allow reactors to be taken to extremes in order to verify predictions about the consequences. A new test reactor, Boiling Reactor Experiment, aka BORAX-I, situated in an open, earth-shielded tank the size of a children's swimming pool, had a relatively small uranium core that was cooled by purified water, which also soaked up some of the neutrons' thermal energy. Would some disaster occur as the power to the reactor was gradually increased when technicians—using remote control—withdrew the control rods? Or would the chain reaction just fade away?

"You need two things to keep a reactor from melting down: a coolant, like water, to carry heat away from the core, and control rods," Rip said.

Control rods absorb neutrons. The precise arrangement of the fuel bundles in a reactor calls for spaces to be left among them or within them for the control rods. They can be inserted or withdrawn to speed up or slow down the nuclear reaction. The insertion in most reactors today is powered by gravity, a fail-safe mechanism that guarantees that even if all the electrical backup systems break down, the control rods can still drop within a few seconds into the reactor to regulate the chain reaction.

In the BORAX-I experiments, through many cautious withdrawals of the control rods, the team gradually brought up the power of the reactor and made many intentional mistakes while trying out different fuel elements and accident scenarios. When control rods were carefully withdrawn, the core would heat up, the water in the open tank would boil up and erupt, and puzzled travelers on the distant highway would be treated to the sight of geysers. According to the official history, "It appeared that boiling-water reactors might therefore be 'inherently' safe; that is, safe because of the way nature took its course, not because automatic controls, machinery, and human judgment operated perfectly one hundred percent of the time."

Because of the admirable record of BORAX-I in the face of so much maltreatment, the design appeared to be ideal for commercial power generation near populated areas. But Samuel Untermyer, director of the project, decided to run the reactor beyond its theoretical limits and find out exactly what the worst possible reactor failure would look like. Just in case contamination ended up spewing out of an exploded core, smoke bombs were set off on the day of the big test to determine the wind direction and velocity. When the air grew still, the last control rod was removed by remote control. The first steam explosion in the history of nuclear power abruptly blew up the reactor with the force of three or four sticks of dynamite, sending a shock wave and some debris hurtling across the desert and causing some visiting dignitaries to rush for cover in the buses that had brought them.

And just what does a steam explosion entail? "Think of a pot of water boiling on the stove," Rip said. "First you see a lot of little nuclei boiling and then you see a big boiling. The neutrons in BORAX-I generated a whole lot more energy to boil a whole lot more water when they took the last rod out and the water came to a boil so rapidly that there was a steam explosion. Only one kind of reactor accident can happen with our reactors in this country. If you can't cool the core because you don't have enough water taking heat away from it, the fuel rods get so hot that they melt. In doing that, they usually incorporate so much crap into the meltdown—melted fuel assemblies, concrete, steel—that the reaction becomes diluted and it goes subcritical. That is, there's not enough enriched uranium to keep the chain reaction going. Of course while the meltdown is occurring, there's a heck of a lot of heat and outgassing, and volatile radionuclides might spread around if there's a breach in the vessel and you don't have the vessel inside a containment dome. But every reactor in the United States by law has to be shielded and contained. I can't emphasize enough that the Chernobyl reactor had no containment structure and that the one at Three Mile Island did."

Bigger and better reactors of the BORAX design showed that various emergency scenarios could be deliberately induced without causing major disasters. The BORAX had built into its system a steam-turbine generator that, in the dark hours of July 17, 1955, sent electricity through the grid to nearby Arco. Idahoans claim it was the world's first town to be illuminated by nuclear power—an honor annually commemorated there—but probably a town in the former USSR has that distinction. Fifty-some years later, electric trains run in Chicago, half of their power coming from nuclear plants, and Times Square in New York City is bathed in radiance around the clock, much of it thanks to nuclear energy. France's 80-percent nuclear-powered grid, and the 436-odd reactors around the world still rely on the breakthroughs made in Idaho and still apply principles of safety and efficiency pioneered there.

The growth in knowledge about reactor safety eventually led to a boiling-water nuclear power plant that supplied all the electricity used by the Argonne lab in Illinois. By 1953, the AEC had created, in tandem with the infant commercial nuclear power industry, a committee that began to study siting, operational safety, risks of fallout, and other safety issues that would have to be addressed in the process of oversight and licensing of commercial plants. The AEC wanted to encourage more people to become nuclear engineers and scientists, and that meant installing small reactors on campuses. In order to determine the safety limits for normal use, the AEC directed the scientists to test reactors all the way to the point at which they destroyed themselves. This led to further programs in which reactors were set up for excursions—"excursion" means "reactor goes wild." The Special Power Excursion Reactor Test (SPERT), built in a secluded spot at INL, used subterranean, earth-shielded reactors and control bunkers to create sudden power surges that took various reactors beyond their capacity, destroying reactor cores. SPERT also developed instrumentation to observe and document the events occurring in the hidden world of water, neutrons, and runaway chain reactions. After analysis and risk assessment, SPERT and its successor programs provided reports and recommendations that became the basis of licensing requirements for university reactors and all American commercial plants. These findings, shared at international Atoms for Peace conferences, helped to establish safety requirements and to guide utilities and university research centers in the United States and around the world in choosing the best designs for nuclear power plants.

As we continued our tour around INL, Summers described many other projects. Over the years, the lab developed other breeder reactors and pressurized water reactors for various purposes. Summers pointed out a group of buildings that belonged to Naval Reactors, the term people usually use to refer to the government office that oversees the navy's nuclear-powered vessels. The first submarine reactor was pioneered here, inside a submarine hull immersed in a tank of water to simulate ocean conditions.

Another building housed an aircraft carrier reactor prototype. All told, the navy has built and operated about 250 reactors.

The Oklo natural reactors bred plutonium from uranium. Although the Oklo phenomenon was unknown to the Argonne scientists and the AEC, they had the same goal in mind. In 1944, Fermi had written a secret paper estimating the total quantity of uranium available in high-grade ores to be twenty thousand tons, half of that occurring in North America. This paltry supply could keep six reactors of his graphite design going for at least seventy-five years—or 140 times longer if plutonium, created in fuel rods by the chain reaction, were also used as fuel. The energy in North America's ten thousand tons of uranium would not be enough to replace the energy being generated by fossil fuels and dams. To do that would require many more reactors and much more uranium.

The postwar government took the position that using uranium for any purpose other than national defense would be foolish. Fermi correctly predicted that methods would be found to exploit more common ores— for example, thorium—but meanwhile he and others thought that any power reactor ought to be designed to recycle its waste into fuel, or even produce additional fuel.

The three practical fissile materials are U-235, found in trace quantities in nature; plutonium-239 and plutonium-240, both reactor products of U-238; and U-233, a reactor product of thorium-232. U-238 and Th-232 are called "fertile" (as opposed to fissile) since they can be coaxed into producing fissile offspring. Although about 99 percent of natural uranium is nonfissile U-238 and can't sustain a chain reaction, a reactor core can be configured to turn U-238 into plutonium-239, the basis of a practically limitless supply of fuel. Nuclear engineers have compared U-238 to wet wood. Like soggy kindling placed around a fire to dry out so it burns more efficiently, U-238 can be similarly transformed. (If U-238 is in a reactor a short time, Pu-239 is made. If U-238 remains in a typical power-plant reactor for eighteen months, Pu-240 is made.) Zinn and his team surrounded the enriched-uranium core of Experimental Breeder Reactor-I (EBR-I) with rods of U-238 in such a way that unimpeded neutrons flying out of the core penetrated the U-238 nuclei, creating plutonium-239. In EBR-I, fast—unmoderated—neutrons brought under human domination had one of two tasks: they had to keep the chain reaction going or they had to make plutonium that could be turned into fuel (a uranium enrichment plant and production reactors at the Hanford facility were already making highly enriched uranium and plutonium for weapons). The scientists needed to maintain a fast flow of neutrons to breed the plutonium, but the thermal energy they released might be so hot that it would melt the fuel unless the right coolant was used—one that, as it circulated around the rods, would neither slow down the flight of the neutrons nor absorb them. An alloy of liquid sodium and potassium could do that job, but it burned when exposed to air, so all the pipes and pumps had to be carefully engineered.

Fermi and Zinn patented a reactor design in 1955 that could run on natural uranium. Further experimentation proved that reactors could operate as effectively on plutonium-239 as they did on uranium-235. Surely, Argonne scientists assumed, nuclear power could supply electricity on a large scale now that fuel scarcity was no longer a concern. And if in hundreds of years uranium in North America and around the world ran out, and if, centuries after that, no more could be bred into plutonium fuel, then, as the lab demonstrated with another reactor, thorium, which is nonfissile, could be used. It's three times more abundant on earth than uranium. With some funding from the Energy Department, a project is now under way in Russia to fabricate and test a thorium fuel that would also burn weapons-grade plutonium. India, which has an abundance of thorium, is also exploring its use in new reactors.

The discovery of the rich Colorado Plateau uranium deposits soothed most official anxiety about a potential shortage of reactor fuel, and the breeder reactor design was eclipsed by that of the pressurized water reactor in wide commercial use today. But in the 1980s, the evolution of the quest Fermi and Zinn had begun resulted in a new, inherently safe reactor prototype that could make more fuel out of its waste in a closed, integrated system.

Early the next morning, we returned to the Idaho National Laboratory complex, this time heading toward Argonne National Laboratory-West and its mosquelike silvery containment dome, set against a backdrop of steep peaks. There, research on a reactor twenty times bigger than EBR-I, Experimental Breeder Reactor-II (EBR-II), had, over a period of thirty years, advanced the research begun with EBR-I while supplying electricity to the whole laboratory complex. EBR-II's spent fuel was rejuvenated at a facility next door and returned to the reactor. In 1983, the EBR-II equipment began to serve a new program: the Integral Fast Reactor (IFR). This design efficiently recycled its fuel, in the process reducing long-lived reactor waste by 99 percent, and was inherently meltdown-proof. Fast reactors could create enough fuel to power the entire country for more than five hundred years while rendering weapons-grade nuclear material into fuel and rejuvenating spent fuel onsite. Congress stopped funding the program three years before its completion. I hoped to get a look at the IFR but didn't yet know if permission would be granted.

William L. Hurt acted as our escort. He had a shy smile, sandy hair, wire-rim glasses, a quiet voice, and a good-natured, thoughtful presence. A nuclear navy veteran of the Vietnam War, he had trained submarine crews at the Idaho prototype before he became a mechanical engineer and a consultant to nuclear utilities. He now was the technical lead in nuclear material engineering and disposition at INL. He had a twenty-year-old son in the National Guard who was later deployed to Iraq.

Hurt mentioned on the drive that in recent years, celebrities, politicians, and assorted rich folks who had built big houses in Jackson Hole, Wyoming, and Sun Valley, Idaho, started to worry about radiation exposure from the lab, an hour or more distant. Some of these moguls do not let their environmental concerns prevent them from burning oil or propane to supply heat for their driveways, which can be miles long, to keep them snow-free all winter. When we pulled into the parking lot, I noticed that a couple of vehicles bore the same bumper sticker as the one Hurt had on his car: "Another Environmentalist for Nuclear Energy."

Inside an administration building, we approached a high counter manned by a posse of burly gun-toting men in camouflage uniforms. One of the guards collected our IDs and went off to check them.

Meanwhile, employees were arriving for work. The men—there were only a few women—mostly sported mustaches and glasses and, with their baseball caps, plaid flannel shirts, and hiking boots or running shoes, seemed prepared to go mountain-climbing on short notice. The complex has 750 employees and each one has to pass through security on a daily basis. As the workers in this area headed toward glass walls enclosing the airportlike metal detector, X-ray machine, and conveyor belt, they passed patriotic emblems and a poster reminding them to watch what they said:

> Threats to National Security
> Weapons of Mass Destruction
> Weapons of Mass Discussion
> **BE OPSEC AWARE**

OPSEC is governmentspeak for *Op*eration *Sec*urity, "a countermeasures program to deny an adversary pieces of the intelligence puzzle."

Paul Pugmire, the lab's public relations chief, appeared. A tall, youthful fellow with a big grin, he was distinctively dressed in a navy blazer, a sky-blue shirt, gray trousers, and a badge on a lanyard.

We were given temporary admittance badges and a dosimeter, which I was assigned to wear and which would announce the presence of any radionuclides picked up by my microfiber pants. I was ready to be shown large metal objects in rooms the size of basketball gyms. But that would have to wait.

After we had passed through security, Pugmire led us along a corridor. "At Argonne, we're all about power reactors, and we have all the facilities here to take the reactor from design to proof," he said, pausing in front of a series of paintings depicting the initiation of Chicago Pile-1 alongside portraits of Argonne's most famous employees, Enrico Fermi and Leo Szilard. "At Argonne we focus on power reactors," Pugmire said. "We're equipped to take a reactor from design to proof. We're not a weapons lab. We originate life-increasing technology for the world." Among other accomplishments, Argonne has developed an inexpensive ceramic building material stronger than concrete and biochips to track toxic chemical and biological agents.

We entered an office with a large polished conference table and were introduced to Bob Benedict, a slender, fair man with a somewhat ethereal presence. He was the deputy associate laboratory director of Argonne-West. We were joined by the former associate director, Michael Lineberry, who was compact, dark, and serious-looking but turned out to have a droll sense of humor. Benedict and Lineberry were research engineers who had worked under Charles Till, the father of the IFR project, and Yoon Chang, who had implemented the program.

After we were seated, Pugmire stood up and launched briskly into a description of Argonne-West's mission to advance safe nuclear power in order to help humanity. "The life span of people in lands with electricity is double that of people in places where there is none," he said. "Over a billion people around the world depend on nuclear energy for some or all of their electricity."

As I was contemplating the degree to which reliable electricity can mean the difference between health and sickness, and between life and death, Pugmire asked what I wanted to know. Rip and Marcia later told me that the body language of the Argonne people indicated that they were weighing whether we would be allowed to visit the facilities.

"Since 9/11, the question of reactor safety has become a big one in New York, where I live," I said. "The Indian Point nuclear plant is twenty-five miles from Times Square. People like me worry about that. You hear all kinds of scary stories."

"It took me three years of tours and talks here before I realized I had to explain to visitors that power plants could not be exploded atomically," Pugmire, who was formerly a congressional aide, said. "This is very obvious to people in the nuclear field and almost unknown outside it. We're working on safe reactors and safe reactor fuel for generating electricity. We're proud of the inherently safe Integral Fast Reactor that was developed here."

"Well, I'd like to see it," I said.

When this elicited no response, Rip intervened and described his background at Sandia in nuclear safety and environmental health. The atmosphere became more welcoming. Cookies and coffee appeared.

After some discussion about the Iraq invasion; about the vast expense of keeping a large military presence abroad to protect our fossil fuel interests, in contrast to the relatively cheaper cost of expanding nuclear power; about the limitations of alternate energy sources; and about how appropriate the IFR design or a kindred one would be for safely meeting electricity needs in the United States while recycling nuclear waste, the meeting ended.

Given the go-ahead, Pugmire led Hurt, Rip, Marcia, and me across a lawn to the silver dome housing the IFR. Like all containment structures for American reactors, this one had been negatively pressurized so that if it were breached, outside air would be sucked into it instead of any potentially contaminated air inside rushing out. In an anteroom, Pugmire paused before a cutaway model of the IFR and explained its innards, and then we passed into an air lock, where we waited for a light to change to green and the second door to open after the first one shut. We then stepped into a sort of basilica that was painted institutional green. As with most containment buildings, the dome wall, three feet thick at the base, tapering off to a foot and a half at the top, was a sandwich of stainless steel, reinforced concrete, and more stainless steel.

About eight yards below the floor, the reactor core had sat in a pool of liquid sodium coolant inside a thick-walled metal vessel. When the reactor was finally shut down in 1994 and the fuel removed, the sodium was pumped out. It had become temporarily radioactive while exposed to the fuel, but within months the radionuclides in it had decayed to natural background level.

"The IFR is a fast-neutron reactor," Pugmire told us. "The moderator, which is usually the same thing as the coolant, determines neutron speed. If the cooling medium is liquid sodium—or a combination of liquid sodium and potassium, or helium gas, or lead-bismuth—it carries heat away from the reactor core, keeping its temperature stable, but doesn't slow down the neutrons flying out of the nuclei and striking other nuclei. Here's a pool table analogy. A pile of feathers between the cue ball and the rack would moderate the energy of the ball. Feathers absorb some of the energy. Some other substance between the cue ball and the rack, like air, might absorb less of the energy. Therefore, fast reactor, fast neutrons: the medium does not slow down neutrons as much as, say, graphite or water."

Those neutrons originated in pure, oxide-free metal that was enriched to about 60 percent—about 55 percent more than typical uranium reactor fuel—and which had superior heat-transfer. Thanks to the fast neutrons, which travel at forty-four million miles per hour, not only could the IFR "burn" existing fuel far more efficiently, it could also make new fuel from spent fuel or even from natural uranium. Slow neutron reactors—that is, the pressurized water reactor (PWR) and the boiling-water reactor (BWR) designs commonly used in the United States and Europe—leave behind 98 percent of the energy in the spent nuclear fuel. After about eighteen months in such a reactor, fuel becomes too cluttered with chain-reaction by-products—plutonium and other actinides (the waste products of neutron capture in fissile material)—to be efficient. In the experimental IFR program, the spent fuel was robotically removed and sent next door to have the actinides and fission fragments removed through electrical and chemical processes and then turned into fresh fuel.

"The way it's done now, our power-plant reactors use the once-through method—the open fuel cycle," Hurt said. "That's like burning the bark off a log in your fireplace and then throwing the log away." The IFR would consume most of the log, using the energy to make electricity and leaving in effect a small pile of radioactive residue that in part would decay to harmless levels in days or dozens of years. The remainder would be longer-lived, radioactively speaking, though at a lower level. This waste could be immobilized in glass or encased in highly impermeable metal or a ceramic material and isolated until, after 400 years, the radioactivity had decayed to the same level as uranium ore.

In the middle section of the core, the fast neutrons from one fission split U-235, and additional neutrons issued forth at random. "Those that didn't hit anything just kept going until stopped by barriers," Pugmire explained. "If you put a blanket of uranium-238 around the reactor, the neutrons hit the U-238 and turned it into plutonium—that's your fuel. But if you removed the U-238, the reactor's stainless steel wall reflected the neutrons back into the core, where they packed a bigger punch. More energy, more electricity made.

"The basic point is this: all the power reactors in commercial plants are water-cooled," he said. "A safe, wonderful technology. But during the chain reaction they're all producing plutonium in their cores. The IFR avoided making plutonium while it generated the heat to make steam to turn the turbine. Another plus was that the IFR avoided the problems of water corrosion, the need for constant attention to water chemistry. The IFR used liquid sodium instead. It's a supremely efficient heat-transfer medium that doesn't slow down neutrons and doesn't react chemically with the stainless steel walls of the reactor." He took us downstairs and showed us the decades-old, thick metal wall, its shiny, neutron-reflecting surface looking brand-new.

"At 700 degrees to 900 degrees Fahrenheit, sodium is well within its liquid phase," he said. "Unlike water, which boils away at 212 degrees, sodium doesn't need to be pressurized to keep it liquid." The sodium was sealed inside the primary vessel, and also isolated from contact with air by a protective barrier of argon gas.

In 1986, Argonne ran tests of the IFR that proved it could protect itself from overheating. "This reactor design is inherently safe, passively safe, in the most beautiful way possible," Pugmire said. Sodium, like any other metal, expands when it's hot—the atoms move away from one another. "The IFR uses a natural physical principle that is immutable: thermal expansion and convection flow, which is the process by which currents form. When the metal, the sodium, gets overheated, it expands. The fissile material has been placed in proper geometry and proximity to work correctly—that is, for a chain reaction to produce the right amount of heat to turn water into steam." What happens if the reactor overheats? Thermal expansion disrupts the fuel's geometry and proximity, and the laws of physics cause the core to go subcritical.

"We demonstrated this to an international audience," Pugmire continued. "First we disabled all of the reactor's safety features and then stopped the sodium pumps. The temperature shot up. Thermal expansion of the fuel occurred. This resulted in increased neutron leakage." Instead of striking nuclei and causing them to fission, neutrons were shooting into the gaps that the expansion opened up among the particles.

This phenomenon slowed the chain reaction. The thermal heat from neutrons caused convection currents in the liquid sodium. The currents carried the heat away from the core. The reactor then went subcritical. "Reactor power and temperatures returned to safe levels. We restarted the reactor an hour later because there was no damage to the core. We *know* this is the safest reactor design."

Some reactor experts consider the IFR design to be more error-proof than those of the pressurized- and boiling-water reactors. The reactor at

Three Mile Island would never have overheated if it had been of a sodium-cooled design. In 1975, a reactor at the Browns Ferry Nuclear Plant in Athens, Alabama, came close to overheating, because of an electrical fire that started accidentally when workers using candles to test for air leaks in a pressurized spreading cable room ignited a polyurethane seal that turned out not to be fireproof. Cables burned and, over several hours, the electrical system that maintained the core temperature was damaged. Emergency repairs enabled the operators to insert the control rods and shut down the chain reaction before the core overheated. With an IFR, the chain reaction would have automatically stopped, no matter what was happening with the electrical system. (The Browns Ferry incident led the Nuclear Regulatory Commission [NRC] to require many changes in fire protection features and plant procedures for responding to a fire.) The power plant, shut down for twenty-two years, reopened in 2007.

Rip now asked Pugmire, "You've talked about the IFR's benefits. What are the detriments?"

"Sodium is chemically reactive with air and with water. Violently so. Sodium in the reactor environment is radioactive, though it decays quickly. A sodium reaction might leak sodium out of the reactor. Some say they're concerned about this. We are not. The sodium would stay inside the containment dome. There are secondary, separate sodium lines that go to the heat exchanger, which heats up the water. The water is in a separate line, which carries the steam to the turbine." Pure sodium isn't toxic, but if it combines with certain other elements it can be poisonous.

Pugmire pointed to a piece of equipment. "We moved the sodium with this electromagnetic pump—it has no moving parts. We've had sodium leaks. Because it's under low pressure, it oozes out, and if enough drips on the floor, the sodium smokes, scabs over, creates ash. But it simply doesn't ignite, so the sodium is not very problematic. The main drawback is the cost of building and operating a system like this."

In 1994, three years before its mission was to be completed, the program was terminated. The people at Argonne told me that the reasons were political rather than technical. Charles Till, lead scientist for the program, has said that the program fell victim to President Clinton's general cancellation of unneeded programs, among them those on advanced nuclear power.

Later, Rip mentioned his thoughts to me. "The IFR offers an elegant way to prevent a meltdown using Mother Nature and to recycle fuel and avoid the enrichment process. The IFR technology could certainly reduce the volume of nuclear waste. The Argonne scientists were arguing that by using the plutonium as fuel to make electricity and by recycling, the IFR put up another barrier to acquisition of bomb-level plutonium. For them, that's a good reason to push the IFR. You burn up the plutonium and make waste that has only about a four-hundred-year half-life, as opposed to waste that has to be isolated for ten thousand years. That can be done with any reactor that has fast neutrons and is sodium-cooled. Actinides that accumulate in fuel during the chain reaction are nasty. The fast neutrons burn them up very, very well. That's good. But this technology needs to be carefully controlled. If you're a no-good son of a bitch and you have this reactor in your country, all you have to do is put depleted uranium—U-238—in a blanket array and you end up with a breeder that makes plutonium that you separate out in a reprocessing plant and use for bombs instead of burning it as fuel. It's really easy to acquire depleted uranium and really hard to acquire plutonium. You wouldn't want a rogue nation to have an IFR and a reprocessing plant, but we don't export this technology anyway." Newer designs based on the IFR can be adjusted to produce no more plutonium than a light-water reactor does.

Another drawback to the IFR was the expense of rejuvenating the spent fuel in an integrated system as long as virgin uranium was cheap and abundant. On the other hand, recycling the fuel onsite would eliminate most of the elaborate handling, transport, and storage of spent fuel. Installed in an American nuclear plant, one IFR could recycle the waste of two or three pressurized water reactors.

Before we left Argonne-West, we were required to pass through radiation detectors, including a metal closet called a whole-body counter. My trousers hadn't picked up any radionuclides and my dosimeter hadn't registered any radiation exposure.

The Idaho prototypes of the 1950s and 1960s constitute the first generation of reactors. The large number of reactors built from the late 1960s through the early 1990s for commercial use when the country was optimistic about nuclear power are members of Generation II. The Japanese launched the first Generation III reactor in 1996; the design has passive safety features and a more efficient fuel technology. Generation III+ reactors, likely to be built in the near term, are more streamlined and have more safety features. All the new designs, like the European boiling-water reactor and the Westinghouse AP1000, which relies on gravity rather than a series of pumps to bring water to the reactor in case of emergency and which the Chinese have chosen to use for new plants they are building, have greatly simplified the piping and controls, making meltdown or radiation leakage virtually impossible. These new systems are cheaper and easier to build and operate.

In 2002, the Generation IV International Forum, searching for new reactor designs for the essential role it expects nuclear energy to play in the coming decades, chose six promising ones. Among them are metal-cooled fast reactors in the IFR mold that can recycle their fuel and the inherently safe and relatively small pebble-bed modular reactor, which South Africa is said to be completing by 2011 for use in developing countries. The Chinese are building one as well.

I asked my friend Chris Crawford, a civil engineer who had in vain insisted to me years ago that nuclear power was safe, and who had gone back to school to get a master's in nuclear engineering, what reactor design he would like to see deployed.

"Of the Generation IV reactor designs, my personal favorite is the Modular High Temperature Gas Reactor (MHTGR), which isn't called modular for nothing; the smaller 200 megawatt units can be installed singly or in groups for flexibility," he wrote me in an e-mail.

It's helium cooled, and has particle-based fuel that can't melt down. Each fuel particle is only a few millimeters in diameter and is surrounded by a containment layer of silicon carbide to prevent release of fission products as well, in effect making each particle a miniature reactor vessel. The particles, demanding to manufacture, are then dispersed in a graphite matrix.

The MHTGR is similar to another design, the pebble-bed reactor, in several ways: they both use graphite for moderation, helium for cooling, uranium enriched to about 20 percent, and fuel in particulate form rather than in the usual pellets. The MHTGR, though, embeds the fuel granules in large blocks of graphite that have cooling ports drilled through them and arranged in an annular core. Both systems can withstand very high heat excursions in the event of an accidental loss of coolant; in fact, they lose reactivity in the event of coolant loss so they can't melt down. The operators can safely take days to shut down the reactor even if 100 percent of the helium coolant is lost. For this reason, neither reactor needs an air-tight containment structure or a wide evacuation zone; that makes them cheaper to build than light water reactors. Since they operate at high temperatures, the heat can be used to make hydrogen gas or to refine oil from oil shales—to name a couple of applications. The spent fuel is more difficult than light water reactor fuel to reprocess, which is good from a proliferation point of view, and is very stable in its graphite matrix so that it can be sent to a repository easily. Because the MHTGR uses fuel that's more highly enriched, it burns the fuel more efficiently, so refuelings are less frequent—meaning the reactor has less downtime.

This stage in the nuclear tour left me assured that every reactor component had been studied and tested for years, that every reactor had undergone simulations and punishing real-world tests, that thanks to decades of experimentation all our power reactors today are sufficiently safe, and that, because of very demanding federal regulations and NRC oversight, newer reactor designs would be even less accident-prone and would be tested as ruthlessly as their predecessors. The technology, so painstakingly elaborated in Idaho over the decades, has been adopted by countries around the world. In 2006, more than 435 reactors in thirty-two countries supplied 16 percent of the world's electricity with a safety record far superior to that of fossil fuel or hydroelectric generation—and that's including the Chernobyl fatalities. The Idaho National Laboratory clearly is dedicated to an important scientific principle: take advantage of mistakes and accidents to increase knowledge.

9 TINY BEADS

EARLY ON, I TOLD RIP that I might be able to accept the idea that nuclear power was beneficial if it were not for my fear—probably just about everyone else's fear, too—that we could have a repeat of Chernobyl here.

"We did have a Chernobyl," he replied softly.

Was he revealing some secret catastrophe? No, he was referring to Three Mile Island.

At INL we were admitted to a site with many, many security barriers where enriched uranium was stockpiled and nuclear waste stored. After touring various buildings, we arrived on foot at a place with two high chain-link fences enclosing rows of thick-walled concrete tombs, thirty of them. They resembled units in a self-storage facility—or perhaps an innovative public monument. Later I was to think of the place as a memorial to certain assumptions.

Within these blocks reposed fragments of the first major reactor accident in the history of civilian nuclear power: the remains of the core and spent fuel of Three Mile Island-2.

One March night in 1979, at a nuclear plant near Harrisburg, Pennsylvania, water-purifying resin beads smaller than mustard seeds blocked a valve in the reactor known as TMI-2. That led to the failure of the main pumps circulating the secondary coolant. The secondary coolant system is a closed loop of pipes wrapped around the primary coolant system, which contains the water carrying heat away from the reactor core. The secondary system absorbs the heat necessary to make the pressurized steam that runs the turbines. The cooling system stopped working, and then the turbine and reactor automatically shut down as they had been designed to do. However, heat released by decaying fission fragments continued to increase the temperature of the water near the reactor core, causing a pressure surge and forcing a pressure relief valve to pop open and emergency pumps to bring water circulation back to the reactor. As the pressure was thus reduced, the valve should have closed, but it stayed open.

Because of faulty indicators, the people in the control room didn't understand what had happened and instead began making a series of suppositions that led to their closing off the flow of water to the reactor core. Without coolant, the core overheated, causing a steam flash in the pressure vessel. In the control room, warning lights blinked and faxes spewed out reports. As the reactor's condition became more precarious, supervisors were overwhelmed with far more information than they could sort out and rapidly respond to. Communications among personnel and with the Nuclear Regulatory Commission (NRC) broke down. As Susan Stacy's *Proving the Principle* sums it up: "One thing after another went wrong with instruments, equipment, computers, and human judgment."

During the next sixteen hours, no one knew for sure what was happening, because there was no way to see inside the reactor. The temperature of the core soared, but at that time no one knew that half of it had melted down into the bottom of the reactor vessel or that the radionuclides flung out of the core had plastered themselves to the inner walls. The containment building flooded with contaminated water and the hot steam around the zirconium cladding of the fuel pellets (the outer layer of metal that encloses the fuel rod, preventing corrosion of the fissile material and the release of its fission products into the coolant) caused a chemical reaction, which formed a bubble of hydrogen gas in the vessel. Inspectors from the NRC finally arrived from Washington, D.C. Some members worried that the hydrogen bubble would not only interfere with coolant water but also might explode, blowing open the containment structure. To prevent that, over several days operators carefully vented gases, which contained only a low level of radionuclides—they were mostly inert and so did not combine with the air—through a tall, charcoal-filtered stack. The wind blew the gases out over the Atlantic.

Meanwhile scientists at the Idaho lab quickly set up the equivalent piping at a test reactor to imitate the conditions inside TMI-2 and soon announced that a hydrogen explosion would not occur. President Carter, formerly a nuclear engineer in the navy, was then able to appear at the plant and offer reassurances. A team from Idaho arrived to help secure the reactor in a cold shutdown position. Although, as in the extreme abuse of reactor prototypes over the years in INL tests, the core had indeed melted down and entombed itself, thereby stopping the nuclear reaction, the Three Mile Island plant officers could not know that.

Meanwhile, misinformation and ongoing communications problems continued to feed a growing crisis. Assumptions flew around like fast neutrons. Reticent plant engineers addressed the public in jargon, and their rather wooden affect made for poor interactions with the press. "What shook the public the most was seeing the men in white coats standing around and scratching their heads because they didn't know what to do," a commissioner from the NRC commented years later. "The result was that accidents were taken seriously in a way they never had been before." Plant representatives made assumptions, and local officials made assumptions, and state and federal agencies made assumptions, and the media made assumptions, and some of those assumptions were heated up by the irresistible hook of *The China Syndrome*, which had come out two weeks earlier: would a meltdown to bedrock cause deadly radioactive gases to come pouring out of cracks in the Pennsylvania earth?

News bulletins were alarming and confusing. Some were highly speculative, especially in the first few days of the crisis. I stood before the TV in my New York apartment staring at the governor of Pennsylvania giving a speech recommending the evacuation of children and pregnant women. How could I protect my young daughter? What had I learned about radiation on those school tours? Was it that the farther you are from the source, the less the exposure? But what if the radiation bursting out of the reactor was—as some commentators were suggesting—*different*: able to travel great distances like bomb fallout and lethal enough to kill hundreds of thousands of people?

Because a helicopter got a high reading from gases in the aerial plume from the containment building, state authorities urged an immediate evacuation of residents nearest the plant. According to a subsequent investigation, that decision was based on erroneous information: at the boundary of the plant complex, the ground level reading was a thousand times lower and so evacuation was completely unnecessary. But by some estimates, two hundred thousand scared people hurried to escape. A Roman Catholic priest offered them the sacrament given to those about to die. After a few days, news anchors began to announce that the accident was under control. Still, assumptions that were wrong and dire predictions of devastation that did not occur have lodged in the public mind.

The seeds of the dread that flourished during the crisis may have in part been sown by newspaper accounts prior to the incident. Bernard Cohen, professor emeritus of the physics department at the University of Pittsburgh, has searched *The New York Times* archives for stories about various types of accidents in the United States between 1974 and 1978—prior to Three Mile Island—and compared their frequency with the annual fatalities caused by these accidents. He writes:

On an average, there were 120 entries per year on motor vehicle accidents, which kill 50,000 Americans each year; 50 entries per year on
industrial accidents, which kill 12,000; and 20 entries per year on asphyxiation accidents, which kill 4,500; note that for these the number of
entries, which represents roughly the amount of newspaper coverage, is
approximately proportional to the death toll they cause. But for accidents
involving radiation, there were something like 200 entries per year, in
spite of there not having been a single fatality from a radiation accident
for over a decade.

Another problem, especially in TV coverage, was use of inflamma-
tory language. We often heard about "deadly radiation" or "lethal
radioactivity," referring to a hazard that hadn't claimed a single victim for
over a decade, and had caused less than five deaths in American history.
But we never heard about "lethal electricity," although 1,200 Americans
were dying each year from electrocution; or about "lethal natural gas,"
which was killing 500 annually with asphyxiation accidents.

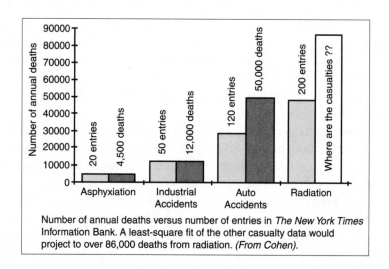

Number of annual deaths versus number of entries in *The New York Times*
Information Bank. A least-square fit of the other casualty data would
project to over 86,000 deaths from radiation. *(From Cohen).*

The crowds who had fled needn't have risked their lives in that heavy
traffic. The containment structures worked. Even if they had been
breached, the radionuclides would have remained close to the reactor, as
investigation would later reveal.

Just months before the TMI-2 failure, the Idaho lab had created
worst-case scenarios in which computer models forecast how different
types of emergency core cooling systems might perform if the core lost
water and became overheated. Then real-world tests were run on a test
reactor designed for the study of loss of coolant. The computer model

turned out to be much more conservative, saying that within ninety seconds the emergency system would restore cooling water to fuel that had heated up to 1,350 degrees Fahrenheit. In actuality, the coolant returned in forty-four seconds and the temperature only rose to about 1,000 degrees. After the Three Mile Island accident, Idaho ran simulations on a test fuel assembly deprived of coolant; these showed temperatures, reactions of the fuel rods, the failure of the fuel-rod cladding, and the movement of escaping fission products. The researchers succeeded in making radiographical images of the melted fuel and rubble at the bottom of the Idaho reactor. Later, after the Three Mile Island core had cooled, a remote camera sent down into the reactor found that the Idaho simulation had been accurate. The footage also revealed that, while the meltdown had been only partial, more of the core had melted than the plant operators had assumed.

Later, the reactor and its debris were shipped to Idaho by rail so that specialists could assess more precisely what had happened. The NRC funded research at the lab to learn more about the consequences of the reactor failure, and the international Organization of Economic Cooperation and Development also contributed to a test program. No country using power reactors could afford to overlook the multiple missteps at Three Mile Island. The Idaho studies involved the deliberate melting of a test reactor's fuel bundle. One aim of this experiment was to find out where fission products went. The scientists and engineers concluded that even if the reactor containment vessel had been breached, the radioactive exposures to the public that previous theoretical scenarios had predicted, based on conservative estimates, were not likely. Fission products that were liquefied at high temperature had been plated on the inside of the vessel and the walls of the containment building.

"The crux was that they shut off the water—the coolant," Rip said. "I still don't see how a group of nuclear engineers could have come up with a reason to shut off the water. At all costs, don't shut off the water! Nobody in his right mind ever thought that some idiot would do that. That's like telling a fireman fighting a fire to shut the water off. The system was programmed to shut itself down if there was a safety problem. If they'd left the machinery to run as it was supposed to, as the designers had intended it to, or if the operators had realized their mistake and added water to the reactor, there would have been no meltdown."

A presidential commission, Congress, the Nuclear Regulatory Commission, the Department of Energy, the Pennsylvania Department of Public

Health, and private groups investigated Three Mile Island. Over seven hundred people living close to the plant were checked for internal contamination and none was found that could be traced to the accident. Internal radioactive contamination isn't uncommon, but it almost always comes from rocks, soil, or nuclear medicine. Radionuclides from a nuclear plant would have a particular signature. Reports about unusual animal deaths or illnesses because of radioactivity proved to be untrue. Through the usual process of gathering and analyzing data and offering up studies for critiques, a scientific consensus concluded that no excess cancers occurred. A state health registry tracked thirty thousand people in the area for two decades without finding any evidence of effects that could be attributed to radiation exposure. The population and the environment showed no detectable signs of harm from radiation. The people living near the facility's perimeter were estimated to have been exposed to a maximum of 100 millirem—about one-seventh of the natural background radiation that they would have received in a year if they had relocated to Denver. The average dose to the local population was a little over 1 millirem, about as much as you'd get from smoking three cigarettes and less than 1 percent of the 100 or so millirem per year all living things in the Harrisburg area receive from nature.

"In Albuquerque, we get a dose from background radiation that's three times as high as the area around Three Mile Island," Leo Gómez told me. "And every day we New Mexicans get a much higher *measured* dose than the average calculated dose people may have received from the small release of radioactivity from Three Mile Island."

The highest dose received by plant personnel was 4,000 millirem—1,000 millirem less than the annual permissible dose for nuclear workers. Over the years, multiple studies of Three Mile Island employees as well as residents of nearby neighborhoods have failed to find any radiation-linked problems such as thyroid cancer or leukemia. The Three Mile Island Public Health Fund engaged independent researchers to determine whether the pattern of cancer in a ten-mile area surrounding the plant had been altered and, if so, whether that was due to the accident. "For accident emissions," the study concluded, "the authors failed to find definite effects of exposure on the cancer types and population subgroups thought to be most susceptible to radiation. No associations were seen for leukemia in adults or for childhood cancers as a group." However, the linear nonthreshold hypothesis led some members of the radiation protection community to make the conservative estimate that even the lowest doses of radiation might result in some cancers, although at such low levels of exposure there is no experimental proof that this would be so.

The notable health effects were psychological. The presidential commission's report identified mental stress as the most important health impact on plant workers and the general population.

The father of Evan Douple, of the Radiation Effects Research Foundation, lived a few miles from Three Mile Island. "When the accident occurred, he called me and said that all the people were locking their houses and driving away," Douple told me. He advised his father to stay home, close the doors and windows, and protect his belongings from looters. "I said that he didn't need to worry. I said that knowing what radionuclides might be emitted, if there were any, knowing that gaseous or wind-borne radionuclides would be emitted if anything at all was, and knowing that at his advanced age a radiation-induced cancer from any exposures he might have would likely take twenty to forty years to become manifest. So the risks of radiation-induced health effects would be very small. And I knew that in any case the doses would be very low and any associated health risk would also be expected to be very low."

Occasionally a newspaper article will appear claiming that people in the Harrisburg area are indeed suffering from radiation-induced illnesses caused by the reactor accident. But such allegations have not been supported by scientific consensus and peer review. Nor are they ever likely to be. In fact, recent research has found that the release of radioactivity into the environment was still less than had been estimated.

Some years later, *The Village Voice* ran a spread of photos of plants growing around Three Mile Island that were said to display signs of mutations. Thanks to the Internet, I was able to find such a photo with the following caption:

> Grossly deformed Gloriosa daisy found in 1989, on the riverbanks of the Susquehanna in Goldsboro, directly across from Three Mile Island. Another stem fasciation; this plant had a very wide flat stem, many deformed flower heads (looking like caterpillars), and a double flower growing back-to-back. This specimen still exists, and retained its color and shape.

I sent along the picture and caption to Gómez. He forwarded them to his son Steve, who has a doctorate in molecular plant biophysics from UCLA, and who is collaborating with his father on a design for a completely sustainable farm that will grow castor beans as biofuel. "Cool, but I bet if you examine any ten-square-mile area of the world you'll find a large number of mutated plants," Steve wrote back. "The trick would be to see if the change is caused by DNA damage, or just by random chance.

That's hard to do. I think the reason they found so many around Three Mile Island is that people were intensively looking for them. If you really want to get people all fired up, then near your favorite nuclear power plant sow some self-pollinated corn from any of the commercial hybrids you can buy. All sorts of weird stuff pops out. Ears where tassels should be, etc." In my organic garden, far from nuclear plants, I'd grown a few of those myself.

"Three Mile Island is the worst accident we've had in commercial nuclear power in this country," Rip said. "No member of the public died or was even physically harmed." He added that it would take a direct hit from a thermonuclear weapon to bring about a total reactor core melt-down or a total dispersion of nuclear fuel. Otherwise, the laws of physics ruled out such scenarios.

The TMI-2 reactor failure led to numerous safety studies, changes in the nuclear plant supervision, and a greatly increased emphasis on human engineering—which meant more intensive training of operators and the initiation of many accident-preventing protocols. Prior to Three Mile Island, NRC emergency procedures required operators to accurately diagnose the initiating event that was causing the problem before they took action. Bill Hurt, the engineer I'd met at INL, described this to me as "like being brought into the emergency room and the doctor worrying about your broken leg instead of addressing the fact that you are not breathing." Thanks to the lessons of TMI-2, operators now follow emergency procedures that restore critical safety functions independent of the events that may have initiated the problem.

Despite the televised images of President Carter walking toward the containment building, and despite local opinion polls that showed support for the way federal and state authorities handled the crisis, most of the American public had been influenced by the unshakable and persistent assumptions that Three Mile Island had been a widespread and probably lethal disaster; that humans were not capable of operating reactors safely or dealing with a reactor emergency; and that evacuation from any future accident occurring in a heavily populated area would be impossible. The actual casualty from the accident was public trust in nuclear power.

The presidential commission that investigated the accident took to task the mind-set of the nuclear industry and the NRC, chastised the companies that designed, made, and operated the reactor, and recommended specific improvements in reactor safety as well as new procedures and practices at nuclear plants. Though the commission found that the multiple containment barriers and the bedrock below it were robust enough to prevent a release of a large amount of radioactivity even in the event of a complete meltdown, the investigators also called for vitally important changes essential to reducing risk: more stringent training for operators, better control rooms and instrumentation, clearer lines of authority, and improved communication both among personnel and with the public.

Scientists and engineers at the Idaho lab began working on better ways to prevent accidents, limit core damage, maintain containment structures, and reduce the scattering of radioactive material in the event of a reactor failure.

Shoreham, a nuclear plant situated on Long Island about a ninety-minute drive from Manhattan, was under construction from 1973 to 1984. After the accident at Three Mile Island, the campaign to prevent Shoreham from opening became popular. Worried about how populous Long Island could ever be evacuated when on holidays traffic was already impossible, I joined the opposition. I followed stories about how corrupt the contractors were and how immense the cost overruns had become— the ultimate price was eighty-five times greater than had been anticipated: $6 billion, most of it footed by the rate-payers. Rumors circulated about steamfitters who labored all day only to find that the night crew had destroyed the pipe work, so that the same job would have to be done again and again. (The way contracts were let to nuclear-plant builders encouraged delays and cost overruns.) Even though Shoreham was fully licensed, Long Islanders, who pay some of the highest electric rates in the country, succeeded in preventing the plant from ever opening. The utility company, under public attack for incompetence in its handling of Shoreham and other matters, met its demise at the hands of the state legislature, which set up a new utility to take over the old one and close Shoreham. It now sits gathering cobwebs, and we customers, who never received one kilowatt from the plant, are still paying the bill for its construction and shutdown.

On that chilly March day in 2003 when I saw TMI-2's multiple tombs through the steel mesh of INL's high fences, I thought I was probably viewing the final resting place of the possibility of commercial nuclear power in the United States, just as on the same day I had seen its birthplace, the EBR-I reactor, developed when optimism for a source of electricity that would help all humankind ran high. No new reactors had been ordered since 1979, mainly for economic reasons. Would anyone ever trust the nuclear industry again? As long as humans operate plants, there will always be risk, yet it can be reduced as new, human-proof, inherently safe reactors come online. Rip maintained that safety procedures could and did minimize the risk at TMI, that the industry had learned from its mistakes, and that the risks from fossil-fuel combustion were by far greater.

By 2007 major newspapers in the country had endorsed a revival of nuclear power in lieu of more coal-fired plants. In a 2005 random survey, commissioned by the industry, of 1,152 people living within ten miles of each of the nation's sixty-four nuclear plants, 83 percent approved of nuclear power, 85 percent gave their local plant a high safety rating, and 76 percent were fine about having a new plant built nearby. The positive attitude toward nuclear power seems more than justified. In over twelve thousand cumulative reactor-years of nuclear plants making electricity in thirty-two countries, there have been only two major accidents in the history of nuclear power, Chernobyl and Three Mile Island-2.

Some environmentalists who once fought nuclear power have softened their stance. In 2007, leading environmental organizations, among them the Pew Center on Global Climate Change and Environmental Defense, acknowledged that nuclear energy has a long-term role to play in mitigating greenhouse gas emissions. Perhaps some of this willingness to consider nuclear power is founded in the recognition that since Three Mile Island the U.S. nuclear fleet has operated safely.

Although many actions have contributed to the current attitudes about transparency and accountability in the industry, the most significant person to shape and enforce them was Admiral Hyman G. Rickover, who founded Naval Reactors and established the first commercial nuclear plant in America. He told the Three Mile Island commission that the nuclear industry could learn from the safety record of the nuclear navy, which has operated 254 reactors and never had a reactor accident. John Deutch, who was acting assistant secretary of energy development at the Department of Energy and later was to head the Central Intelligence Agency and to coauthor *The Future of Nuclear Power*, also urged the commission to study the naval nuclear propulsion program's technical depth in regard to "training and education, continuity, certification of operators, exercises, component testing, and quality assurance" and to contrast "the archetypal commercial power system" with Rickover's organization, which was "built on integral engineering and technical competence throughout its whole pattern."

Part 4 THE KINGDOM OF ELECTRICITY

At a touch, a whole room was lit; hundreds of rooms were lit; and one was precisely the same as the other. One could see everything in the little square-shaped boxes; there was no privacy; none of those lingering shadows and odd corners that there used to be; none of those women in aprons carrying wobbly lamps which they put down carefully on this table and on that. At a touch, the whole room was bright. And the sky was bright all night long; and the pavements were bright; everything was bright.

— VIRGINIA WOOLF, *Orlando*

Responsibility is a unique concept. It can only reside in a single individual. You may share it with others, but your portion is not diminished. You may delegate it, but it is still with you. If responsibility is rightfully yours, no evasion or ignorance or passing the blame can shift the burden to someone else. Unless you can point your finger at the one who is responsible when something goes wrong, then you never had anyone really responsible.

— ADMIRAL HYMAN G. RICKOVER

10 MAN'S SMUDGE

"TO UNDERSTAND THE NEED FOR nuclear energy, you have to understand what the alternatives are," Rip said. To observe where uranium did its work and how it compared with other resources, we headed to the domain of Duke Energy Carolinas, formerly known as Duke Power. Its headquarters is in Charlotte, North Carolina. To serve more than two million customers in the region, the company obtains less than 1 percent of its electricity from gas and oil combustion turbines, about 2 percent from hydroelectric plants, 51 percent from nuclear power, and 47 percent from coal combustion.

In China people began using coal as an energy source three thousand years ago, but it became popular in the West only about two centuries back, after the Industrial Revolution began. Coal smoke became commonplace, causing the nineteenth-century poet Gerard Manley Hopkins to mourn:

> *And all is seared with trade; bleared, smeared with toil;*
> *And wears man's smudge and shares man's smell*

Today coal provides about 40 percent of the world's electricity and over 50 percent of America's.

Riverbend Steam Station, built in 1929, is located on a bend of the Catawba River in a treeless industrial zone. The old brown brick building with six tall white stacks resembles illustrations in schoolbooks from the days when American achievement and prosperity were symbolized by factories belching smoke. However, today nothing appeared to be coming out of these stacks: a person might believe that those ads about "clean coal" on public television and public radio were true.

For two decades, Duke's fleet of coal stations has been rated among the most efficient in the country. That is, their equipment is such that it burns less coal per kilowatt hour, resulting in fewer emissions and lower costs than average. Having already reduced emissions of particulates and gases to below federal requirements, the utility is spending $1.7 billion on equipment to bring emissions into compliance with North Carolina's 2002 Clean Smokestacks law. Tougher than the federal regulations, which allow utilities to buy pollution credits from other states instead of cutting air pollution locally, the legislation has also made North Carolina, which gets 60 percent of its electricity from coal and 32 percent from nuclear power, a leader in air-quality control. When implemented fully over the next decade, Clean Smokestacks will reduce by about three-quarters the emissions of nitrogen oxide and sulfur dioxide and at least two-thirds of those of mercury.

Duke personnel had been gracious about our wish to visit the company's various energy-generating sites and for the fossil-fuel leg of our journey recommended Riverbend. Duke's efforts to reduce pollution suggested that Riverbend would be a reasonably modernized plant. In the entrance foyer were certificates and plaques. For eight years in a row this station had won the Edison Award, a top prize among electric utilities; an EPA Environmental Leadership Program certificate of recognition; and praise for "safety achievement" from the North Carolina Department of Labor.

Our guide was Jerry Kinley, a friendly Tarheel who wore blue coveralls, a Beatles T-shirt, and a hard hat and lives just five miles away. After providing us with hard hats and earplugs, he swung open a big door and we were promptly engulfed in the roar, teeth-rattling vibration, and high-pitched shriek that precedes the wresting of electricity out of matter. We stepped into a cavernous space with ducts snaking around and blackened pieces of giant equipment: conveyor belts, hoppers, crushers, blowers, furnaces, water and steam pipes, turbines, and generators. Tall, glaucous windows provided some light and a little ventilation. A hot wind blew in our faces and swirled the coal dust and feathery gray ash that collected in the corners and crannies. Paint curled off the walls. Fans pumped the heat and acrid-smelling exhaust up the stacks and into the atmosphere. There was a strong blend of odors: Marcia detected steam on hot metal, acrid sulfur, and that peculiar ozone smell of electricity. My earplugs kept falling out, perhaps shaken loose by the loudest din I've ever encountered. They sometimes hit the floor. I'd pick them up, try to wipe them off with my fingers, already grimy from the omnipresent soot, and reinsert them. It was hopeless. Sometimes, when no one was shouting, I just put my fingers in my ears. For the next few days a ringing in my head persisted, and for two weeks all normal sound seemed wrapped in cotton batting.

"We use a coal blend!" Kinley hollered. "It's a mix of different grades!

Watch out! The handrails are real dirty!" He coughed deeply from time to time, and eventually we did too.

To make about 500 megawatts of electricity, during periods of increased demand, Riverbend runs continuously, day and night, seven days a week, with eight operators, a supervisor, four people to do manual labor, and a fireman for each twelve-hour shift.

We tiptoed through puddles—Rip stopped to stare at a rusty overhead pipe that was leaking—and around oil spills. We scaled trembling metal stairs and crept along narrow, twitching catwalks that passed alarmingly close to huge, sooty vessels radiating intense heat. On top of one lay a pigeon that must have flown in through an open window and been roasted on the spot.

Kinley had us step out onto the roof, where we took in cooling breaths and a view of green hills and a rail line that brought in low-sulfur coal. A mountain of it towered behind the plant.

A coal shipment to feed just one plant the size of Riverbend that supplies baseload power for a year requires 14,300 train cars pulled by diesel locomotives. A mile-long coal train is not unusual. The most important single commodity hauled by rail is coal, which makes up about 44 percent of the tonnage for Class I railroads and 21 percent of their revenue.

Coal industry employees tend to live shorter lives than, say, judges. Miners work in unsafe, poorly ventilated galleries where coal dust hangs in the air, causing black lung and other serious ailments. From 2002 to 2006 there were a total of 154 coal-mining fatalities in the United States alone.

The enormous mound of crumbled blackness that rose before us had, hundreds of millions of years ago, been alive: it was composed of organisms that had thrived in a very oxygen-rich environment. Their submerged remains became bogs that were fossilized and compacted by geological processes into coal. Today the coal industry shaves off mountaintops, strip-mines open lands, and has a long history of casting the rubble into streams and valleys. Coal waste, which contains a variety of toxins, is stored in slurry dams that sometimes break and flood creeks and rivers. In addition to lives lost, hundreds of mountains, plains, and watersheds have been ravaged, and more will be. The mountain of coal dwarfed a toylike D-9 bulldozer pushing one ripple of black after another in a hypnotic rhythm toward a tower where a chute took the coal into the plant. At full power, the Riverbend station devours 4,500 tons daily.

The ash from burned coal collects at the bottom of the plant chimneys or is propelled up through them by hot gases from the combustion. Riverbend traps the solid waste with precipitators; these employ an electrostatic charge to capture almost all the particulates and are 99.5 percent efficient. This waste is mixed with water, and the resulting slurry, which contains a host of toxic heavy metals like arsenic, lead, molybdenum, cadmium, and chromium as well as uranium, thorium, and their daughter decay particles, is pumped to an ash pond, a bleak, gray, liquid expanse behind the plant, to dry. Eventually some will be sold for use as landfill and to make bowling balls and golf balls. Waste from other plants is turned into wallboard and paving materials.

"The environmental standards here are strict," Kinley said. Duke was now installing scrubbers and other equipment to catch even more of the particulates and opacities. "Opacity" means visible pollution. Coal combustion produces oxides of sulfur and of nitrogen. As they move through the atmosphere and combine with water, they can form tiny particles that settle in our lungs as well as making smog and acid rain. The acidity can decimate forests, kill fish, and, because of tremendous worldwide emissions, is also changing the chemistry of the ocean and affecting its organisms.

Then there are transparent gases, released from coal when it is burned. Those "clean smokestacks" look good because you can't see mercury rising as vapor from them into the sky. The mercury returns to earth in raindrops. In bodies of water, this toxin, which changes chemically to methylmercury, is consumed by algae, becomes increasingly concentrated as it goes up the aquatic food chain, and can cause ailments in humans who frequently consume affected fish. Around the Great Lakes, people are urged to limit their consumption of local fish and pregnant women are warned to avoid it entirely. A comprehensive assessment of methylmercury from the National Academy of Sciences (NAS) in 2000 estimated that each year more than sixty thousand children are born at risk for neurodevelopmental problems associated with in-utero mercury exposure. The NAS recommended a major reduction of anthropogenic mercury pollution (natural mercury also erodes out of rocks into the water supply). Government scientists have found methylmercury in every river and lake in the United States; its level has also increased in the ocean, and *Consumer Reports* now advises pregnant women to avoid canned tuna altogether. Coal-fired plants are the biggest producers of mercury emissions in the country, spouting fifty unregulated tons per year. President Bush's Clear Skies initiative now calls for gradual reduction of this output to fifteen tons a year by 2018. The problem, however, is global.

A 1,000-megawatt coal plant also freely disperses about twenty-seven metric tons of radiological material a year, exposing people to much more low-level radiation than a nuclear plant would. But it is the nuclear industry that by regulation must track and isolate the smallest actual or estimated quantities of radioactive substances and foot the gigantic bill for doing so. Like mercury, radon rises invisibly from coal-fired plant smokestacks—scrubbers and precipitators can't catch these vapors—and eventually decays into daughters that can damage the pulmonary lining, especially in people who are also inhaling tobacco smoke and fine particulates. On average, every year fossil fuels expose the American population to about a hundred times more low-level radiation than nuclear plants do.

If you live within fifty miles of a coal-fired plant, you're exposed to 0.03 millirem a year. Living near a nuclear plant exposes you to 0.009 millirem a year. Anthracite coal is cleaner, but it's scarce now; plants rely on soft coal, which is dirtier. The big ones in the Four Corners area burn sub-bituminous coal, which is even worse. Those plants give off four hundred times more radionuclides a year than a nuclear plant—one to four millirem. The Nuclear Regulatory Commission would close down a nuclear plant with that emissions rate. In either case, emissions are low-level and the EPA does not consider them harmful. But those who believe even a tiny radiation dose is risky might consider refocusing their concerns on coal-fired plants.

According to EPA statistics, a plant the size of Riverbend burning about 1.6 million tons of coal a year would concentrate in its fly ash about two tons of uranium and five tons of thorium. In the United States in 1999, coal combustion produced over 1,000 tons of uranium and over 2,500 tons of thorium. This is enough fissile material to exceed the amount consumed by all the nuclear power reactors in the country in a year. After World War II, when scientists believed uranium to be rare, they considered extracting it from fly ash.

"Remember the tailings piles at Ambrosia Lake that were being capped with clay to prevent the release of radon?" Rip asked. "That was slurry after the uranium had been *removed*. And people get their shorts in a twist about the radioactivity of those piles. Well, an ash pond at a coal plant would have more thorium and uranium and release more radon than those depleted tailings. It's not the radiation that concerns me. It's more than we get from nuclear, but it's still low-dose. The problem is all the other crap—toxic heavy metals, smog that causes acid rain, and greenhouse gases. Since the 1800s, ash piles have been left to weather away or to leach into the soil and the water table. Toxic heavy metals wind up in the rivers and lakes. Riverbend has been running since 1929. That's a bunch of fly ash, and it's not enclosed. In regions with a lot of coal-fired plants, you can see mountains of fly ash. They're not capped. And you see people living in houses near them. The radiation exposure isn't significant, but the toxic heavy metals in the ash may be." The Environmental Protection Agency exempts large-volume fossil-fuel combustion wastes—all ash, slag, and particulates removed from flue gas—from federal hazardous waste regulations. If placed in landfills, the wastes are subjected to some regulation.

Every year a single 500-megawatt coal-fired plant alone sends up into the sky the same amount of carbon dioxide as 750,000 cars do. Coal combustion is responsible for a major share of the world's man-made carbon dioxide, a significant cause of global warming.

At one end of the plant we got a look at the bend in the Catawba, where water was sucked in from the river through filters. "We took one cove of the river for our use," Kinley said, pointing to a small dam. "We block the fish and we filter them out, and tree branches, too. It's one big cycle."

The river water is purified, then heated by coal furnaces to make superheated steam. It's pumped through a system of pipes and blown through a high-pressure turbine, and then on through other turbines until as much energy as possible has been wrung out of it. Then the steam cools into liquid, falls into a well, and high-pressure coal-fired heaters bring the temperature up until the water becomes steam again. After several cycles, the hot water discharged from the turbines is mixed with cool water before being returned to the river.

On we went, following the black stream of coal as it flowed along a conveyor belt to a noisy crusher. "Forced drafts take outside air and suck it in through a preheater!" Kinley shouted. "The air gets real hot! Hits the coal to heat it up so that it crushes and grinds a lot easier!" The crushed coal was then dumped into huge spinning hoppers and milled using centrifugal force. "Now it's like baby powder and it's sucked up and pushed along by forced air into a heated vessel!" Kinley said. "There it ignites automatically in the hot air!" Through a small open hatch in a big furnace we glimpsed the flames that made our eyes tear up and that heated the steam. "It's a tornado of coal dust and hot air!" he said. "We have alarms on everything in case there's a problem!"

The electricity from the generator goes to the substations next to the plant, where transformers step it up and pass it on to the grid. In a quiet, air-conditioned control room lined with computer screens—incongruous in this nearly nineteenth-century setting—a couple of men were monitoring electricity production and communicating with headquarters in Charlotte, where decisions were being made about where to channel the stream of electrons. They always have to go some place, because they can't be stored. The average American city-dweller today is responsible for about four tons of coal a year going up in smoke. Since electricity generation accounts for 92 percent, or 1.039 billion tons, of the coal we burn, it's our reliance on it that helps make our nation the biggest single per capita contributor to the earth's burden of anthropogenic greenhouse gases. Our nation's 626 coal-fired plants, over 500 of them quite old, are major offenders. American coal production reached a record 1.133 billion tons in 2005, while consumption reached a record 1.128 billion tons. Utilities plan to burn even more coal as demand increases.

Tour completed, Kinley led us to back to the foyer. We paused in front of an old photo of Riverbend with all stacks spewing thick smoke.

"If we did that now we'd be shut down," Kinley said. "My dad used to live nearby, and he'd get fly ash in his hair when he went outdoors."

Although federal environmental policies have made standards stricter for new coal facilities, the old plants have been grandfathered in—they aren't held to the stricter standards of newer ones. Even with improvements, coal combustion is still doing a lot of damage and there is little motivation for utilities to make changes. (In 2007 the Supreme Court decided against Duke in regard to a lawsuit brought by environmentalists claiming that the corporation used legal loopholes to get around emissions regulations.) Nevertheless, other states besides North Carolina are taking action to lower emissions, and states in the Northeast have banded together to cap greenhouse emissions caused by electricity production. The continued operation of nuclear power in the region will be essential to meeting this goal.

The nonprofit Clean Air Task Force commissioned an independent firm, Abt Associates, to research the health effects of coal combustion in the United States. The most recent findings indicate that it causes an estimated twenty-four thousand premature deaths a year. Mortality is greatest in the states that burn the most coal. Higher rates of certain respiratory and cardiac illnesses also occur around coal-fired plants. Pollution from the plants is responsible for 38,200 nonfatal heart attacks a year.

Better coal technology exists. Coal can be turned into a gas that is cleaned of noxious substances before being burned. Toxins in coal waste can be removed and the residue burned. But utilities have had little or no regulatory incentive to spend money to obtain lower emissions. Of the 150 or so new coal-fired plants on the way, only a handful of them will install technology that could be modified to trap carbon dioxide. Coal-fired plants, which are not required to isolate their waste, are profitable and expected to remain so, particularly if they don't have to upgrade to meet stricter pollution standards.

If all the existing nuclear plants around the world were replaced by non-nuclear sources in the same proportions provided by the present mix of energy sources, annual carbon emissions would rise by 600 million metric tons. The total amount that experts estimate will be avoided in 2010 by the Kyoto Protocol is approximately 300 million tons. In any case, at the present burn rate, we may have only 200 more years of coal left in the United States, which possesses one-quarter of the world's supply—and fewer years if coal is converted to a liquid fuel to replace oil, as is already being done in South Africa.

Robert Socolow, a mechanical engineer, and Stephen W. Pacala, an environmental scientist, head the Carbon Mitigation Initiative at Princeton University, and they have put together some possible future scenarios by applying a variety of solutions that they believe would make a considerable difference worldwide by mid-century. Among other things, they recommend improving efficiency at 1,600 large coal-fired plants so that they can get from 40 to 60 percent more energy out of fuel than they now do—and more cleanly; installing carbon capture and storage at 800 large coal-fired plants; replacing 1,400 large coal-fired plants with gas-fired plants; and doubling today's nuclear output in order to displace coal. As it happens, they base their recipes for carbon mitigation not only on recycling carbon and using nuclear power and renewables, which Rip finds appropriate, but also on the assumption that the rate of carbon dioxide absorption by the ocean remains constant.

He considers that assumption unreasonably optimistic, given present data. The rate of increase in fossil-fuel combustion now strongly suggests that the ocean's carbon absorption rate will only increase as well. In the long run, something terrible will occur. "As the ocean continues to sorb carbon, and acidification of the upper layers of seawater continues, the rate of carbon dioxide absorption will drop to zero," he said. "At present the ocean is absorbing massive amounts of carbon, but if it ceases to do that, then CO_2 in the atmosphere will spike astronomically."

The industry is planning about 154 new American coal-fired plants.

The world's largest private-sector coal company, Peabody Energy, which is based in St. Louis, fuels 10 percent of all the electricity in the United States and 3 percent of the world's electricity. Gregory H. Boyce, Peabody's president and chief executive officer, and one of the biggest donors to the Republican Party, served as chairman of a Department of Energy advisory panel that recommended exemptions to the Clean Air

Act that boost coal's clout over the next two decades. Peabody has launched a campaign celebrating clean coal technologies that one day will produce synthetic natural gas, automobile fuel, and hydrogen.

Surely laws will compel new plants to be cleaner than Riverbend—but perhaps not, if the industry sets and enforces pollution limits. As of 2007, the White House and Congress were still refusing to regulate carbon dioxide release, claiming that stricter emissions control would harm the economy.

Over the past thirty years our air has become cleaner because of regulations curbing smokestack and tailpipe emissions, yet these recent decades have seen plenty of economic growth. Some utilities, in particular Duke and others that own nuclear plants, have begun to ask for a clear federal policy to control carbon emissions, probably a carbon tax, so that companies with coal-fired plants will be able to plan ahead and minimize economic risk. "Climate change is real, and we clearly believe we are on a route to mandatory controls on carbon dioxide," James E. Rogers, chief executive of Duke Energy, said in 2006.

Two truckloads of uranium ore contain the same energy to make electricity as two million tons of coal. Damon Woodson, a mechanical engineer formerly employed at a nuclear plant, remarked to the writer John McPhee that he had never understood nuclear power until he came to work at a 12,000-acre coal-fired plant. "The way to go is nuclear if you want to have power. To get a million BTUs, fuel oil costs nine dollars, natural gas six dollars, coal a dollar-eighty-five, nuclear fifty cents. We'll see how it all turns out."

11 FROM ARROWHEADS TO ATOMS

RIP, MARCIA, AND I WERE THE first visitors to step inside the Oconee Nuclear Station since September 11, 2001, and we'd had to prove our bona fides well in advance.

Oconee, Duke's first nuclear plant, began producing power in 1973 near Greenville, South Carolina, and after eight years had paid for its start-up costs; in 2000 the Nuclear Regulatory Commission (NRC) re-licensed it to operate for twenty more years. We drove toward the Blue Ridge Mountain Escarpment along a hilly road that wound through pine forest and past the 300-mile shoreline of Lake Keowee. It was created by Duke's damming of the Savannah River's headwaters for a hydroelectric plant that is part of a grand energy scheme to make electricity from falling water, burning coal, and nuclear fission. The Oconee plant motto is "From Arrowheads to Atoms" because "Oconee" was the Cherokee name for the area. When the dam was built, Duke relocated tribal arti-facts and graves. The forest at the site was clear-cut—such habitat destruction usually accompanies dam-building—and the lumber from this vast harvest fed the state's furniture industry. In the early 1970s, this was an isolated backwater (the movie *Deliverance* had been filmed nearby), but over time the lake attracted developers, and big houses with white-columned verandas have gone up, the most expensive ones rather near the nuclear plant. Today we passed campgrounds, boat docks, and new retirement communities with names like Waterford Estates.

The day was misty, with some rain. "It rains a lot in these mountains," said Tom Shiel, our escort during our visit to the Piedmont. He handles public relations and corporate communications at Duke Energy. Previ-ously, he'd been a newspaperman and sports columnist in his native Con-necticut and a spokesman for Northeast Utilities there. He remains a dedicated Yankee fan and writes a humorous sports column for his local paper. Today he wore a blue denim shirt with the Duke Energy logo embroidered on the pocket.

"We take rainfall into consideration when we have emergency drills at the nuclear station," he said. "We have scenarios for the road getting flooded when an evacuation might occur. We have an emergency operations center away from all this. See the towers with security cameras mounted on them? We're being watched. Every vehicle that comes along this road is observed."

The nuclear plant, situated between two arms of Lake Keowee, has a tastefully landscaped timber-and-stone visitor center, The World of Energy, with a soaring atrium and an immense glass wall facing a butterfly garden and three cylindrical, windowless reactor buildings resembling silos. As is the case with Duke's other nuclear plants, much is made of the successful environmental stewardship of the nearly three hundred unpopulated acres surrounding the plant.

Since September 11, 2001, public tours have been restricted to the visitor center, and it had taken several months and many calls and e-mails for us to obtain authorization to enter the business end of the plant.

Shiel turned us over to our Oconee guide, Dayle Stewart, a kindly, poised woman with prematurely graying dark hair who lives with her husband and young children nearby. "We have thirteen hundred Duke employees here," she said. "Three generations have worked at Oconee. It was the first big boon to employment in the area and has made a big contribution to the economy."

Duke's system of hydroelectric dams, pumped storage (an elegant technique in which excess electricity generated during off-peak hours is used to pump water uphill to an elevated reservoir where it can be released as needed to turn turbines), and nuclear reactors on the lake here comprises the largest power generation source of its kind in the world; it has produced more electricity than any other plant in the United States. "When there's not a drought like there is now, the hydroelectric power comes from the Jocassee Dam, up at a higher elevation, and Keowee Dam," Stewart said. "We're the only nuclear plant in the country with a hydroelectric backup. We also have a dedicated line to a coal-fired plant, and we have a battery backup as well." All nuclear plants must have multiple systems of backup electricity.

In the visitor center, various displays demonstrated how electricity is made and how to conserve it. There was of course a Geiger counter alongside examples of radioactive materials that might set it off: a chunk of potash, a rock from the building foundation, a pebble of uranium ore, and, inevitably, tangerine-hued FiestaWare—an ashtray.

"I look for FiestaWare at yard sales—it's hard to find anymore," Stewart said. "All the nuclear facilities like to buy it up."

As we drove downhill into a natural bowl containing the tight configuration of big buildings that make up the plant, I noticed that Stewart was wearing a dosimeter, and I asked if she'd ever gotten an unusual reading.

"Only once. While I was taking a certain gentleman on a tour, I noticed that a dose was registering. I thought my dosimeter must be broken, and I made a mental note to have it looked at. But at the exit, radiation alarms went off. The station always has a radiation protection person on duty, and he checked the gentleman out and sure enough, he was more radioactive than the normal human body usually is—and so was I! He got asked a lot of questions. It turned out he'd had a strong dose of a radioisotope to treat a thyroid ailment—and a month later he was still giving off 2 millirem."

The alarm went off one other time when she was conducting a high school tour. "During my introductory talk about safety, the teacher started heckling me about the dangers of radiation exposure and kept it up the rest of the time. When we were leaving afterward, he set the portal alarms off. Someone from radiation protection came and wanded him. The teacher was wearing an old watch with a radium dial. He could not believe that was the problem, but when he removed it, he was able to pass through without setting off the alarm. The radiation-protection person talked to him about not wearing the watch anymore." If the teacher wore it twenty-four hours a day for a year he would receive up to 130 millirem—thousands of times more exposure than a nuclear plant could give him, even if he lived on-site. The teacher decided to keep wearing the watch, an heirloom, but on subsequent tours did not heckle Stewart.

After passing through security checkpoints and along polished corridors inside a building where we were scrutinized and screened by various guards and detectors, we stepped outside and got a closer look at the reactor containment structures. Each one had nineteen stories, twelve of them underground. The reactors were anchored in bedrock a hundred feet below the surface. A typical American power reactor core is twelve feet tall and about twelve feet in diameter; it could fit in a home kitchen. The core is enclosed by a large carbon steel pressure vessel with a thick stainless steel liner and walls of dense concrete about five inches thick. This containment vessel in turn sits within an immense containment building made of a shell of steel covered with four to six feet of concrete reinforced with steel bars and negatively pressurized to contain any explosion or leak. Anyone entering has to pass through an air lock. Any escape of radionuclides would be so small that it wouldn't cause a health or safety problem.

Stewart pointed to a massive door that's opened only when refueling takes place. "This building has been closed for eighteen months. To run the reactor from the control room, the operators use cameras and sensors. If workers do have to enter the building, they wear radiation protection clothing and a chill vest—the temperature is high in there. And of course they're checked afterward for exposure. The reactor cooling system also operates at a lower pressure than outside. So water from the lake would leak *in*, not *out*. And we're constantly testing the lake, the steam, and the environment in the reactor building for radiation."

Each Oconee reactor unit generates over 900 megawatts (some of that electricity is used to operate the facility), and the station always keeps two of its three reactor units running. Each one has to be shut down for about a month when they're refueled. Every eighteen months or so, a traveling team of about two thousand experienced nuclear refueling specialists operates heavy-duty cranes and other equipment to transfer the old fuel assemblies to an underground pool of water to cool until the hottest fission products have decayed. Eventually the assemblies go into interim storage in dry casks, which are widely spaced, thick-walled concrete cylinders cooled by naturally circulating air. Already about a quarter of American plants have begun using this method to reduce the accumulation in spent fuel pools, and more will follow. The NRC says that the fuel can be safely stored in dry casks for at least fifty and probably for one hundred years.

After NRC inspectors discovered that boron in the coolant water had corroded welds on the reactor head at the Davis-Besse nuclear plant in Ohio, they examined reactor heads at all the other plants of similar design and found that a unit at Oconee also had microscopic cracks in the weld material. The reactor-vessel head is unbolted during refueling. "The cracks are not a safety problem at this point," Stewart said, "but it's cheaper and safer, dose-wise, for the workers if we purchase new heads and replace the old ones before anything happens." The new seal would be protected against corrosion by a film of polymer.

And what was the worst that would happen if the reactor head completely failed? In America and in most other countries, the negatively pressurized containment structure, built to withstand pressure of the primary steam should it escape from the pipes, would fill with steam containing radionuclides. The contamination would not affect the public or the workers, because the building normally remains off-limits to everyone when the reactor is in operation, but the station would have to conduct a major clean-up operation inside the building.

A company that builds and services reactors was coming to Oconee soon to replace the damaged reactor head, and that would require the use of special heavy equipment to cut a temporary opening in the side of the containment building. "It will be quite a job," Stewart said. "The walls are made of high-density concrete and contain steel tendons the thickness of a man's arm that can be tightened to make the building withstand more pressure. We keep them lubricated and torqued. The crew will also have to cut through an inner lining made up of prestressed concrete and steel sheathing." Any radiation from the core would be shielded by the pressure vessel, so workers would not receive a dose from the fuel rods. Nevertheless, appropriate protective garb must be worn.

As I gazed up at the three structures, which reminded me of turrets in a medieval castle, I remembered that within their formidable ramparts something rather simple was occurring: water was being heated up. Such is human progress since the days when we stuck hot rocks in a fire and then dropped them into vessels of water to accomplish this chore. Water in pipes wrapped around the reactor core—basically composed of rods filled with hot rocks called uranium pellets—carries away heat. As the primary coolant, in its closed system, leaves the core, it's about 600 degrees Fahrenheit, but it doesn't boil because it's kept under pressure in a steam generator while flowing through thousands of tubes that are intertwined with the pipes of the secondary coolant system. That system, also closed, in turn picks up the heat and, because the pressure is lower, creates the steam that flows through big conduits wrapped in silvery insulation out of each unit's containment building and into the nearby turbine building. Meanwhile, the cooled water in the primary system is returning to the reactor core.

The steam from the turbines eventually passes over a third, independent system of pipes containing cool water from Lake Keowee. The steam in that way is condensed into water within its closed system, and thus cooled, circulates back to the reactor. The water in the tertiary system is allowed to cool down and then is returned to the lake as slightly warmed water. This discharge is always being tested to make sure it's unchanged and that it's not affecting lake organisms. Oconee and all other nuclear plants also monitor the air, water, aquatic life, and local dairy milk to make sure that no radionuclides or toxic chemicals get into the biosphere. Keowee and Jocassee are considered by the South Carolina Department of Health and Environmental Control to have good water quality. Gina Kirkland of that department told me that if there is a pollution concern associated with energy generation, it's the deposit of airborne, coal-sourced mercury from the worldwide load.

Other plants use cooling towers to make the steam condense into water. Some steam issues out of the tops of the towers, and people wrongly assume that this is radioactive. It's just pure water vapor that has never been near a reactor; it's from the tertiary system that condenses the steam. Rip, always concerned about the world's growing lack of clean water—about a billion people have no access to it—wants to see nuclear power replace the fossil fuel combustion now in use to distill pure drinking water from the ocean. Japan has already coupled nuclear plants to desalination, and India and Pakistan have such projects under way.

We followed the silvery conduits to Oconee's turbine building, put in earplugs, which fortunately stayed in this time, and, as we entered, were blanketed by the piercing whine of steam flowing through ever tinier pipes and forced against blades in a row of turbines whose generators were spinning at 3,600 revolutions per minute. The noise seemed somewhat less maddening than that at Riverbend, which runs a lot of heavy, coal-moving and -pulverizing machinery in addition to the turbines. We crossed a wide, spotless, pale terrazzo floor. As a reactor operator later told me, "You can eat off the floor in a nuclear plant. I just wish hospitals were as clean." In a relatively dust-free setting radionuclides are easier to track.

We passed pieces of equipment that were cordoned off and posted with radiation warning signs. How could this be? How could radioactivity be brought into the turbine building if the primary and secondary systems were completely separate? After we exited the turbine room and conversation became possible, Stewart explained. "Our steam generators are getting old and it's cheaper to replace them than to fix them. It's a big job that will take five hundred million dollars and require twenty-four hundred workers. As we set up to do that, as an extra precaution we make the assumption that any secondary water system is contaminated—just in case maybe there's a pinhole leak or something."

Now when I see high-tension wires silhouetted against the sky, I remember that overwhelming scream that accompanies the birth of electricity. The electrons from Oconee flow in cables from the generator out of the building to a collection of transformers that, smelling of ozone, step up the voltage to match that of the grid that they're feeding, and from them the electrons proceed outward, notes in a vast, complex fugue, along transmission lines on the shoulders of tall pylons that march away into the mist. People at computer terminals in Charlotte guide the electrons through the grid, providing more at peak hours of use, less in the deep night.

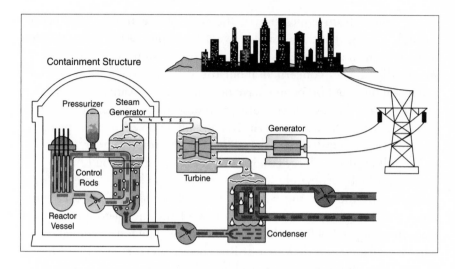

Oconee supplies baseload electricity around the clock. When I woke up in the night in my hotel room in the Piedmont and turned on the lamp, electrons coming from Oconee and other power stations made the filament in the bulb glow. Some huge, rapidly spinning dynamo perhaps hundreds of miles distant slowed down infinitesimally as it adjusted to the load. The work done on one end is compensated for at the other.

When my nuclear tour began, I didn't know what baseload electricity was or that this necessary, steady, reliable supply of electricity in our country comes almost exclusively from fossil fuels and nuclear power, along with a small fraction from hydroelectric dams. Those of us who cherish hopes that some other, better energy might take over the running of our nation twenty-four hours a day, seven days a week, may imagine simply replacing coal or nuclear or both with some pleasantly alternative resource like wind. But in the real world, that's unfortunately not how things would happen. A 2006 National Academy of Sciences study of the Indian Point nuclear plant near New York City found that if it were to be shut down, the state would lose the source of about 10 percent of its electricity. That would have to be made up by burning fossil fuels—probably carbon-dioxide- and methane-emitting natural gas—and customers would have higher utility bills. To replace the power generated by Indian Point with a wind farm would require three hundred thousand acres.

The amount of electricity generated in a day, week, or year is measured in kilowatt-hours or megawatt-hours. Oconee recently became the first nuclear station in the country to generate 500 million megawatt-hours of electricity—enough power to supply electricity for twenty years to every house in South Carolina—nearly two million households. And how much uranium did Oconee use each year to pump out all these electrons?

I asked Tom Shiel.

"I can only give you a ballpark estimate," he replied. He reminded me that one pellet of uranium the size of the tip of my little finger produces as much energy as 1,780 pounds of coal. Each Oconee reactor is fueled with 8.3 million of these pellets, packed inside 208 fuel rods. Each fuel assembly, nine inches square and twelve feet long, weighs about sixteen hundred pounds (that includes the weight of the pellets' cladding and the fuel assemblies). There are 177 assemblies per reactor. The total weight of a fuel core is 103.7 tons, and 34.57 tons of fuel are replaced every eighteen months for each unit. The average American power reactor produces twenty tons of spent fuel a year, and the annual total of spent fuel from all of our 104 power reactors comes to about 2,000 tons.

And how does this compare with coal combustion? A pound of pure uranium contains the energy equivalent of over two million pounds of coal and produces about 400,000 kilowatt-hours of electricity. One pound of coal when burned produces 1.2 kilowatt-hours. Per capita, Americans burn twenty pounds of coal a day.

Using a very small volume of fuel that is always monitored, shielded, and sequestered from the environment, a 2,000-megawatt station like Oconee saves the equivalent of twenty-seven thousand tons of coal a day from being burned and therefore keeps its dangerous waste from entering the biosphere. Only 27 percent of America's electricity comes from emission-free sources, and about three-quarters of that is made by nuclear energy.

Back in North Carolina, we had lunch with Tom Shiel at a nautically themed restaurant on Lake Norman, which is about twenty-five miles from Charlotte, and which Duke created in 1963 by damming the Catawba River to run a hydroelectric station. The lake and the river also provide water for steam for Riverbend and another coal-fired station as well as for the McGuire Nuclear Station on the lake. Duke also maintains public recreation areas around the lake and has built housing developments as well.

Duke initially thought people would never move here to the sticks. But the surge in population in the hot, damp Carolinas, attributable in part to technological breakthroughs in dehumidifying climate control units and cheap electricity, has naturally created more consumers. The newish communities nearby, built on Duke real estate, were pricey. White frame houses with front porches lined quaint streets. Picturesque buildings had been made to look weathered. As with developments like this in other rapidly growing parts of the Piedmont, the aura of Disney World–like artificiality will no doubt blur and fade as the trees grow and the buildings age.

"Our customers have a lower electric bill because we diversify," Shiel said. "Essentially, Duke base rates have remained the same for over fifteen years. We control costs by running whatever is the least expensive. The cost of nuclear fuel is cheap and stable."

"What about renewable resources, like solar and wind?" I asked.

"When the NRC reviewed our recent application for license renewal for the McGuire Nuclear Station here on the lake, we had to analyze the alternative power sources and the impact of each one on the area as opposed to nuclear. To replace the electricity provided by a single reactor with solar, the entire lake, thirty-two thousand acres, would have to be covered with photovoltaics. There's not enough wind here to turn turbines. Anyway, the environmental impact of replacing a nuclear plant with renewables would be big." After looking into government statistics on that topic, I found myself in agreement with Shiel.

But nuclear utility companies are scarcely perfect. "Aren't energy corporations that own nuclear plants out to make money, just like any other business?" I asked. I was thinking of news reports about releases of mildly radioactive water or gases that utilities have covered up, even though the radiation levels did not violate EPA regulations. "Aren't utility owners likely to cut corners? You know, cheat on safety and security maintenance? Fail to report problems to the NRC?"

"Even though we're competitive with other energy companies, when it comes to nuclear, one plant's problem is everybody's problem," Shiel said. "Take as an example Davis-Besse in Ohio, the worst-run plant in the country. It had the problem of a corroded reactor head. Contract workers doing refueling there had performed some maintenance on it. When they were finished, they came to Oconee to do refueling. As we always do, we tested the crew for radioactivity before they entered the plant and found that they were wearing slightly contaminated clothing. We called Davis-Besse to inform them. Davis-Besse responded by saying defensively, How do you know it came from us? They knew that the crew was triggering the detectors but had attributed it to an internal dose." Atomic workers can accumulate a small lifetime permissible dose that's strictly regulated and that might be detectible in their bones and organs.

"The contamination on the clothing was very low-dose," Shiel continued. "But it should not have happened. They should have investigated. Davis-Besse was still playing the game of 'let's ignore it and maybe it will go away.' People were fired after the reactor-head incident. The company cleaned house. But Davis-Besse lost confidence with the public and with the industry and had to pay a 5.45-million-dollar fine. Every nuclear station has to be well run or we all suffer. Think of Three Mile Island. Those guys made a big mistake and we're all still paying for it."

After Three Mile Island, the nuclear industry established the Institute of Nuclear Power Operations, which has reactor operators with seniority follow a practice Rickover instituted in the nuclear navy: inspection visits at other plants to check on how well their colleagues are doing their jobs. "It's important to have outsiders come in and quiz you on what you're doing," Theodore Rockwell, a nuclear engineer who worked with Rickover for many decades, had told me. "They see stuff that the insiders overlook." He added, "Nuclear plants are hostages of one another. Each of us in the industry has a stake in seeing that nobody else screws up."

I asked Shiel how problems like the reactor-head failure could be better prevented.

"The next generation of nuclear is going to be far better, because of standardized designs," he replied. "The same equipment and the same design will be in place at every plant. Inspectors will know what they're looking at. Now each plant is different, so inspectors have to learn all the designs. There's an evolution in any industry. There was a time in this country when the railroad tracks became standardized in size. Standardization, having all the replacement parts the same, will save money in building nuclear stations and in operating them. We need new plants. Ours are running at about ninety-five percent capacity right now and will run more." Capacity factor is the percentage of electricity actually produced, compared with the total potential electricity that the plant is capable of producing. By contrast, wind power operates at about 30 percent and can't improve, because the capacity is limited by the amount of wind that blows.

Rip asked Shiel about output.

"Duke has seven nuclear units putting out electricity at a cost per kilowatt-hour that's *lower* than coal."

Over time, nuclear electricity has grown steadily cheaper. In 2005, the production cost of electricity from nuclear power on average cost 1.72 cents per kilowatt-hour; from coal-fired plants 2.21; from natural gas 7.5, and from oil 8.09. American nuclear power reactors operated that year around the clock at about 90 percent capacity, whereas coal-fired plants operated at about 73 percent, hydroelectric plants at 29 percent, natural gas from 16 to 38 percent, wind at 27 percent, solar at 19 percent, and geothermal at 75 percent.

Using past performance of existing nuclear plants as the basis of their calculations, John Deutch and Ernest Moniz, the cochairs of the panel that wrote the MIT-Harvard study *The Future of Nuclear Power,* go beyond simple production cost to arrive at a much higher cost per kilowatt-hour than the industry does. They write in an article in *Scientific American,* "The 2003 M.I.T. study estimated that new light-water reactors would produce electricity at a cost of 6.7 cents per kilowatt-hour. That figure includes all the costs of a plant, spread over its life span, and includes items such as an acceptable return to investors."

By comparison, the study determined, a new coal plant kilowatt-hour would cost 4.2 cents, and a new gas-fired plant kilowatt hour, 5.8 cents. Cutting back plant construction costs and time could reduce the cost of new nuclear electricity, as could improvements in operation and maintenance and the way a plant is financed. "All these reductions in the cost of nuclear power are plausible—particularly if the industry builds a large number of just a few standardized designs—but not yet proved," Deutch and Moniz say. (This is the plan the NRC now fosters.) If carbon emissions from fossil-fuel plants are ever priced, whether taxed or traded, then nuclear power will have a distinct economic advantage over fossil fuels.

A comparative study of nuclear and fossil fuel resources done by Finland in 2000 indicates that if the price of uranium doubled, the cost of nuclear power would increase by 9 percent. If coal and natural gas costs doubled, the cost of electricity would rise by 31 percent and 66 percent, respectively. If carbon emissions trading, a scheme that has been instituted in the European Union, were to be used in the United States, the cost in pennies per kilowatt-hour of fossil-fuel-generated power would rise considerably above that of nuclear-generated.

Emissions trading is supposed to ensure that reductions—of greenhouse gases, for example—take place where they cost the least. In this way the costs of addressing climate change are lowered, as the U.K.'s environmental department explains:

> Emissions trading is particularly suited to the emissions of greenhouse gases, the gases responsible for global warming, which have the same effect wherever they are emitted. This allows the Government to regulate the amount of emissions produced in aggregate by setting the overall cap for the scheme but gives companies the flexibility of determining how and where the emissions reductions will be achieved. By allowing

participants the flexibility to trade allowances, the overall emissions reductions are achieved in the most cost-effective way possible.

Participating companies are allocated allowances, each allowance representing a tonne of the relevant emission, in this case carbon dioxide equivalent. Emissions trading allows companies to emit in excess of their allocation of allowances by purchasing allowances from the market. Similarly, a company that emits less than its allocation of allowances can sell its surplus allowances. In contrast to regulation which imposes emission limit values on particular facilities, emissions trading gives companies the flexibility to meet emission reduction targets according to their own strategy; for example by reducing emissions on site or by buying allowances from other companies who have excess allowances. The environmental outcome is not affected because the amount of allowances allocated is fixed.

In 2003 nine states in the Northeast made a commitment to apply this scheme to carbon dioxide emissions. The following international projections indicate the decline in the cost of nuclear power that Deutch and Moniz predict if fossil-fuel plants have to start paying for carbon emissions.

Some comparative electricity-generating cost projections for 2010			
	nuclear	**coal**	**gas**
Finland	2.76	3.64	—
France	2.54	3.33	3.92
Germany	2.86	3.52	4.90
Switzerland	2.88	—	4.36
Netherlands	3.58	—	6.04
Czech Rep	2.30	2.94	4.97
Slovakia	3.13	4.78	5.59
Romania	3.06	4.55	—
Japan	4.80	4.95	5.21
Korea	2.34	2.16	4.65
USA	3.01	2.71	4.67
Canada	2.60	3.11	4.00

US 2003 cents/kWh, Discount rate 5 percent, 40 year lifetime, 85 percent load factor.

Source: OECD/IEA NEA 2005.

Antinuclear advocates say that external costs such as liability insurance and effects on health and the environment must be included in calculations of the cost of nuclear power, and that therefore it's very expensive. These opponents often grumble about insurance for the nuclear industry, which they assume is paid for with tax dollars. In fact the industry is self-insured. Utilities deposit money into what is basically a savings account that is rarely drawn upon. More than $10 billion in liability insurance protection has accumulated for use in the event of a reactor incident. This insurance is governed by legislation enacted in 1957, the Price-Anderson Act, which has worked so well that it stands as a template for other similar legislation. The Nuclear Energy Institute defines the arrangement as follows:

> Insurance pools have paid more than $200 million in claims and litigation costs since the act went into effect. They disbursed approximately $71 million of that total in claims and litigation costs related to the 1979 accident at Three Mile Island 2. . . . The Price-Anderson Act provides no-fault insurance to benefit the public in the event of a nuclear power plant accident the Nuclear Regulatory Commission deems to be an "extraordinary nuclear occurrence." The costs of this insurance, like all the costs of nuclear-generated electricity, are borne by the industry, unlike the corresponding costs of some other power sources. Risks from hydropower mishaps (dam failure and resultant flooding), for example, are borne directly by the public. The 1977 failure of the Teton Dam in Idaho caused $500 million in property damage, but the only compensation provided to those affected was about $200 million in low-cost government loans.
>
> Under the Price-Anderson framework, the public has paid nothing. . . . The act has proven so successful that Congress has used it as a model for legislation to protect the public against potential losses or harm from other hazards, including faulty vaccinations, medical malpractice and toxic waste.

The European Union has found nuclear power's external costs regarding the environment and public health care lower than those of coal, gas, and solar power. In terms of avoidance of damages related to ecosystems and global warming that cost money, nuclear power is on a par with renewables.

The U.S. Energy Department's Information Administration predicts that global electricity production will more than double by 2030, with developing countries expected to ramp up consumption. But the leaders will remain the United States and China. American demand will increase 45 percent and require construction of hundreds of new power plants to generate 350,000 additional megawatts.

When people ask me about nuclear plants now, I say, "If you could go to the visitor center at Oconee, and you had time to see only one display, it should be the one of a little metal rod less than an inch long, identical to one I first saw in a uranium mill. It's a pellet of uranium dioxide that weighs about as much as three pennies and produces the energy equivalent of nearly a ton of coal."

12 BARRIERS

SINCE 9/11, SPECULATION HAS BEEN rampant about how terrorists could take over a nuclear plant or fly a plane into one. Links to websites, articles, and videos sent to me by well-meaning friends before my visit to nuclear plants informed me that terrorists could crash a small plane or a jetliner into the spent fuel pool or into the reactor or both, resulting in fire and an explosion like the one at Chernobyl. Or the collision could cause the core to melt deeply into the earth while giving off radioactive gases that would bubble up through fissures as the reactor fire or burning jet fuel led to the dispersal of all the spent nuclear fuel over hundreds of square miles, causing cancer deaths in half a million people as far as five hundred miles away and making the surrounding land forever uninhabitable. Security, I was advised, was notoriously poor at nuclear plants: at any moment suicidal commandos could hop over the fence and overwhelm a tired, unfit guard and then, assault weapons blazing away, race directly to the reactor and blow it up, causing an atomic explosion that would dwarf Hiroshima. Or the marauders could grab the uranium out of the reactor to make a nuclear bomb, maybe even assembling it and detonating it on the spot. Or attackers might cut off the coolant water, causing the reactor to overheat and melt down.

Because of concerns about terrorism, these days most people will never experience firsthand what security at a nuclear plant is like. Oconee, hidden away in its mountains, must—like all other nuclear plants in America—meet strict standards in this regard. We visited McGuire Nuclear Station, about a forty-minute drive from Charlotte, which has a population of more than six hundred thousand, and now that I understood more clearly how a nuclear plant worked, I focused on security measures.

McGuire is set among a few hundred acres of forest on the shores of Lake Norman. Because of advance arrangements made by Tom Shiel, the guards at the first gate expected us; they waved us onward for a few miles to the Explorium, as the visitor center was called. There our guide, Rita Sipe, ushered us into an auditorium and seated us in the front row.

Behind her petite and slender form was a huge window and a lakeside landscape of loblolly pines, flowering dogwoods, and lawn, as well as the reactor and turbine buildings. "This used to be farmland in the middle of nowhere," Sipe told us in the soft accent of the Piedmont. "Now it's a high-traffic area, with people who work in Charlotte living out here. They bought million-dollar houses on the lake and didn't even know there was a nuclear station here—we don't have those big cooling towers. Since 9/11, the community has noticed us. Now they're asking questions about security."

"They're worried?" I asked.

"The longtime residents are not. And neither are the newcomers once they understand how a nuclear plant works. It's unfortunate that we can't do open-plant tours anymore. If you drive by the property, you don't see the security. But we've always had it—long before 9/11. We built the Explorium here at the beginning, and put in displays about how electricity is made and about radiation safety, and we gave tours. We wanted everyone to understand and see what we do. I showed one couple around. The woman was a skeptic. By the end, she was telling her husband that she wanted to become a reactor operator. Visitors—mainly school tours—can still come to the Explorium, though not to the plant. Our engineers, radiation-protection staff, and others go speak at the Rotary and the Lions Club. Everyone living within a ten-mile radius gets mailed updates about McGuire. People here just aren't hostile about the plant."

Sipe lives nearby with her husband and two young children. "We know the station is secure—we never give it a thought. An independent agency does annual customer surveys for us that ask: do you feel Duke is operating the plant safely? And people say it is. The people who don't know how it works are the ones who fear it. Our industry is the most heavily regulated in the world. We have two NRC inspectors here all the time. It's like having a highway patrolman in your car to make sure you don't speed. We also have self-reporting, and there are small things that come up that we have to report to the NRC. The equivalent of saying to the patrolman in your car, 'I forgot to signal.' "

McGuire has won environmental stewardship awards for the care of its 586 acres, most of which remain undeveloped. Sipe described sunset concerts on the greensward here, given by the Charlotte Symphony. Crowds come, boaters anchor nearby, people bring picnics. Perhaps the environmental angle was hatched by a public-relations expert, but the result has been the preservation of wildlife habitat as well as a demonstration that nuclear power can be easier on the environment than other kinds of energy production. We strolled down to an inlet to inspect water-monitoring stations, picnic tables, the Backyard Habitat, the "fish-friendly" dock. "Here fifth grade children study electricity, and in the eighth grade nuclear power, so we take classes through the exhibits and we make our informational material correspond to the class curriculums and also to Scout badge requirements," Sipe said. "We have a nature trail, a bluebird trail, a foraging garden for animals, and an herbivore pond."

We spotted a pair of mallards in flight and an osprey with a fish in its talons alighting in a nest on top of a pole with a security camera. It overlooked a favorite spot of fishermen that was adjacent to an enclosed rectangular pond holding clean water discharged from the steam turbines. After the water from the turbines cools to air temperature, it's released from the pond into the lake. "Lake Norman is one hundred and twenty feet deep and can take the heat," Sipe said. "In the summer the water temperature is in the upper eighties naturally. The operating limit is ninety-nine degrees Fahrenheit. Fish come to feed here. Each time Homeland Security declares a higher alert nationwide, people phone to ask if they can still come here to fish." Of course, we, the fishermen, and any other visitors were being watched.

The fish are not radioactive. "We're required by law to think about it this way," Sipe said, turning to the example of our old friend, Fencepost Man. "If a person lived right at the plant and gardened, and ate all his food from the garden and drank the water only from here, he might pick up one additional millirem a year. Natural background here is 360 millirem. Of course nobody spends twenty-four hours a day at the plant, and no one grows food here, so no employee is going to get as much as a millirem. We also have radiological monitoring stations. We test—within a ten-mile radius—the fish and cows, the water, soil, air, trees, and grass. We drill test wells. We have a sanitary landfill for all the waste that isn't nuclear in origin. It's lined, grassed over, and surrounded by test wells. We follow EPA and NRC regulations about all this."

The nuclear industry has fewer worker accidents than other utilities. According to statistics from the U.S. Bureau of Labor, it's safer to work at a nuclear power plant than in the manufacturing sector—or even in a real estate office or a bank.

Employees of nuclear plants are healthier and less likely to die prematurely of cancer and other diseases than the rest of the general population. In 2004, the Mailman School of Public Health at Columbia University issued the results of a study it had conducted on fifty-three thousand workers from fifteen American nuclear utilities for periods of

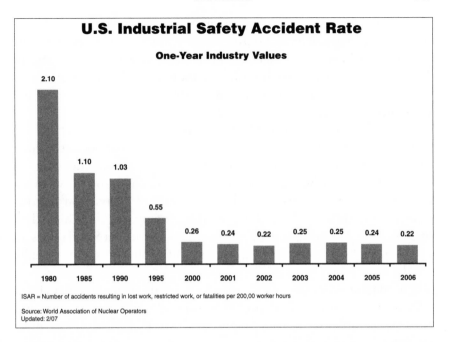

U.S. Industrial Safety Accident Rate

One-Year Industry Values

ISAR = Number of accidents resulting in lost work, restricted work, or fatalities per 200,00 worker hours

Source: World Association of Nuclear Operators
Updated: 2/07

up to eighteen years between 1979 and 1997. "Mortality rates of these workers showed that they were 60 percent lower than cause-specific U.S. mortality rates for a population similar in terms of gender, age and calendar year," the study concluded, supplying the reason for this rosy outcome—the healthy-worker effect: "In order to work in the nuclear industry, workers have to be healthy and are usually required to have annual medical check-ups."

The security at all nuclear plants, always high, has been greatly stepped up. After 9/11, McGuire's main entrance was closed and barricades were erected. Everywhere we went at both plants, before and after we had gone through entrance gates, we saw hydraulic pop-up barriers. Duke security officials refused to discuss them except to say that they are designed to stop vehicles. I obtained a general idea of what an understatement that was from a description of such a device on the U.S. Army Corps of Engineers website: "The hydraulic barrier has the capability to stop a tractor-trailer truck going 55 to 65 miles per hour dead in its tracks." Even when you've received authorization, getting inside a plant is a complicated affair, as we were about to find out.

Our rental car was permitted no farther than the Explorium parking lot. Sipe drove us in a company van toward the plant complex and stopped at the first security checkpoint, which was under a sort of car-port. Three-quarters of McGuire's security force comes from the military and they look it. A muscular man wearing a uniform and carrying an M-16 collected our identification and took it inside the guardhouse. Meanwhile a uniformed woman used a mirror on a pole to examine the van's undercarriage.

The man returned and held a private discussion with Sipe. The NRC had to approve every visitor, and a form she'd filled out did not include the name of a company, since we weren't here as representatives of one. This had caused a bureaucratic glitch. After it was resolved, we were asked to leave the van and walk about a quarter of a block and wait for Sipe to drive through. I asked Rip what was going on. Until his retirement in 2002, as director of nuclear waste management at Sandia Labs he had been involved in counterterrorism measures. His terse reply: "Sandia has the same thing."

Continuing on in the van, we passed a fire department and a clinic. "There are always medical people on-site as well as emergency and hazmat—hazardous material—responders," Sipe said. "We're like a mini-city here."

After passing through another checkpoint, she escorted us into a red brick building, where we came up against a squad of armed guards in flak jackets who looked fit and alert. They wore combat boots, pagers, and headsets. We were waved on to a big room with a smaller room projecting into it, the walls made of mirrors—or, rather, one-way glass.

A door opened and a man stepped out of the mirror world, which was dimly lit, with people seated in front of video monitors connected to cameras. They'd no doubt followed our progress from the main road, through the first gate, the Explorium, the nature center, and through all the checkpoints thus far.

We gave our IDs to the man, who took them into an office and reappeared with a badge and a dosimeter for each of us. We then had to put whatever we were carrying onto a conveyor belt to be X-rayed while we walked, beltless and shoeless, through a metal detector. The whole time, more guards armed with automatic assault weapons who were positioned on the other side observed us attentively. These men seemed even bigger and tougher than the previous group, and there were more of them.

Sipe then escorted us to a bank of floor-to-ceiling turnstiles with wide, thick, flat bars of case-hardened steel. She swiped her ID card on a machine. A formerly invisible window in the mirror wall suddenly opened and the head of a new security officer appeared. Sipe explained who we were. Eventually, she was permitted to take us a little farther.

Now she put her hand into an open metal box with a few metal pegs sticking up as guides. In less than a second, ninety different measurements had been taken of her palm and compared with those already keyed to her employee number in a database. Every palm is unique; the results had to match the information on her badge. Our badges were attached electronically to her badge. One after another, we each swiped them and put our hands in the box to be scanned. Most nuclear plants now use this method, called palm geometry. (We learned the macabre fact that an intruder could not get access by cutting off an employee's hand and placing it on the reader, because it can tell the difference between a severed hand, whose measurements have necessarily changed owing to blood loss, and one still attached.)

The turnstile unlocked but only briefly. One by one we passed through, and it automatically relocked after each of us.

Once we made it to the other side, Sipe warned us that at all times we had to stay near her, remaining within her vision. We were to go through doorways ahead of her. If one of us needed to use the restroom, the others would wait outside the door. She then unlocked a heavy door and let us outside, where we encountered an impressive series of defenses that included three parallel chain-link fences that were perhaps twenty feet high; spirals of concertina wire, its razor-sharp edges glinting in the soft Carolina light; berms of gravel between the fences where motion sensors, sirens, and cameras were planted; and Jersey barriers—long, pyramidal, precast concrete blockades—arrayed to prevent the approach of vehicles. Periodically an armed guard on patrol strode by.

We looked up at the reactor buildings, two brick-faced, windowless structures, each about five stories high, that were separated from the rest of the plant by still more of those graveled berms with their broad-spectrum detection systems—just in case trespassers somehow managed to get past the cameras and pop-up barriers on the approach to the plant, dodge cadres of guards with major weapons, sneak through the many checkpoints, and figure out a way to trick the palm sensors and persuade the steel turnstiles to unlock. I guessed that there were other security devices we would never learn about. To reach a reactor, terrorists would have to get through a tornadoproof, locked-nine-ways-from-Sunday, heavy steel door in the containment building wall, and then to negotiate an air lock.

If they were to set off a blast inside a negatively pressurized containment building, not much would happen. The design for such structures has its origins in the bunkers engineered to withstand the ferocious shock wave from atomic bombs during tests. As at every U.S. nuclear plant, the walls have been made robust enough to withstand powerful hydrogen and steam explosions. Not much would ever happen to this structure unless it took a direct hit from a thermonuclear weapon. For terrorists to overcome every obstacle while carrying an undetected conventional explosive large enough to cause any serious damage would be very problematic. And why waste an extremely valuable and hard-to-obtain nuclear suitcase bomb on a single rural reactor deep underground when terrorists could instead take out, say, lower Manhattan?

But what if attackers managed to get inside a containment building without being stopped? What if somehow they got inside the reactor containment vessel and pulled the fuel rods out of the reactor? Of course those rods weigh tons, and would have to be moved robotically, with cranes, so the terrorists would have to be adept at that. Could they cobble together a weapon using the fuel? The radiation from the assemblies would kill anyone within a minute, so special, thick shielding and remote manipulators of the kind we'd seen in the hot cells at Argonne-West would be required. Even if somehow the terrorists could penetrate that far and get their hands on the necessary equipment, the bomb they assembled would be a dirty bomb, not an atomic one. First of all, putting together an atomic bomb and its triggering mechanism takes a lot of time and effort and special expertise, as I've said, and, second, it's not possible to make an atomic bomb out of low-enriched nuclear power plant fuel without putting huge quantities of it through a reprocessing plant. So the rods would have to be combined with conventional explosives. If the terrorists set off any kind of bomb in a negatively pressurized containment dome, the result would be only some contamination restricted to the interior of the building. To affect a population, the group would have to carry the heavy fuel assemblies out of the plant without falling dead, surreptitiously load them into a large, specially reinforced truck that could carry the weight, and stay alive long enough to drive undetected out of the McGuire complex and to a metropolitan area without being tracked. This would require an army. Crews of two thousand are needed, along with special cranes and other equipment, when fuel assemblies have to be swapped out of the reactor. In any case, at the first hint of a breach of security, every alarm on the grounds would be sounding as well as alarms at Duke headquarters and no doubt in police stations and government agencies.

In fact terrorists would never leave the plant. It's just as hard to get out of one of these places as it is to get in. Believe me, I know. When we had been about to exit the Oconee Nuclear Station, we had to pass through radiation detectors and yet another security barrier. Rip and Marcia breezed through, swiping their passes to unlock floor-to-ceiling turn-stiles there. I was to follow. But when I swiped my card, nothing happened. Dayle Stewart, our guide, tried swiping it. Nothing. She didn't care to swipe too many times, because it would set off an alarm. While she was phoning security, a young man had appeared. He was armed with a pump-action assault shotgun and I thought I saw a pistol in a holster, too. He was using a headset to explain the problem to the chief of security. Finally, the message went up the chain of command, and it was verified that no security breach had occurred. My pass's magnetic strip had gotten too worn to work. At last another security officer buzzed open the turnstile. The security chief, a lean, leathery, well-armed fellow, was waiting there. Assured that we were the much-anticipated nuclear tourists, he bid us a warm farewell. "We don't get many visitors nowa-days. Y'all come back now, you hear?"

If terrorists managed to cut off the electricity and stop the supply of water to the reactor buildings at a nuclear plant, that still wouldn't cause a meltdown because there are multiple systems.

"We have an automatic containment spray system to keep the reactor cool, if something happens to the primary coolant system," Sipe explained. "We have multiple safety systems in case the power fails. We have combustion turbines—fuel oil in the winter and natural gas in the summer. We have diesel generators we test frequently. We've done some experimenting with photovoltaic cells. We have four redundancies when it comes to electricity. We have a very large maintenance program, and whenever we refuel a unit, we do thirty days of maintenance."

Rip and many others in the technical community told me that nuclear disaster scenarios that are popularly envisioned contradict the laws of nature. A paper by the American Nuclear Society (ANS), a professional organization, states, "Peer-reviewed analysis shows that physical science precludes creation of a serious public health hazard after any realistic series of mishaps." But suppose that somehow, defying all risk-assessment, a meltdown actually occurred—whether through intention or human error. We know from the partial meltdown at Three Mile Island that the process takes days, that radionuclides stick to internal walls, and that the containment building would prevent the release of all significant radioactivity. There would be time to bring a great deal of attention and manpower to bear on the problem.

OK, but what if the terrorists fiddled with reactor controls and produced a steam explosion like the one that occurred at Chernobyl? I think finding the control room in this labyrinth would be a major feat; doing so without being stopped and killed would be virtually impossible. Furthermore, very few people have access to the control room. Presumably more palm geometry and other restrictive measures would come into play. And contrary to many TV and movie scripts, it's just not possible to push a button that would produce a meltdown or an explosion. Finally, most employees have worked at McGuire and other plants for years without ever getting near the control room or the reactors.

But let's say the terrorists, or some colossal chain of mistakes, produced a steam explosion. The building has been engineered to contain it. At McGuire, a backup technology for preventing overheating of the core and a release of radioactive materials would kick in: the ice-condenser containment system. Nine ice-condenser nuclear plants operate in the United States and just a few others abroad. An insulated, refrigerated compartment that runs around the inner perimeter of each reactor building contains about two thousand tubular steel baskets, each forty-eight feet tall and weighing about a ton, each filled with shaved ice. If cooling water from the reactor, under pressure, were accidentally released from its closed system, it would quickly flash to steam and, being hot, rise from the deep subterranean room where the reactor is located to the upper part of the containment building. This event would cause the compartment doors to fly open and the steam would hit the ice and condense into water, thus protecting the walls and equipment from overheating. In containment buildings without this technology, the same safety measures are obtained by erecting a larger containment structure of greater volume and by walls that are even more reinforced than those at McGuire.

The area is not known for its earthquakes, but what if one occurred under McGuire? Or a tornado struck? Or someone fired a rocket-powered grenade at a containment building? "The reactor can be shut down in 2.6 seconds by dropping the control rods," Sipe replied. "There's automatic shutdown and manual shutdown. All nuclear plants in the United States have this automatic shutdown response to a sudden jolt."

Even prior to September 11, 2001, when all American nuclear facilities were put on highest alert indefinitely, nuclear plants were less vulnerable to attack than any other type of commercial facility. The NRC, harshly criticized after the events at Three Mile Island, had already instituted many improvements in plant security. After the terrorist truck-bomb explosion at the World Trade Center in 1993 and the crash of a station wagon driven by a mentally ill intruder into the turbine building (*not* the reactor building) at Three Mile Island, every nuclear plant added more vehicle barriers and other defenses, stepped up detection systems, access controls, and alarm stations, and enhanced its response strategies, which are periodically tested in mock raids by commandos intimately familiar with plant layouts.

These staged intrusions have occasionally been successful, leading to further corrections. Critics have argued that the make-believe assaults demonstrate how easy it is to slip into nuclear plants and do damage. In 2001, NRC commissioner Edward R. McGaffigan Jr. rebutted this charge. He pointed out that in spite of receiving beforehand detailed, normally closely-held information about plant layout and defenses—a major advantage—along with the equipment necessary to damage the reactor core, the hired attack squad reached its target only 15 percent of the time. "They are credited with very substantial capabilities to penetrate barriers in short periods of time," he said, and compared the plant's cooperation with the mock attackers to a coach giving his defensive playbook to the opposing team before the Super Bowl game. Reaching a target does not "equate to core damage, for operators could well still recover the plant. And core damage does not equate to a radiological disaster, as Three Mile Island showed." He concluded, "Nuclear plants are hard targets by any conceivable definition." The NRC's analysis of the results of these force-on-force exercises, observations derived from how security orders have been carried out, a review of plant security plans, and an improved baseline inspection program have all contributed to further upgrades of security requirements.

Antinuclear groups cite the threat of terrorism as a reason to shut down plants, although a closed nuclear plant still would have spent nuclear fuel on the premises. "We have found diagrams of American nuclear power plants," President George W. Bush claimed in his 2002 State of the Union speech; he asserted that they'd been discovered in Al Qaeda hideouts in Afghanistan. Greenpeace asked the NRC what it was doing about this troubling state of affairs. Commissioner McGaffigan replied that he'd already testified in a closed congressional hearing that President Bush was probably wrong in his State of the Union speech. Nevertheless, the commissioner said he believed Al Qaeda was interested in nuclear plants and pointed out that since 9/11 the NRC had changed plant security rules five different times. After a thorough review of all levels of plant security, the agency mandated additional access controls and personnel screening to reduce the chance of insider collaboration and ordered nuclear plants to cooperate more closely with local law enforcement agencies.

Because Rip, as director of Energy Programs at Sandia National Laboratories, was involved in counterterrorism efforts, especially after September 11, 2001, I kept phoning him that terrible fall with my worries about the Indian Point nuclear plant two dozen miles from Manhattan. He maintained that all of our nuclear plants were well protected and that there were many softer targets that would produce bigger bangs and cause far more widespread harm—natural gas storage terminals, oil refineries, electrical stations and the grid, and chemical factories. He and other chemists have pointed out, for example, that the dispersal of chlorine from a water-purification facility would be likely to kill more people than the worst event at a nuclear plant. Even if damage to the core occurred, the accident at Three Mile Island had shown that radioactive materials would be contained, whereas deadly chemicals from a factory explosion or release, as accidents each year demonstrate, become easily airborne and dispersed.

The disaster in Bhopal killed thousands and caused illnesses and birth defects when methyl isocyanate, an extremely toxic compound, contaminated the area. There's a nearly identical plant in the United States. Transportation mishaps and accidents at large chemical plants and refineries with tanks of gaseous and poisonous material are common. According to the U.S. Chemical Safety and Hazard Investigation Board, over a thirty-nine-month period from 2003 to 2006, plant and refinery explosions and contact with cyclohexane, butyl acrylate, methanol, propylene, vinyl chloride, and other toxic chemicals caused 25 deaths, injured or compromised the health of 199 people, and resulted in evacuations in some cases and in significant property damage in others. Chemical plants continue to remain vulnerable to attack. Despite repeated requests to the industry by Congress to improve security, the chemical industry has declined to do so, on the grounds of cost. Voluntary measures have been poor. Five years after 9/11 Congress gave Homeland Security $10 million "to build a capacity to police what has been recognized as about 15,000 facilities that have the means to injure or threaten the lives of up to 100,000 people around them," says Stephen E. Flynn, homeland security expert for the Council on Foreign Relations. "This is totally unsatisfactory in light of the threat that some very deadly chemicals can pose." He rates the vulnerability of chemical plants as far greater than that of nuclear plants. Senator Susan M. Collins of Maine, hoping for stricter regulations, has said, "This vulnerability has been raised time and again by terrorism experts. It is a glaring one that invites attack."

"If we misallocate funds to gold-plate nuclear power plants, but we leave chemical plants and public transportation unprotected," my friend Chris Crawford, the nuclear engineering graduate student, remarked, "people may die."

In 2002 the Center for Strategic and International Studies, in partnership with two other independent institutes dedicated to counterterrorism, conducted a large-scale, two-day national security exercise called Silent Vector to test and analyze the nation's greatest vulnerabilities to surprise attack. Government officials, politicians, defense personnel, and various experts were asked to respond to a credible but unspecific threat. Which potential targets were most vulnerable? The panel reviewed a list of refineries, large liquefied-natural-gas or liquefied-petroleum-gas storage facilities, pipeline infrastructure, petroleum terminals, nuclear power plants, chemical plants, and dams. At first glance, the targets that the nonexperts gave top priority to defending were nuclear plants, but once the facts about their sturdy construction, multiple barriers, and security procedures were explained, the panel concluded that terrorists would choose more vulnerable sites.

The containment structures for power reactors, which are subject to both federal and international regulation, are among the most durable and impenetrable structures on the planet: they're constructed to withstand 200-mile-per-hour hurricanes, tornadoes, earthquakes, and floods, all of which can provide a more energetic impact than anything terrorists would have at their disposal apart from a hydrogen bomb. The NRC requires all plants to demonstrate that they can withstand flooding. When Hurricane Katrina devastated the Gulf Coast in 2005, in the nation's worst natural disaster, petrochemical facilities and chemical plants were breached, releasing toxic substances. Three nuclear plants in Mississippi and Louisiana—one near New Orleans—lay directly in the storm's path but remained intact and unharmed.

According to an Electrical Power Research Institute study commissioned by the Nuclear Energy Institute, nuclear plants are highly likely to withstand a full-force impact from a wide-bodied jet fully loaded with fuel. Other studies suggest that airliners, which have a thin metal skin over a lightweight frame, would crumple on contact at high speed with a containment building. Before our trip to the Piedmont, Rip had driven me through the grounds of Sandia National Laboratories. He showed me the ramp where a pilotless F-4 fighter jet had been catapulted at a speed of 480 miles per hour against a big, thick block of concrete. Later I saw a Department of Energy (DOE) film of the experiment. The plane was smashed as if made of foil and fell apart on contact. The wall was penetrated to a depth of a little over two inches.

The findings can be applied to what would happen if a plane somehow managed to overcome many likely physical obstacles and succeeded in crashing at high speed into a containment building. A fighter jet like the one in the test has heavy engines and its concentrated mass would make a

more powerful, more focused impact than a big airliner. James Muckerheide, a nuclear engineer who is codirector of the Center for Nuclear
Technology and Society at Worcester Polytechnic Institute, told *The
New York Times* in 2002 that, compared with the F-4, "a large passenger
aircraft is a slow, empty, tin can." He added, "The mass of the aircraft can
put a heavy compression load on the containment structure, but it has
negligible penetrating ability."

"In the Sandia test, the energy of the impact disintegrated the fighter
jet but not the concrete wall," Rip said. "It's the laws of physics. What
holds true in the experiment would hold true for a containment building,
which also contains steel reinforcement bars and other steel barriers.
The concrete block at Sandia contained some rebars but not as many as a
containment building."

Once again Rip brought up the problem of mistaking risk for consequence. He now applied this observation to threats to nuclear plants, saying that the probability of an individual or a group penetrating a nuclear
plant all the way to the reactor and causing harm is extremely low, and
even if somehow terrorists managed to cause a breach, the physical consequences would be relatively minor.

"Some people make this mistake when they talk about the consequences of a jet flying into a reactor or a spent-fuel pool," Rip said.
"They're going for a worst-case scenario, an extreme set of circumstances that don't exist in the real world or exist so rarely that the probability of occurrence is very, very small. Or the circumstances *can't* exist
simply because of the laws of nature. This idea of a plane ramming into a
reactor building assumes that the pilot would get a running head start at
the building, or at the spent-fuel pool. But a nuclear plant—Oconee, for
example, which is in a hollow—is made up of a tight cluster of buildings
not easily approached at a flat angle from the air. All aviators know about
a phenomenon called ground effect. As a plane is coming in for a land-
ing, it's compressing the air down, but the closer the aircraft gets to the
ground, the more the air molecules are pushing back. The air actually
gets denser under the plane. If it flew at high speed close to the ground it
would skip like a flat stone does when you throw it low across a pond.
You'd have to get way, way back from the containment building and fly
very low—twenty feet off the ground—on a flat approach. I don't think
any airline pilot could do that. If he approached from an angle, diving
down at the plant, the plane would ricochet off the containment building
the way bullets do. A pilot flying that low, where the air is thicker, could
not get up to anywhere near the speed of the planes hitting the Twin
Towers—and they were almost a quarter of a mile high. And the walls of

both the World Trade Center and the Pentagon are like paper compared to the walls of a containment building. Also, relative to the Pentagon and the World Trade Center, two of the biggest buildings in the world, the structures of a nuclear plant are narrow, small, and low to the ground— extremely difficult targets. There's just no comparison when it comes to structure. Remember, uranium is dense, so the fuel at even a big plant takes up only a small space. Very different from gigantic office buildings where thousands of people work. The other thing about a nuclear plant to keep in mind is that the reactors are located belowground, and so are the majority of spent-fuel pools. Usually the buildings housing spent-fuel pools are located next to containment and turbine buildings, which would have the effect of shielding them as well. To do any damage you could not come in at an angle. But even if you could approach a reactor building or a building containing a spent-fuel pool directly, the laws of physics are going to make most of the airliner pass over the top of the structure. Even if a collision could somehow break open the roof so that the rest of the plane penetrated the structure, the I-beams and eighty-ton cranes inside would shred it on contact. The probability of damage to the reactor or the spent-fuel pool is minuscule. The pool would remain intact because of the stainless steel that lines it and the thick concrete with rebars. Pieces of the plane would not be able to crack the pool walls, but even if somehow the wall was cracked, it would take a long time—days—for the water to leak out and for the fuel to get hot enough to melt down. You had better believe crews would replace that water immediately."

An improbable chain of events would have to occur in order to bring about the meltdown of a commercial spent-fuel pool and the dispersal of its radioactive contents. To begin with, a number of barriers surround the ceramic fuel pellets themselves, which are stacked inside long hollow rods made of Zircaloy, a zirconium alloy originally developed for the nuclear submarine program. Zircaloy doesn't readily absorb neutrons, so it doesn't interfere with the chain reaction; its melting point is nearly 2,000 degrees Centigrade (3,660 degrees Fahrenheit), whereas reactor temperatures reach only around 600 degrees Centigrade (1,112 degrees Fahrenheit). The fuel rods are bundled together in assemblies usually made of stainless steel. Even if repair crews couldn't get to the damaged pool and all the water leaked out after many days and the spent fuel gradually heated up sufficiently, it would slowly melt through the bottom of the steel and concrete pool and fuse with the soil and rock underneath. (At Three Mile Island-2 the meltdown over several days penetrated only a fraction of an inch of the containment vessel.) As for the spent-fuel pool water being evaporated by a fire from a portion of the jet fuel leaking into it from a crash of a plane much bigger than the pool, the water is forty feet deep. A jet fuel fire could not burn up and disperse nuclear materials in the pool, because jet fuel burns in air at a temperature much lower than the melting point of the Zircaloy fuel rods—at 550 degrees Centigrade (1,022 degrees Fahrenheit). The jet fuel could not melt the fuel rods, or the fuel pellets for that matter. The uranium fuel itself might be able to build up enough heat over the course of days, once its coolant was gone, to melt the fuel assemblies and settle to the bottom of the dry pool. The point is that nuclear fuel poses no danger unless it comes in contact with the public, which it cannot do under any realistic scenario.

Larry R. Foulke, formerly president of the American Nuclear Society, is an engineering professor at the University of Pittsburgh. He gave an address at the society's annual meeting in 2004 in which he referred to sensationalistic reports about terrorist threats to spent-fuel pools and reactors and made a plea for "realism":

> Starting about thirty years ago, extensive research by industry, academia and government, including theory, experiment, large-scale tests and epidemiological studies, provided a basis for establishing realistic premises and inputs. And these realistic inputs are supported by known effects of several real casualties.
>
> This body of work showed that physical properties of materials severely limit the release of radioactivity, even from molten fuel, and dispersion from such fuel, especially in a water environment, and transport of radioactivity even with compromised containment. The basic nature of the risk is different than previously portrayed.
>
> Instead of arguing that any ultimate casualty is highly improbable, we can now show that there is a nature-imposed limit to the consequences of even the worst *realistic* casualty.
>
> From the examinations of Ted Rockwell and other senior greybeards (see their *Science* article of September 20, 2002), we can conclude that for any dispersion of radioactive materials, the risk to the public is in the vicinity of the release where there are high concentrations. For a ground-level release, this would be downwind to less than a mile. It appears that mass evacuation (e.g. everyone within a five- or ten-mile circle) is probably counterproductive in most circumstances.

The nineteen authors of the *Science* article Foulke mentioned were members of the National Academy of Engineering and were from universities, national laboratories, consulting firms, and private industry.

They stated that they were using "engineering principles and long, practical experience in nuclear technology" to rebut "a large number of outrageous public statements" that had appeared after 9/11 and claimed any attack on a nuclear plant or its spent fuel "would be catastrophic." The paper documented the conclusions of engineering tests and analyses of radioactivity from molten nuclear fuels that showed that failed containment, "under realistic worst-case assumptions, would produce few, if any, casualties."

The probabilistic risk assessments of potential reactor catastrophes conducted decades ago assumed the most extreme scenarios for the purposes of doing calculations, not because they were feasible in the real world. These scenarios, now inapplicable because a number of findings over the years have invalidated many of those assumptions, have, as I mentioned earlier, been disavowed by the NRC as a valid basis for calculating damage to nuclear plants or radiological harm to surrounding populations because those hypothetical events relied on extremely conservative estimates.

However, some opponents of nuclear energy keep using these outdated documents and old hypothetical worst-case scenarios to craft up various outcomes. One, involving a fiery jet crash into a spent-fuel pool, among other mishaps, appeared in *Science and Global Security*, a political science journal published by a program of the same name at the Woodrow Wilson School at Princeton University. The research group seeks "to provide the technical basis for policy initiatives in nuclear arms control, disarmament, and nonproliferation." The article, by eight authors (environmental and energy consultants, theoretical physicists, political scientists, nuclear engineers, and a geologist; several of them are active in antinuclear causes and some have dual professions), argued that the risks and costs to society of terrorist attacks on spent-fuel pools were so great that serious modifications had to be made in order to ensure safety. The predictions made in the article continue to be repeated by various antinuclear groups. Herschel Specter, once the head of a DOE emergency planning committee and a former federal regulator of nuclear plants, provided a response in an op-ed piece in *The New York Times*. He began with the assumption, however unlikely, that terrorists would be able to invade a nuclear plant, destroy every safety system, and thus damage the plant enough to cause a release. He then pointed out the weaknesses of this assumption: the finite amount of radioactive material at any given plant and the physical laws governing behavior of radionuclides that would prevent spent fuel from being widely dispersed into the environment.

Natural forces cause radioactive material to stick to various surfaces, to fall quickly to the ground, or to form soluble salts that remain largely in place. Terrorists can do nothing to eliminate these natural forces. Other natural forces come into play once the remaining radioactive material is carried from the damaged plant by air currents. The resultant plume is narrow, covering but a small percentage of the surrounding area.

People outside the radioactive plume would not be at risk for exposure, he went on, and most emergency workers would therefore not be affected. In addition, the plume would naturally weaken as it traveled, and after a few miles—even in the worst possible release—would be insufficiently radioactive to cause early fatalities or injuries. Evacuation would be limited, realistically, to about 4 percent of the ten-mile zone: "Beyond this area, people would be well served by staying inside until the very weak plume had passed. People in New York City, some 35 miles south, would never have to evacuate promptly from any terrorist attack at Indian Point."

In another essay for the *Times*, Specter pointed out miscalculations that had been made by the Princeton group, writing, "Even if the [spent-fuel] pools drained, new research from the Nuclear Regulatory Commission indicates that it may be physically impossible for all the radioactive material from the spent rods to escape. A 10 percent or smaller release seems more plausible." Specter went on to note:

> The Princeton study's authors have also responded to criticisms that they made serious errors, including miscalculating the size of the population at risk in an improbable terrorist attack. Correcting this error resulted in a reduction in their calculated worst-case health predictions, to 5,600 from 250,000 long-term cancer fatalities. But even the 5,600 figure is highly inflated. By comparison, cancer fatalities from non-nuclear causes that are certain to occur in the same population would be about 1.2 million.

The NRC reviewed the paper at the request of one of the directors of *Science and Global Security* and, citing many inconsistencies and inappropriate assumptions, noted that the argument rested on excessively conservative estimates and relied on NRC studies that were inapplicable to terrorist attacks; the agency made a public disavowal of such notions and concluded that the paper did not "address such events in a realistic manner," that its use of overly conservative studies provided misleading results, and that the authors' recommendation that all spent fuel more than five years old be transferred to dry cask storage at a cost of many billions of dollars was unjustified.

The Princeton authors defended its methods and findings in another rebuttal.

When I asked Rip about this controversy, he was already well into his analysis of the conservative estimates the Yucca Mountain Project had been making for its probabilistic risk assessment about the long-term storage of spent nuclear fuel. His reply was to paint a surrealistic portrait of what can occur when people misapply data based on extremely conservative, subjective assumptions and take worst-case risk assessments for reality.

"What engineers will do when they design a bridge is make an estimate about the heaviest vehicle that they think will go over the bridge and then double that estimate just to be safe. That's called 'being conservative.' But, as we've talked about before, the conservative approach doesn't rely on our knowledge of the real world and usually winds up relying instead on wild guesses. Suppose you want to design a car that takes into consideration every possible contingency about who is going to drive it. If you take the conservative approach, you need to look at the most extreme scenario. So you go to the record books and find the weight of the heaviest person in the world—let's say nine hundred pounds—and the height of the tallest person in the world: eight feet eleven inches. And the worst-case scenario you could anticipate would be a nine-hundred-pound person who's nine feet tall, and that's who you design cars for. But if you want a real-world scenario, you place this data on a bell curve, and these extreme examples of the human form would be at the far end."

Sketching his usual bell curve, he indicated the flat, shallow, trailing right-hand side. "Using probabilistic analysis based on real-world data rather than conservative estimates, you'd find that there was nearly zero likelihood that someone that heavy and that tall would be able to get out of bed and walk through the door of his house in order to drive a car. The chance of that is so tiny that you can rule out that person when you design a car. And in fact that's just what car companies do. But here's what happens if you assume that worst-case scenario to be highly probable and then make conservative estimates about your car design. You assume that there must be people even *bigger* than those mentioned in the record books and you assume that somewhere in the world there is or will be a person who weighs one thousand pounds and a person who's ten feet tall. You say, OK, this means we need to design a car that a person ten feet tall and weighing one thousand pounds can drive. The internal dimensions of the car would have to be huge. The seats would have to be five feet wide and five feet high. The location of the gas and brake pedals and their height would have to be changed. The doors would have to be much bigger. And you'd want a car that could carry four passengers, so the structure would have to be reinforced to support a load almost three times what my standard pickup can haul. The external dimensions would have to be gigantic—twelve feet wide by thirty feet long. So now, using the conservative approach, you're stuck with a lot of problems of your own making just by relying on the most extreme scenario. What happens if you move to the middle of the bell curve? How does your worst-case design work for drivers with sizes in the mean and median ranges? Well, their feet won't reach the pedals or touch the floor, because the seats will be five feet away from the dashboard to give the people on the extreme end of the bell curve room for their long legs. And the vehicle will be too wide to fit into a normal traffic lane. How is this worst-case scenario, calculated by conservatively including the extremes, going to work for people in the normal range of the curve? You are not in the real world here with such suppositions."

Rip then returned to the fundamental problem of the argument. "When scientists—or political scientists—do an analysis, and they don't do it in a probabilistic mode, they end up using worst-case values. And that does tell you a lot, because if you can't make a facility or a plan fail with worst-case values, it must be superstrong. But the drawback is this: you don't know *how* strong. So you have uncertainty. Opponents of nuclear power quote worst-case analyses as real ones, but the worst-case ones are not realistic. And this is what all those who have an adversarial agenda, who are anti whatever, do to make a potential outcome sound so bad. And some of them have learned to do this really well. The result is that the public loses out, because we don't always have a way to tell what's real and what's not. In the case of the car analogy, we can understand the reason not to design cars for drivers ten feet tall and weighing half a ton because we know what the real world is like—we know about cars and driving and the average human form. But nuclear plants are mysterious to most people."

At the behest of Congress, a National Academy of Sciences (NAS) panel—which included Frank von Hippel, a physicist who has made his career in political science and is one of the authors of the *Science and Global Security* article—recommended to the NRC some upgrades on securing spent fuel from aircraft attack. The NRC announced that it was giving the recommendations serious consideration and was continuing to work with the NAS on the matter. In March 2005, the NRC provided Congress with a report responding to the recommendations and addressing a number of issues raised.

The NRC has conducted tests about plant safety and security; it noted:

> These studies assessed the capabilities of these plants to withstand deliberate attacks involving large commercial aircraft. The NRC studies included national experts from Department of Energy laboratories, who used state-of-the-art experiments, structural analyses, and fire analyses. The studies at the specific facilities confirmed that the plants are robust. In addition, the studies found that even in the unlikely event of a radiological release due to a terrorist attack, there would be time to implement the required offsite planning strategies already in place to protect public health and safety.

Rip thinks the controversy is actually an argument about conservative assumptions versus real-world data. "Until someone sits down with an independent outfit, a national lab like Battelle Northwest that does research using real-world projections, we won't have a realistic resolution."

In 2007 the commissioners of the NRC ruled that America's nuclear plants, being inherently robust structures, did not need to build special defenses to protect themselves from suicidal aircraft attacks. The commissioners also recommended that new designs include practical schemes to reduce the effects of such an emergency, and in fact most recent plans already include such features as a water tank above the reactor that would automatically flood it with additional coolant in the event that the external water supply were cut off.

At the McGuire Nuclear Station, we proceeded to "the war room": the Technical Support Center, from which emergency drills are run. The plant's district managers work with mayors and other elected officials in the five counties and diverse local governments in the ten-mile radius around McGuire that constitutes the emergency planning zone mandated by the NRC. Lessons from TMI-2 have been applied. "Emergency planning officials around here are well-versed about our plant operations and have taught the community that in case of a big release of radioactivity everyone has to evacuate," Rita Sipe, our guide, said. "Really, we don't foresee *ever* having a big release."

All nuclear plants must by law have safety drills several times a year under the supervision of the Federal Emergency Management Agency and the NRC and are graded on performance. The drills include simulations of scary events. "It's very realistic," Sipe said. "We have a group that develops scenarios. For example: What if a plane crashed into the plant at the same time as a tornado struck? We also practice what to do if the hydroelectric or fossil-fuel plants are threatened. The dam here is guarded."

The drills, practiced frequently, are coordinated by McGuire's radiation protection manager along with the person in charge of assembling personnel on-site and evacuating them. Sirens that are tested weekly announce each drill; they're operated by the county and the state. Potassium iodide pills have been distributed to the public living within the zone, and people have been taught to take them to protect their thyroid glands in case of a release, and only then.

The war room was glass-walled and in it stood a conference table; and, at every seat, like a place mat, there was a binder for each drill participant that listed his or her designated tasks in case of an emergency. As they're accomplished the person must sign off on them. An adjoining room was a communications center, its phones and computer monitors in constant contact with different parts of the plant and with Duke headquarters—yet another innovation thanks to Three Mile Island. "The group who works here keeps everything up-to-date and also interfaces with the county and the state," Sipe said.

I'd wondered how reactor operators here could possibly make the kinds of mistakes that occurred in the control room at Three Mile Island—and had no doubt inspired the creators of the popular satirical television cartoon *The Simpsons* to cast Homer, the daffy, loutish protagonist, as a reactor operator. Since TMI, the NRC and the nuclear industry have invested considerably in the education of the few who are permitted entry to a plant's control room. Just as today no pilot would ever be allowed to fly an airliner unless he had trained on a simulator, reactor operators must spend one week in five training at the facility's simulator, to keep their responses honed. It comes up with various unpredictable chains of failure that demand accurate responses. For example, an operator can practice responding to the warning of coolant loss to the reactor by shutting down the reactor.

"The operators are tested and videotaped as various make-believe scenarios are run, then critiqued," Sipe said. "A lot of our operators come from the nuclear navy, and I want you to know, these folks are *very* dedicated."

Recruits for the nuclear navy are selected from the top 2 percent who enlist; veterans who go into private industry take with them a relentless approach to safety and accountability that originated with Admiral Rickover. A small, straight-spined, human dynamo who was born in Russia to Jewish parents and was fond of literature and philosophy, he began his career in the navy as an electrical engineer. A farsighted maverick, he went from designing submarines in the 1930s to leading a program to invent and test the first nuclear submarine and he oversaw the first nuclear aircraft carriers. When the citizens of Shippingport, Pennsylvania, became fed up with the pollution from their local coal-fired plant and when the federal government saw that the Soviets were developing nuclear-plant technology to sell abroad, Rickover and his team established the first American plant at Shippingport. His bluntness and obsession with taking every precaution and inculcating personal responsibility were legendary and probably kept him from being promoted. Submarines could travel faster with less shielding, but the crew might be exposed to slightly more radiation; therefore, Rickover insisted on slower, well-shielded submarines and asked the engineers designing them to imagine that their own sons would be serving on them. He liked to say that "any one detail, followed through to its source, will usually reveal the general state of readiness of the whole organization." He literally knew every nut and bolt on a nuclear submarine. Once he got a late-night call from a skipper about to depart on a long voyage who was frustrated because a tiny screw had fallen into the turbine, which would have been a huge job to disassemble. Rickover directed him to remove a small inspection plate at the bottom of the turbine casing that had been installed for just such emergencies. The screw was sitting on it. Theodore Rockwell, who was technical director of Rickover's programs to create the nuclear navy as well as the world's first commercial atomic power plant, writes that since these were the first large-scale industrial operations involving radiation and radioactivity, Rickover's team "took the first steps to define radiation and safety standards for industry and to write model state laws. . . . Our methods of presenting plant designs and procedures for outside safeguards review, and our manuals, checklists, and other techniques, were adapted for commercial use by Rickover alumni, and these in turn became exemplars for the rest of the world."

Rickover insisted on broad and deep technical training for everyone, from the top down, and strict accountability. He thought that instead of spending a lot of time preparing for events with a low probability of occurrence, the nuclear industry should concentrate more on the likeliest high-risk scenarios. Nearly all the world's nuclear power plants are direct technological descendants of Rickover's programs, and most of them rely as well on his system for the educating and training of reactor operators and plant workers.

Nowhere else has human engineering vis-à-vis reactor operation been more thoroughly studied and executed than in the nuclear navy, where operators not only run reactors—the navy has 103 of them in operation today—they also live for months at a time within meters of them. And the power of naval reactors is significant: the two aboard the aircraft carrier USS *Harry S. Truman* produce enough electricity to supply 250,000 homes. Rip, Marcia, and I had visited the submarine base at Groton, Connecticut, and gone aboard the USS *Connecticut*, a state-of-the-art, fast-attack nuclear submarine. Crew members are virtually bonded with the machine they inhabit. Several large studies have shown that submariners are healthier and longer-lived than their counterparts in non-nuclear jobs. When the men are on duty, shielded by the ocean from cosmic and terrestrial radiation, their dosimeters register less exposure than when they are on shore leave. The nuclear navy has educated over a hundred thousand people in reactor operation and safety and has demonstrated that it is possible to avoid reactor accidents. As of 2006, nuclear-powered submarines and ships had safely traveled a total of 134 million miles, and registered 5,700 naval reactor-years of safe operation of a total of 254 reactors.

The nuclear industry imitates the naval reactors program in having an independent safety specialist on-site. As a result, on the grounds of every nuclear plant there must by law be an NRC representative who has the authority to shut down the reactor. Before they get too cozy with the staff, they're rotated to other plants.

We expect to have light and power around the clock at the flick of a switch as if it's our right, and many of us are indignant about the pollution from fossil fuels or troubled about the dangers we've heard about concerning nuclear energy. I've felt anger and concern about both issues. But now I found myself thinking differently. Electricity from any source has a price; everybody wants more of those helpful electrons. We're trained from early childhood to expect our toast crisp, our homes properly heated or cooled, to want more things, to consume more, to get the newest toy. To make do in a city or a suburb without electricity, or to get by on a few watts a week, as is the case in many parts of the world, would be unthinkable.

As we left McGuire, Rip remarked, "The nuclear industry is far from perfect, but there's nothing that can happen to a nuclear plant, or even to a whole bunch of them, that can ever begin to come close to the harm that's already been caused and will continue to be caused by burning fossil fuels."

Since the trip to Ambrosia Lake, I balance things differently: on the one hand, mountains of coal and their prodigious quantities of waste; on the other hand, that little uranium pellet.

Throughout the tour I'd been nagged by an illusion of familiarity about the Riverbend coal-fired plant, even though I'd never been to any place remotely like it. Was it the noise, the heat, the grime, the massive pile of fuel, and that lake of waste like the sullen tarn in Edgar Allan Poe's tale "The Fall of the House of Usher"? Suddenly I realized that a real coal-fired plant strongly resembled my old fantasy of a nuclear power plant.

I asked Rip and Marcia what they'd found the most surprising during our days in the Carolinas.

"That coal-fired plant," Marcia said.

Rip chuckled and shook his head. "I couldn't believe it." He found the corroded pipe dribbling water the most worrisome. "Fatal steam explosions happen in coal-fired plants," he said. "If that sucker blew, we could have been killed, scalded to death. People pay no attention if someone dies in a coal-plant explosion. But if you have a steam accident at a nuclear plant—man! Big headlines. And then there's the health impact from the fly ash and slurry and gases, and the fact that coal combustion kills thousands of people every day around the world and is a major factor in the biggest ecological disaster in human history."

These are the risks we take—or arrange to have someone take for us—in order to enjoy all that brightness that came to us with the dawn of the twentieth century and stays with us through the night and enhances our lives in countless ways. I thought of a luxurious ocean liner where, down in the hold amid a clanging din, diesel or coal is burned to propel the ship, and men toil, faces blackened by soot, to keep the power coming, while above in a hushed, immaculate, grand salon with chandeliers and white linen napery, violins play and people consume their food in ignorance of the smoke and heat and sweat and shrieking below.

"We have to get rid of coal-fired plants, then," I said to Rip. "Just from an ethical point of view."

"We can't do without them, at least not in the short run," he said. Today America would be brought to a standstill if coal-fired plants were shut down overnight. Along with nuclear power, they supply baseload— the steady low hum in the electron fugue, reliable around the clock whether the sun shines or not, whether the wind blows or not, whether rainfall replenishes shrinking dams or not. "People don't want fossil fuels, don't want nuclear," he said, "but nobody is willing to give up electricity."

13 UNOBTAINIUM

MY HUSBAND AND I LIVE in a two-hundred-year-old cedar house, formerly a boat-building shed, on Long Island. One day our old furnace started belching black smoke. The fuel oil company repaired it, and in so doing blew greasy black dust particles through the heating registers and all throughout the house. I've always hated having to rely on fuel oil, having grown up with natural gas. But there's no gas pipeline where we live. We have to buy propane in big bottles for our kitchen stove. Using electric baseboard heating is too expensive. Our utility, the Long Island Power Authority (LIPA), charges 96 percent more than the national average for electricity, and since 2001 has increased the average household bill by 35 percent.

While I was vacuuming and sponging up soot and reflecting on how particulates from oil and wood combustion had to be accumulating in our lungs even when the furnace was working properly and the fireplace damper was open, I heard an energy guru promoting his new book on public radio.

He spoke of a cheap, sustainable power source that emitted no greenhouse gases. Hydrogen, he said, was the most abundant element in the universe, and it produced no toxic or radioactive waste. Fossil-fuel plants and nuclear plants would soon become obsolete, because each household would have a device the size of an air conditioner designed to take the energy from sunlight and store it for future use in a hydrogen fuel cell, which would provide electricity to heat and light your home cleanly. The only waste would be water and heat. And nothing could be easier than making your own hydrogen from solar power, by using its electricity to perform electrolysis. Through this simple process taught in high school chemistry labs you separate the hydrogen from H_2O. That same convenient, straightforward technology would run your electric car. Hello decentralization and good-bye to exploitative utilities and carbon emissions! Furthermore, jets and ships could run on hydrogen as well. Good-bye to dependence on foreign oil! I was ready to sign up.

"When we've depleted most of the oil, we'll spend a lot of money and energy extracting it from tar sands and shales," Rip told me in the late 1990s; in the 1980s he'd worked on an investigation by Sandia into the feasibility of oil extraction from tar sands. "Oil, gas, coal, wood, uranium, thorium, and water behind a dam are all ways of storing energy for long periods for later use. Most people don't think of these as energy storage mechanisms, like a battery, but they should. Sunlight, wind, and falling water can be used to generate electricity, but they don't store energy."

Petroleum is a stunningly efficient, portable container of energy. As rumors of depletion grow and oil prices increase, Canada has decided that it's worthwhile to extract oil from tar sands and refine it using energy that perhaps will come from nuclear power but more likely will come from fossil-fuel combustion. Others are planning to make fuel from coal, but estimates suggest that any such program would deplete our rather abundant coal resources within decades and increase output of greenhouse gases. Biofuels have limitations as well. "In the end, I expect we'll switch to hydrogen," Rip said. "It isn't an energy source—it's an efficient carrier of energy, and it's also portable. But there are some difficulties."

About five years later, when we visited the Idaho lab and chatted with the Argonne-West reactor engineers, the subject of hydrogen as an energy panacea came up. Antinuclear people had told me to forget nuclear plants—hydrogen could provide all the energy we needed. Michael Lineberry asked rhetorically where that hydrogen might come from. "There are no hydrogen wells or hydrogen mines. People talk about just going and getting hydrogen to power our cars with as if that were no problem. But it takes a *lot* of energy to make it."

"What folks don't understand is that you only get out of hydrogen a portion of the energy you put into it, because at each step of the process you have inefficiencies and energy is lost," Rip said.

Hydrogen is invariably bonded with other elements—locked up in H_2O molecules, for example. To break the bonds between the hydrogen and the oxygen requires energy that today comes from burning fossil fuels. "You always have to invest more energy in making hydrogen than you're going to get from it," Rip said to me later. "Nothing can change that physical law. But the waste from burning hydrogen is pure water—so the energy investment might be worth it. Nuclear power is the only clean, large-scale, relatively cheap way to make hydrogen."

To find a hydrogen factory, you have to go to an oil refinery, where hydrogen is extracted from fossil-fuel derivatives (usually methane) by using heat from combustion of still more fossil fuels. The hydrogen is bottled and mostly used to refine crude oil and make carbon dioxide and fertilizer. Hydrogen can also be derived by passing steam over coal, or by combining natural gas and steam; both methods release carbon dioxide, however. Despite discussions about sequestering the CO_2 under the earth or under the ocean floor, considerable uncertainty remains about how to do that on the necessarily vast scale we need; about how long the gas would remain isolated before leaking into the atmosphere; and about what the carbon, which would produce acidic reactions, might do to the rock or soil.

That morning at Argonne-West, Lineberry had said, "People imagine that you can use hydrogen without any cost, without any pollution. You hear all this speculation. It's easy to say, Hey, let's go get some *unobtainium* and use it to make hydrogen."

Unobtainium, magic dust, handwavium—these are terms used by scientists, engineers, and science fiction fans to describe that mysterious energy source or substance that's supposed to drive a space ship faster than the speed of light, or bridge a gap in an invention in progress, or fill in the blanks of a scheme that looks great on paper or sounds good when touted by an alternative-energy oracle.

One pound of nuclear fuel could provide hydrogen in a quantity equivalent to the energy in 250,000 gallons of gasoline without the carbon emissions, according to Idaho National Laboratory (INL). It has begun researching reactors that combine heat and chemical processes to split water into hydrogen and oxygen—an efficient, rapid way to produce large amounts of hydrogen. This method would take two times more energy value of the hydrogen produced rather than the approximately four times more that using fossil fuels would. To take the methodology to the necessary commercial scale would be costly. But as a replacement for gasoline, reactor-bred hydrogen would be more energy-efficient than our present arrangements, which involve the purchase and shipment of oil from distant, usually troubled locales as well as the investment of staggeringly huge military expenditures in some of those lands to protect oil fields and on the seas to protect shipping routes. To maintain a military presence in the Middle East and prosecute wars there costs taxpayers about $100 billion a year. That's roughly the cost of thirty or forty new large-scale reactors.

Burning fossil fuels to make hydrogen is like driving fifty miles to a fill up your tank at a station that sells gasoline for four times what it costs per gallon at home. It makes no sense to extract hydrogen using gas or coal. Although antinuclear advocates say that wind power and other renewables can make all the hydrogen fuel we need, energy experts say the arithmetic does not add up. Steve Herring, a scientist at INL working on the reactor method, says that we'd have to triple the number of reactors we now have in order to meet the country's current hydrogen needs and we'd require 4,000 new reactors to make enough hydrogen to replace all the gasoline being used today.

Once the hydrogen is liberated, it must be captured. It's the lightest of all elements and must be strictly isolated from all the other elements it likes to bind with—otherwise it will rush back into their embrace. Storing and transporting this fickle gas is an expensive, major undertaking requiring special valves and seals, because a hydrogen molecule is so tiny that it easily escapes. It also acts on container walls in a way that makes them brittle. Existing oil trucks, tanks, and pipelines would be useless. An entire new infrastructure would have to be built to get hydrogen from a plant into your car.

How much hydrogen would it take to replace gasoline? Paul M. Grant, a Science Fellow at the Electric Power Research Institute, wrote in *Nature* in 2003 that in America we burn twenty million barrels of oil a day—twelve million in automobiles and other forms of transportation (the rest goes to home heating, power generation, and industrial uses). Because the combustion engine is inherently wasteful, in that only a third of the fuel is used to turn the drive shaft that makes the wheels go, nine million barrels are burned each day without doing any work at all. A hydrogen fuel cell is basically a battery—an electrochemical energy conversion device that combines hydrogen with oxygen to make water, in the process making electricity. A fuel cell would operate three times more efficiently than a car engine, so hydrogen would need to replace the energy content of only four million barrels. "Hence, we need to generate daily some 230,000 tonnes of hydrogen," he writes, "enough in liquid form to fill 2,200 space shuttle booster rockets, or, as a gas, to lift a total of 13,000 Hindenburg dirigibles." To get hydrogen from factory to the consumer would mean not only a new fuel infrastructure but also outlets at over one hundred thousand filling stations and the like. Hydrogen, unlike oil or gasoline, is odorless and invisible, and when it burns you can barely see the flame; it also can blow up. Static discharge from a cell phone or a lightning storm could spark explosions if hydrogen leaked. Advocates reply that you simply chill hydrogen into a liquid. But for liquid hydrogen to work on the scale of gasoline, a complicated technology would have to be developed. INL is working on this as well.

"OK—supposing you overcome all the obstacles to getting hydrogen to us consumers," Rip said. "We put that hydrogen in a fuel cell. A fuel cell is one of those things people believe in without knowing what they're talking about, as if it's a magical solution. But it's not. It's one more unobtainium type of problem. If any big breakthrough were going to happen, by now it would have. We don't have a Manhattan Project or a Silicon Valley for fuel cells. It's going to take an all-out effort bigger than even the Apollo moon program and a humongous expenditure of capital. I don't see the government taking this very seriously. So the technology is going to evolve slowly, in increments. Real-world applications are what count. In order to have hydrogen fuel cells good enough to, say, store energy from solar power arrays in order to provide electricity for Albuquerque, well, at this point you'd need unobtainium, not to mention millions of gallons of water required for electrolysis—and we're already in a water deficit here. Burned hydrogen makes heat that produces electricity through a chemical process in a battery, or hydrogen can be put in an engine cylinder the way gasoline is, ignited, and the explosion drives the piston down. Problem is, it's less efficient, because hydrogen can't be stored easily in liquid form—you have to chill it to hundreds of degrees below zero. If you don't keep it that cold, it boils away. If it's stored in gas form, you don't have as many calories per cubic foot of hydrogen in a storage tank. Gasoline is denser and contains thousands of times more energy than its equivalent in hydrogen, so you can have a relatively small gasoline tank in your car. For hydrogen, you'd have to have a much bigger storage tank, even if the hydrogen is compressed, or the car won't go as far." Researchers have come up with million-dollar hydrogen-powered prototypes, and, if certain breakthroughs occur, they might one day be mass-produced more cheaply.

Geoffrey Ballard, inventor of the modern fuel cell, has called for the clean production of hydrogen by using nuclear power, emphasizing that it is "extremely important, unless we see some other major breakthrough that none of us has envisioned." The MIT-Harvard study *The Future of Nuclear Power* concluded that hydrogen produced by electrolysis of water depends on low-cost nuclear power. The availability of water itself for the project is another obstacle to the hydrogen economy. The nonprofit World Resources Institute calculates that electrolysis would require over four trillion gallons of water a year and that American water consumption would increase by 10 percent if we went to hydrogen.

"We'd need unobtainium to create all that water without resorting to nuclear-powered desalination," Rip said. "And not only do you need unobtainium to make an affordable fuel cell, there's also the unobtainium materials problem when it comes to shipping hydrogen in leak-free containers. But when you discuss hydrogen power with true believers, sooner or later they say that there's something out there: a miracle is going to happen. Something will be found or invented that will get us to the next step. If they just had that unobtainium it would all come together. Fans don't want to hear about the real-world stumbling blocks. People in the nuclear world who cling to certain arguments, like the near-term application of fusion power, don't want to hear about real-world stumbling blocks either. As far as cars are concerned, why not skip the hydrogen and just build nuclear plants that make cheap electricity and build electric cars? And electrify our mass transit the way France has? And why not improve battery technology?"

Ever since the 1970s, when I worked on an issue of *Harper's* magazine about renewable energy, I've believed in solar and wind power. They've been successfully introduced in developing countries to provide a modicum of electricity to households unconnected to the grid. About two billion people have no electricity at all. Even a few watts from time to time have been found to make a difference in health and life expectancy. A wind turbine, a solar panel or two, a micro-hydroelectric plant can make a big difference. Girls who formerly spent many hours a day walking long distances to fetch drinking water for their families can instead go to school while a solar-powered pump does their work.

In 1979, around the time I planted my first organic garden, the second energy crisis in a decade was under way. The steady flow of cheap oil was threatened. Imported oil could become too expensive or scarce and Americans might not control it. In 1973, as I remember it, *Harper's* sent me to an upbeat State Department press conference at which appointees of President Richard Nixon announced that it was in our best interests to control the Middle East, because we needed the oil. But those men, who I now realize were harbingers of the neoconservative agenda that was to culminate in the invasion of Iraq three decades later, seemed captives of an obsolete vision. I felt that the revolution that had begun in the counterculture was poised to burst forth and show the world how alternative energy resources could solve our problems. Whatever was natural was good. Humanity had to get back to the garden . . . to nature as it had been before we fell from grace into industrial, hydrocarbon hell. Among other things, that meant using a bike instead of a car, wearing homespun cloth imported from India, recycling, and getting all our energy from sunshine and wind. Treading this gentle path, we'd bring ourselves back into harmony with Mother Earth, referred to by her ancient Greek name, Gaia.

In 1979, James Lovelock, a British scientist with a background in chemistry, oceanography, and medicine, published *The Ages of Gaia*. "Living organisms and their material environment are tightly coupled," he wrote. ". . . I see the Earth and the life it bears as a system, a system that has the capacity to regulate the temperature and the composition of the Earth's surface and to keep it comfortable for living organisms." His book became a bible of the environmental movement. The iconic photograph, taken from space, of our uniquely endowed planet, with its luminous ice caps, sapphire oceans, and translucent shell of air, captured the vision of the whole earth as a solitary breathing organism suspended in a black void.

Beautifully small, simple technologies relying on clean, natural resources gave me hope, much in the way that in Long Island, where I started spending time, my garden did, by turning sunshine, soil, compost, water, and seeds into vegetables that were tastier and cheaper than store-bought. Each year my biodynamic-raised beds repeated the same miracle. I kept harvesting them, but the forecast by energy gurus that within a decade solar and wind power would be lighting up most of America did not happen. Environmentalists complained that the fossil-fuel and nuclear industries were gobbling up federal subsidies and suppressing the development of marvelous breakthroughs for renewable energy. For whatever reason, it made only a tiny contribution to the grid, and fossil-fuel combustion continued to increase—all the more so when no new nuclear plants were ordered after 1973. (Worldwide, all new solar power installed in 2006 generated fewer megawatts than the Oconee nuclear station did that year.)

In my backyard, the campaign of "split wood, not atoms" worked, and in 1987, after numerous court battles, the Shoreham nuclear plant was prevented from opening. Sometimes I happen to fly over its dormant reactor building. Long Island continues to suffer electricity shortages. Outages in the summer are common, and—as is the case throughout the Northeast—diesel generators, with their large carbon output, meet peak demand, usually during heat waves. Had Shoreham gone into operation, it would have prevented three million tons of carbon dioxide a year from being emitted. A cable—vociferously opposed by environmental groups claiming that its electromagnetic field could harm marine life—has been laid across the floor of Long Island Sound so that more power can be brought in from Connecticut, which gets 53 percent of its electricity from nuclear plants.

My husband and I try to be sensible about energy expenditures. And we compost and recycle and get our food whenever possible from local sources. We've heeded Bill McKibben, who in 1989 called attention to human causes of global warming in *The End of Nature*, as well as other environmentalists who have inspired us to be more conscientious about conservation. We've added insulation, weather-stripping, and thermal-paned windows to our house. We use energy-saving lightbulbs and appliances. In winter, we close off rooms, keep the thermostat low (even lower at night), and wear extra layers of clothing. In the summer, we rely on a ceiling fan, roof vents, and big old shade trees to keep the house cool.

Recently, as oil prices reached a new high and winter temperatures plunged, people in upstate New York began stoking up their old coal furnaces, but we could never countenance that. We replaced our antique hulk with an Energy Star model that is cleaner and more efficient. We don't have a lot of electronic gadgets and we don't live extravagantly. Some of our heat in winter comes from burning our fallen trees instead of imported oil, and we often take public transportation. I like to think our carbon footprint is smaller than average, but nevertheless our participation in modern life makes us polluters: although we plan to get a hybrid car—preferably a plug-in—we drive gasoline-fueled cars (reasonably small, elderly ones with good gas mileage); shop at the supermarket; buy clothes, electronics, and plastic household objects; take diesel-powered train and jet-fuel-powered plane trips. The list could go on for pages and would have to include things such as the purchase of an organic cotton skirt brought from Asia in a diesel-powered ship and then trucked to a big, brightly lit, climate-controlled health food emporium that brags about its participation in wind farms. Our hands are not clean.

Rip and the scientists we met who were thinking about future energy resources didn't oppose the renewable ones—I never met anyone in the nuclear world who did, and in fact encountered some enthusiastic fans. But these experts considered expectations about wind and solar power as unrealistic as some of the early claims made on behalf of nuclear energy. The scientists who were accustomed to looking at the entire life cycle of energy resources debunked the assumption that renewables were environmentally immaculate.

Maybe, as people had told me, on a large scale, wind and solar power would not be practical, because they relied on weak, diffuse, inefficient sunlight and wind. But couldn't I use them at home to get out of the fossil-fuel trap and to become more independent of the grid—especially in case of a hurricane or some other disaster? I began looking into alternative sources.

Sunlight in the often cloudy Northeast is an iffy matter. In the winter the overcast skies can last for weeks. Fog is common where I live. However, our house is passive solar, in that it has some big south-facing windows so that it warms up nicely on clear days. What if photovoltaic panels on the roof could replace fossil-fuel consumption?

Solar advocates envision the deserts (seemingly lifeless to some renewables fans) covered with photovoltaics that would pipe their electricity to distant cities. Celebrities champion solar energy as a fabulous new way of life. Barbra Streisand's website recommended hanging out laundry on a clothesline rather than using a dryer; she explained to the *New York Post*'s gossip columnist that "she never meant that it necessarily applied to her." Christie Brinkley, the former supermodel, seeking to close all nuclear plants, has testified at Congressional hearings. She's especially troubled about the plants near her Hamptons residence—Millstone in Connecticut, across Long Island Sound, and Indian Point. A reporter asked her where electricity would come from if not from nuclear power or fossil fuel, and she replied that it would come from the sun. Did she use solar power in her home? No, it was impractical on her forested property. I've noticed that few of the people who say that solar energy is the answer actually rely on it.

The manufacture of photovoltaic panels requires highly toxic heavy metals, gases, and solvents that are carcinogenic. Brookhaven National Laboratory's department of environmental sciences, which has amassed one hundred fifty studies on the topic, lists poisonous and flammable substances as well as hazardous chemicals that go into making the panels. Some of the gases are lethal and others are explosive. Scrubbers to control accidental releases are not in place at all such factories. Workers in such plants must be strictly protected. If a residential fire burns a solar panel, people would be at risk for exposure to toxic vapors and smoke, though significant health impacts would probably occur only in the heat of an industrial fire. When solar panels are decommissioned, after twenty to thirty years of useful life, they must be disposed of in special toxic waste dumps. If they wind up in a municipal waste incinerator, the heavy metals—such as cadmium and lead-based solder—will partially vaporize into the atmosphere. The ash must be disposed of in a controlled landfill. If modules are dumped into municipal landfills, then heavy metals such as arsenic and lead can leach into the soil and water table. Hundreds of thousands of years from now, some of those substances will still not have decayed: their life spans are essentially eternal. Newer designs are lower in toxicity but still must pass a DOE test about whether their dangerous components will be released after they're decommissioned. Scaling up the technology would present difficulties. Solar farms big enough to supply 1,000 megawatts per year or more would cover over fifty square miles and produce a quantity of toxic waste that would be significant. Controlling it would be costly, perhaps prohibitively so.

When we compare energy sources, we tend not to think about the comprehensive cost—the fuel, metals, and plastics required to fabricate, transport, and install solar panels and wind turbines and, twenty-five years later, to dispose of them. Solar and wind construction on a large enough scale to make a significant contribution to the grid would require maintenance, roads, platforms, transport, and cables, and would add hidden costs in their environmental impacts and waste problems to an already expensive way to make electricity. For the 70 to 80 percent of the time when nature isn't cooperating, you need the grid or a fossil-fuel generator. Until the technology for storing energy improves, photovoltaic panels will not be able to produce electricity as efficiently, cheaply, and reliably as coal combustion or nuclear power. Innovations—solar towers with lenses that concentrate sunlight; mirror-lined dishes that collect the sun's heat to power engines; solar roof tiles; and solar film, as well as new designs for industrial buildings that efficiently employ sunlight—suggest that one day a bigger portion of our energy than the present fraction of 1 percent may come from this resource.

Although since 1979 solar power has become 80 percent cheaper, Solarbuzz, an industry website, says that a residential system costs $8 to $10 per watt, but government incentive programs and large purchases can reduce the installed cost to as low as $3 to $4 per watt. The largest systems of unsubsidized solar energy in a sunny place range from 22 to 40 cents per kilowatt-hour. In other words, solar is the costliest alternative energy of all.

LIPA, my local utility, is offering rebates to customers who go solar. A small system producing about 600 watts of power would cost about $10 per watt ($6,000). "These small systems will offset only a small fraction of your electricity bill," the utility's website warns. The cost includes the system and the installation. A 2-kilowatt system that will meet almost all the needs of household that is very energy-efficient ". . . may cost $8 to $10 per watt ($16,000–$20,000). At the high end, a 5-kilowatt system that will completely offset the energy needs of many conventional homes may cost $7 to $8 per watt ($35,000–$40,000)." The higher-efficiency modules were even costlier.

A thousand square feet of roof space would be required, and our cedar shakes would have to be replaced by asphalt shingles. Our venerable trees, which have been sequestering carbon dioxide for sixty years and providing not only shelter and food for wildlife but also our shade in the summer, would have to go. We'd have to use an air conditioner or a lot of electric fans. If we were to build a new, energy-efficient house on a tree-less lot and incorporate solar equipment, over the long run maybe we could lower our electric bills, as all the promotion promises. But perhaps the cost in maintenance and eventual replacement of the panels would overtake any savings and make solar power an indulgence we could not afford, even after the rebates, tax breaks, and assurances that we'd be sell-ing excess electricity we didn't use back to the grid. I've heard testimoni-als from people around the country who are happy with their panels and take pleasure in seeing their electric meters run backward.

What about a home wind turbine? Thanks to big federal and corpo-rate subsidies for wind power in the past few years, a boom has begun. Because I belong to a cooperative organic farm, I'm on a mailing list of the Renewable Energy for Long Island organization. RELI says that a residential turbine would require an eighty-foot tower, and the turbine, batteries, and inverter to create electricity would cost around $40,000. (The American Wind Energy Association, a lobbying group, claims a home turbine is a lot cheaper than its solar counterpart, which would cost $80,000.) But because of the deforestation and development occurring apace around us, our little glade has become a sanctuary to birds, and I wouldn't want to see them or our bat population disoriented by the sound or sliced up by what the Audubon Society once called, in its oppo-sition to a California wind farm, a "condor Cuisinart." Wind proponents say that newer designs are quieter and more bird-friendly and that in any case many more birds die when they fly into windows. Wind tends to fail during heat waves. A severe heat wave in California in the summer of 2006 challenged the grid. Consumers prevented rolling blackouts by cut-ting back on electricity use. Wind power turned out to be highly unreli-able, with capacity plunging from its usual 33 percent to 4 percent during the time of peak demand.

To meet growing needs, LIPA must add 100 megawatts a year through 2011. Supported by a coalition of environmental and stakeholder groups, the utility wants to build a $300-million, 140-megawatt "wind park" about four miles off scenic ocean beaches. Fishermen and other residents wonder about the big footprint (eight square nautical miles) and the row of tall towers (260 feet high and with blades 182 feet long) as well as the installation process itself, which would involve pile drivers sinking steel monopile towers about nineteen feet in diameter deep into the seabed. (The U.S. Army Corps of Engineers is planning to test monopile installation and has announced that four species of endangered turtles, four species of endangered whales, two species of endangered seabirds, and a threatened beach plant may be affected by such an experiment.) Warning lights for aircraft and boats would light up the wind park at night and foghorns would bellow as needed. An underwater cable would connect the turbines to the grid. (Curiously, environmental groups don't appear to mind *that* cable—just the one across Long Island Sound.) Though Greenpeace has attained fame for its fierce defense of oceans from pollutants, the organization supports the project and has no problem with what is in fact the establishment of heavy industry offshore. But turbines and other equipment need lubrication, and fishermen worry about what oil leaks will do to marine habitat. Others are concerned about migratory flocks whose flyway includes the south shore of Long Island. LIPA says that the site is not on their flight path and that the wind park would meet the needs of about forty-four thousand homes. The turbines would operate only intermittently and, optimally, at about 34 percent capacity— which means that most of the time any given turbine will be idle. Still, wattage obtained would be cleaner than that from fossil fuels.

Ardent advocates of wind power have a motto: "It's not the vista, it's the vision." The vision includes wind farms on rooftops in industrial zones, on mountain ridges, and on vast tracts covering hundreds of square miles of the Great Plains, where a steady wind pours down from Canada. Wind power requires 200 to 300 square miles to make 1,000 megawatts. Coal-fired cement plants would be needed to make millions of cubic yards of concrete for the bases, each of which would form a platform of 300 cubic yards of concrete forty-seven feet in diameter and rooted forty feet deep in the ground. Rocks might have to be blasted away, or trees cut down to make room for the bases, towers, and the wind itself. On the other hand, in the spaces between the platforms, farming and grazing can continue.

While LIPA works on an environmental impact study of the wind park, opposition is arising from civic groups along the south shore who object to "view-shed" destruction and the high cost to the consumer; they recommend upgrading existing power plants and cite the attack the noted environmental activist Robert F. Kennedy Jr. mounted on the proposal to build a wind farm in Nantucket Sound near his summer home. He worries about, among other things, the impact on fish. Others point out that if the ocean continues to acidify because of carbon emissions and to heat up because of heat-trapping greenhouse gases, there won't be any fish. Ironically, the debate becomes similar to the one about nuclear power once risk-benefit ratios have been put on the table. Like nuclear energy, wind is also held to a higher standard than other forms of power production. The wind-farm controversy is dividing the environmental community as people weigh the urgency of acting to stop emissions on the one hand and, on the other, the prospect of formerly pristine settings and wildlife habitat being radically altered.

Residents near scenic mountain ridges, like those of the Berkshires in Massachusetts, shudder at the thought of "ridgelines industrialized with 34-story turbines," as Eleanor Tillinghast, head of Green Berkshires and vice-chair of the Massachusetts League of Environmental Voters, wrote in 2006 in an op-ed piece in the *Berkshire Eagle*. That state has a quota for renewable energy that involves the construction of hundreds of wind turbines in the next four years. The state has "defaulted to a bumbling strategy of spending extraordinary amounts of public funds to entice developers into the Berkshires and hoping that something happens," Tillinghast wrote. "There's no plan, no guidelines for where turbines should be erected to minimize impacts. . . . There are, however, many unintended and costly consequences," among them lucrative subsidies for renewable-energy producers. Wind power supporters disagree with her. Whatever the merit of these claims, they indicate how much hostility toward wind power there might be among people who consider themselves preservationists.

At least the average price of wind power has gone down, from 10 cents per kilowatt hour to about 5 cents. Some big projects are generating electricity in areas with strong prevailing winds for 3.5 cents per kilowatt-hour, which is competitive with natural gas generation—but that's because of the rise in the cost of natural gas.

The Long Island Power Authority is trying to sell me on renewable energy because New York State, like Massachusetts and ten other states, has mandated its growth. Most of the 6 percent of the renewable energy utilities supply to the grid is generated by hydropower, and the rivers that provide that energy have already been dammed. There is also no environmental plan to get rid of dams. Thus, as a practical matter, the mandate applies mainly to wind and solar. To ramp up these expensive technologies from supplying less than 1 percent of our electricity to meeting the DOE's goal of 6 percent in twenty years will be expensive indeed, even if the government intends to help out the utilities. And the environmental costs could be big. To produce 1,000 megawatts of installed capacity, a wind farm would require about 60,000 acres, or about ninety-four square miles, according to the American Wind Energy Association. Others say that two hundred square miles are required. John Ryskamp, a nuclear engineer at INL, has compared the impact on ecosystems of generating 1,000 megawatts of electricity from renewables and from nuclear power. A nuclear plant would need only about a third of a square mile. Bio-alcohol would take sixty-two hundred square miles of cornfields; bio-oil nine thousand square miles of rapeseed fields; and bio-mass in the form of wood twelve thousand square miles. Rip considers ethanol an energy-draining fad that will enrich entrepreneurs and investors. "That bubble will burst," he said. "You have huge tracts given over to growing fuel feedstock instead of food. Ethanol made from corn-stalks and other agricultural waste would deprive the soil of nutrient-rich humus." The United States exports food to more than ninety countries. In a world where millions are starving and farmland is shrinking because of development and desertification, use of crops to power our automobiles is ethically troubling to me.

For now I'll have to put up with an oil furnace. An expert on home heating came to my house to inspect the furnace and the chimneys. I begged him to recommend a clean resource to replace oil. He replied, "You're not going to like this, but there is an answer. Nuclear power." He turned out to be a retired reactor operator. If nuclear power ever becomes widely available in the Northeast and reduces our bill to the low, stable level paid by nuclear utility customers in the Southeast, we could heat our home with electricity and one day use it to recharge an electric car. Our carbon footprint would dramatically shrink.

The DOE asked in its 2007 research and development budget for greater expenditure on renewables and fossil fuels than on nuclear power. Despite the sums spent on wind and solar power in the past thirty years, and investments on the part of big energy corporations, fossil fuels remain dominant, and the small reduction in their use is almost totally due to the increased efficiency of nuclear power. Many a promoter of renewable resources insists that their use is growing so rapidly that it will soon overtake other forms of energy generation, and it's true that there are more wind turbines and solar panels at work than ever. But we're burning 400 million more tons of coal a year than we were when I planted my first organic garden. That increase far outstrips the contributions of all other sources of energy generation.

While we Americans each contribute on average about twenty tons a year to the carbon load of the earth, by contrast Sweden, which relies on nuclear and hydroelectric power for most of its electricity, contributes less than a third of that per person, as does France, with most of its electricity coming from nuclear power. Denmark, which has thousands of subsidized wind turbines providing over 20 percent of its electricity, has per capita carbon emissions of eleven tons, thanks to fossil fuels. Electricity always has to go somewhere. It can't just be deposited in a tank and pumped out when needed. According to a study by a U.K. group, ABS Energy Research, Denmark "was forced in 2004 to export some 85 percent of the wind-energy its mammoth turbines generated, often at a loss. . . . Worse, the carbon-emissions reducing potential of that power was compromised because two of the countries it exported to—Norway and Sweden—reduced their hydropower production sources to accommodate cheaper wind power." Meanwhile, Denmark's primary power was still delivered by fossil-fuel plants, ABS said, effectively "nullifying" wind power's chief benefit. ABS concluded, "Wind power has been promoted for politico/environmental reasons and wind developers have benefited from substantial subsidies, leading to exaggerated claims. A reality check is needed." Germany presently gets a quarter of its electricity from nuclear plants, but the Green Party has chosen to phase them out by 2020. To meet rising electricity demand, the country is building one carbon-sequestering coal-fired plant as a demonstration model and twenty-five traditional carbon-emitting plants that will burn coal and, worse, lignite. Nuclear plants emit almost no carbon, a natural gas-fired plant emits 425 grams per kilowatt-hour, black coal over double that amount, and lignite almost three times as much. The result of the Green Party's dream of eliminating nuclear power: Germany will fail to meet the European Commission's target for carbon dioxide reduction, and by 2020 carbon emissions will shoot up still further.

Paul Lorenzini is a nuclear engineer, an attorney, a former general manager of contract operations at DOE's nuclear defense facilities in Hanford, Washington, and a retired executive of PacifiCorp, a utility that derives over four-fifths of its electricity from coal and none from nuclear power. In "A Second Look at Nuclear Power," an article he published in the National Academy of Sciences' journal, *Issues in Science and Technology*, Lorenzini notes that renewables haven't lived up to their promise, causing Denmark and Germany, among other countries, to mandate their use by law the way some American states have. He goes on to analyze the environmental lobby's rigid ideological opposition to technology as an explanation for its focus on the supposed ills of nuclear power even as it has ignored—until quite recently—the vast and rapid expansion of fossil-fuel-power generation. Polls indicate that although the majority of scientists favor nuclear power, as do about three-quarters of college graduates, the majority of environmentalists oppose it and have succeeded in sowing skepticism among policymakers. Furthermore, the lobby has used "the legal system of reviews intended to protect the public . . . as a vehicle for blocking nuclear power."

Meanwhile, fossil-fuel plants have continued to thrive and multiply. The International Energy Agency has indicated that in 2000 about 90 percent of global energy came from coal, oil, and natural gas. Lorenzini says that current estimates of future use continue to favor fossil fuels, which will probably provide 90 percent of all new energy through 2030. These dismal predictions bring to mind that dark plume over the splendor of the Colorado Plateau, the open-pit coal mines visible from space, the endless trains of coal cars, the mountains of coal and the lakes of slurry.

President George H. W. Bush once commented, in regard to an international environmental summit he was obliged to attend, "The American way of life is not up for negotiation." Bill McKibben wrote in *The End of Nature*, "If that's true, if we can't imagine living any differently, then all else is mere commentary."

Rip repeated the sentiment when I asked him what was to be done about the future of energy. "Some event has to trigger a big change in our way of thinking. We have to start doing something about greenhouse gases pronto, and so far the United States doesn't have a real energy plan. We've *never* had a real one. Nobody thinks about the big picture, about the long term. Things have just happened this way and that way, depending on the politics of the moment, on various lobbies and their contributions to campaign funds."

Since the 1990s, Congress has lacked a scientific review board to evaluate the claims lobbyists make and to ensure that policy is science-based. Politicians focus on the next two, four, or six years and pass along the big problems to the next officeholder. The last big national initiative to benefit the American people was President Eisenhower's interstate highway system, and one of its purposes turns out to have been to depopulate cities, supposedly the targets of Soviet nuclear warheads, and to encourage us to scatter into suburbs. Now most of the American population is hostage to those suburbs and their fossil-fuel-demanding energy-inefficient individual houses and private automobiles. Meanwhile, since 1980 public and private funding of research and development in the energy sector has plummeted.

In 2006 the government spent over $70 billion on military research and development and less than $5 billion on civilian energy research and development. Corporations spent even less than that. Perhaps it's not entirely relevant, but I am struck by the fact that Americans choose to part with $10 billion a year for bottled water.

In recent decades, most politicians have been publicly opposed to nuclear power. Some who privately understand that we can't do without it studiously avoid the appearance of supporting it. Until very recently, few in office have considered what might happen when fossil resources become scarce or much costlier to extract. Those who decide energy policy may make optimistic remarks and seemingly heartfelt, voter-pleasing promises about clean energy. But the campaigns of most elected officials of both parties have received fossil-fuel funding, and legislation rarely has taken into account uncomfortable facts such as the depletion of resources, accelerated global climate disruption, the premature deaths and the incidence of disease caused by pollutants, or the bloody, treasury-draining foreign entanglements that result from our need to import oil and gas.

"Hardly anyone is looking at the whole structure across time," Rip said. "It's all stopgap and living in the moment and grabbing all you can. We're more dependent on foreign oil now than we were during the energy crises in the 1970s and more inclined to use military force to make sure nothing interferes with our importation of that oil and gas."

World CO_2 emissions from electricity generation come to 9,500 million metric tons a year. Using a small footprint, hundreds of nuclear plants in more than thirty countries cut carbon emissions by 600 million metric tons every year. But that fact has been dismissed over the years by most major environmental lobbying groups. Of course they're worried about global warming and wilderness areas being ravaged to extract gas and oil, but they keep repeating that the only solution is renewable energy. This policy seems to arise from entrenched assumptions about nuclear power and from a belief in, well, unobtainium.

"We should not be sidetracked into talking about nuclear," Greenpeace International's spokesperson, Damon Moglen, says. "It's irrelevant in the climate change debate." Debbie Boger, an energy expert with the Sierra Club, acknowledges that it's "absolutely true" that the nuclear industry has improved its safety record, but she remains concerned about "the potential for catastrophic accidents. What kind of risks are you willing to take for that?"

Certainly the nuclear industry has evolved, but minor mishaps still occur, and cover-ups as well, and since most Americans appear ignorant of how radiation works, and of what comparable exposures from nature are, they assume any leaks to be dangerous. In 2006 word got out about tritium leaks at nuclear plants in Illinois and New York that the utilities had tried to hide. An NRC task force investigated these and similar leaks over the past decade, concluded that the public hadn't suffered any risk, and recommended more candor on the part of the utilities.

"This story points out a really big part of the problem faced by the resurrection of the nuclear power option," Leo Gómez, the radiation biologist, said to me. "Not only is the public's perception a concern, but also the companies haven't disclosed past releases or spills. Their credibility is lost, and when that happens it's nearly impossible to regain it. The detectable releases are low dose and may not exceed permissible releases, but the view is that if a radioactive leak is detectable, it's bad." So technical and social factors remain important considerations. But what are these compared with the looming consequences of global warming?

Other environmental spokespeople also emphasize the risk of nuclear power, its expense, the potential scarcity of uranium, and the carbon emissions associated with its extraction and refinement. And the antis claim that, most important of all, safe disposal of nuclear waste is impossible. None of these claims turn out to be correct. And as choices become more obvious a pro-nuclear movement is growing among environmentalists.

A French engineer and author, Bruno Comby, founded the international organization Environmentalists for Nuclear Energy. He travels extensively, giving impassioned lectures about how nuclear power uses a million times less raw material than do fossil energies and therefore produces a million times less waste. "It's important to note that renewable energies will never be able to fulfill the energy needs of the planet in the coming century," he says. "Windmills produce kilowatts, but the demand for energy is in gigawatts." That demand will grow even if citizens of prosperous, developed countries conserve energy, because the world's population is increasing and developing countries want more electricity. Comby's assertions are supported by a paper published in *Science* in 2002 by a multidisciplinary panel of scientists:

Nuclear power can play a significant role in mitigating climate change. There are no insurmountable technical barriers to nuclear expansion, but the expansion must be performed under very high safety standards. Additionally, capital cost reductions from advanced designs and production methods will be required. It is therefore important to maintain and intensify current programs of research and development on power reactors, waste disposal, and nuclear safeguards to assure that safe nuclear power is available when it is needed.

James Lovelock, the patriarch of the environmental movement, defines himself as a Green. His Gaia hypothesis, initially dismissed by many scientists, is now widely accepted. Lovelock, among the first to warn of global climate change, thanks to his research across several disciplines, thinks its risks surpass those of war, pandemics, and terrorism. With six billion humans and more to come, he says, fossil-fuel combustion must end, and renewables can't provide enough energy in time:

If we had 50 years or more we might make these our main sources. But . . . the Earth is already so disabled by the insidious poison of greenhouse gases that even if we stop all fossil fuel burning immediately, the consequences of what we have already done will last for 1,000 years. Every year that we continue burning carbon makes it worse for our descendants and for civilization.

Lovelock wants to awaken environmentalists everywhere to the reality of the situation:

Opposition to nuclear energy is based on irrational fear fed by Hollywood-style fiction, the Green lobbies and the media. These fears are unjustified, and nuclear energy from its start in 1952 has proved to be the safest of all energy sources. . . . I entreat my friends in the movement to drop their wrongheaded objection to nuclear energy. Even if they were right about its dangers, and they are not, its worldwide use as our main source of energy would pose an insignificant threat compared with the dangers of intolerable and lethal heat waves and sea levels rising to drown every coastal city of the world. We have no time to experiment with visionary energy sources; civilization is in imminent danger and has to use nuclear—the one safe, available, energy source—now or suffer the pain soon to be inflicted by our outraged planet.

Other environmentalists have made the same case. Patrick Moore has accused his former colleagues at Greenpeace of being ideologically blinkered, out of touch with the facts, and prone to fearmongering when it comes to nuclear power. (He once compared nuclear plants to atomic bombs; he's now co-chair of a nuclear-industry-funded initiative called the Clean and Safe Energy Coalition.)

"Renewable energy might be a good idea in the long term," Lovelock said in a 2004 address titled "The Selfish Greens," "and is a showy way for politicians to prove that they are doing something, but it is already too late to expect it to play a significant role":

> [T]he magical appearance of a completely clean, safe and economic source of energy would do little to stop the burning of fossil fuels; such is the human tendency to over-consume. Indeed, the widespread use of wind turbines in Germany has been accompanied by an increase in coal combustion, the dirtiest of all fuels. The Earth system, Gaia, functions because within it are powerful restraints to growth, as we will soon discover. We have to make our own restraints if we are to avoid those the Earth will surely apply.

He closed by appealing to what he considers best in human beings:

> Perhaps when the catastrophes of the intensifying greenhouse become frequent enough we will pull together as a global unit with the self-restraint to stop burning fossil fuel and abusing the natural world. . . . Greens, let us put aside our baseless fears and be brave enough to see that the real threat is to the living Earth, which is our nation and our home.

There was one last possibility that I'd heard was promising. What about nuclear fusion, as opposed to nuclear fission? The sun fuses atomic nuclei in a thermonuclear process that makes heat and light, so why can't we do the same? For fifty years humans have been experimenting with various methods of tapping into this fundamental stellar power and have made many advances. However, it takes a tremendous amount of energy—the kind found in the heart of a star—to overcome the force that repels atomic nuclei from one another, and there are also many other practical barriers. While some fusion enthusiasts say that the quantity of radioactive waste would be reduced and would decay much more rapidly than the waste from today's fission reactors, others remain skeptical, saying fusion would produce plenty of waste. Fusion also faces many other challenges—for instance, how to economically harness the necessary 100-million-degree-Centigrade plasma and create components that would last in such an environment. Still, fusion remains promising enough in the very long term for nations to have joined together to continue experimentation.

Rip and the others at the conference table at Argonne-West had agreed that probably the obstacles to fusion would eventually be worked out, but that many decades would pass before it could replace nuclear fission. Lineberry said, "The saying is, 'Fusion is fifty years out no matter what date you make this statement.' "

Forced by the scientific method to do their work in the real world, scientists do not have the luxury of relying on miracles. But many of the rest of us live so comfortably in an enclosed sphere, in which a single supermarket offers dozens of different kinds of pet food and shampoo, that we've come to imagine we have choices that really we do not.

"It's good to have renewables and we need to grow them," Rip said. "I completely support them. But to make clean baseload energy, to make hydrogen efficiently, to stop the carbon cycle, we have no choice but to rely on the nuclear fuel cycle. Nuclear plants are never idle and they work at top capacity. The cost of the electricity from them is low once the start-up expenses have been earned back. The fuel and the waste are small in volume. We already have the equivalent of unobtainium, except it's obtainable. It's called uranium."

Part 5 **CLOSING THE CIRCLE**

An outstanding advantage of nuclear over fossil fuel energy is how easy it is to deal with the waste it produces. Fossil fuel burning produces twenty-seven thousand million tons of carbon dioxide yearly. This is enough if solidified to make a mountain nearly two kilometres high and with a base ten kilometres in circumference. The same quantity of energy if it came from nuclear reactions would make fourteen thousand tons of high level waste. A quantity that occupies a sixteen metre sided cube.

—JAMES LOVELOCK

14 TEN THOUSAND YEARS

UNTIL THE PUBLIC CAN BE SHOWN that nuclear waste is safer than fossil-fuel waste and that it is reliably controlled and of low risk today and will be in the future, nuclear power in the United States will be constrained. California is aggressively curbing greenhouse gases and plans to cut back on reliance on coal-fired plants. Nuclear plants supply 17 percent of its emissions-free energy, but the state has forbidden construction of new ones until there is proof of adequate disposal of spent fuel. The Nuclear Regulatory Commission (NRC) says that spent fuel can be safely stored in dry casks for fifty to a hundred years, but that "as a nation it would be beneficial to move toward a permanent solution."

During the last leg of the nuclear tour—to nuclear waste interim storage sites and long-term repositories—I had many questions about whether the radioactive products of the nuclear fuel cycle could be safely isolated during the centuries it would take for them to decay to the level of natural background radiation. Without a doubt, this was my biggest concern. So far, not one country has put spent reactor fuel in permanent storage.

In the United States, the fate of uranium after it has finished making electricity, enabling medical diagnoses and treatments, and serving in the national arsenal is determined by the open fuel cycle, also called the once-through cycle. Fuel assemblies that have made one eighteen-month trip through the reactor and there accumulated enough radioactive clutter to render them inefficient are placed in spent-fuel pools or concrete casks to await permanent disposal in an as yet unbuilt central repository. The closed cycle, practiced in France, Russia, and Japan, separates out that clutter and then sends the uranium, or a mixture of plutonium and uranium oxides, back through the reactor again. This reprocessing method is repeated, reducing the ultimate volume of waste considerably, but it's costlier than using fresh uranium.

At present the Department of Energy (DOE) is researching extraction of the reusable uranium and plutonium from spent fuel and elimination of the useless but long-lived radioactive isotopes by burning them up in a fast reactor similar to the model we saw in Idaho. The Integral Fast Reactor (IRF) design and its offspring, the Advanced Fast Reactor, can provide not only electricity but also a way to keep nuclear materials from becoming suitable for bomb production. Two fast reactors can make electricity by getting energy from consuming all the plutonium and 99 percent of the actinides created in fuel burned by five pressurized water reactors (the design used in most American plants). The basic fast-reactor design is efficient, recovering almost all of the energy remaining in the spent fuel. Because this closed fuel cycle takes place at one site, the hazards involved in transporting and handling fissile and radioactive material multiple times are obviated. In addition, the very high radioactivity of the type of spent fuel used in a fast reactor would present an obstacle to anyone wanting to divert it to acquire plutonium. The volume of waste that would eventually go to a permanent repository would be small. The radioactive decay of the plutonium-free final residue is much more rapid than that of conventional fuel; isolated and immobilized in metal or ceramic, the waste, after four hundred years, would have a radioactivity level lower than that of uranium ore.

The IFR project pioneered the method, called pyro-processing. The spent fuel, after being cooled in dry cask storage, was moved to an adjacent remote-handling facility and then chopped up and subjected to a series of chemical and electrical processes that separated out the non-fissile components and turned the still-fissile material into fresh fuel. Annually about 100 kilograms of spent fuel from Experimental Breeder Reactor-II and the IFR is still being processed there at the Fuel Cycle Facility.

To accomplish this, technicians use manipulators while looking through windows into "hot cells" of various sizes. On our tour of Argonne-West, Rip, Marcia, and I visited the building, which was dedicated to dealing with highly radioactive material and equipment. The hot-cell walls were three feet thick and the windows composed of glass slabs with transparent mineral oil between them that gave a pale orange cast to the scene within the hot cells themselves, which were filled with inert gas. The technicians maneuvered various materials and hardware on the other side of the glass by moving their hands, attached to sensors, in slow motion. A man remotely opened a canister of fuel, his hands raised and slowly miming an unbolting motion, something like modern dance. Inside the hot cell, obedient robot arms and hands that held tools mimicked his human movements. Outside another hot cell a woman was teaching a few trainees to use manipulators. Through the window, we observed a robot hand meditatively wiping a tabletop with a rag in the eerie light. She told us that women leave a hot cell in better order than men do. The work is often delicate. Technicians must pass a test that involves being able to sign their names legibly using manipulators.

After spent nuclear fuel leaves a reactor, it's usually placed in a spent-fuel pool. When we visited the Duke nuclear plants, we never even knew where theirs were, but at the Idaho Nuclear Technology and Engineering Center (INTEC), we were escorted to a corrugated steel building that housed a spent-fuel pool. We paused outside to watch spent nuclear fuel in a dumbbell-shaped transportation cask being lifted from a truck bed with a crane. The opening of the cask, removal of the waste, and transfer of the contents to the spent-fuel pool were all done robotically because of the weight and the radioactive and thermal heat.

We then entered the building, which was reinforced with huge steel I-beams connected by cross struts with gigantic bolts—I was reminded of the architecture of an old railroad bridge. Rip pointed out the structural sturdiness—necessary in order to support eighty-ton-capacity cranes that glided noiselessly overhead on horizontal tracks and certainly could shred an airplane crashing into the facility. Resembling a giant swimming pool and bigger than commercial spent-fuel pools, the INTEC one was twenty feet deep except for a section that was sixty feet deep; the design was state-of-the-art and included a leak-detection system. The water, deionized, demineralized, and continually cycled through a purification system—and a beautiful, limpid blue, thanks to the presence of neutron-absorbing boron—was the purest I'll ever see. Storage racks lay shimmering ten feet below the surface. They held fuel rods hanging down into the depths under numbered hatches. The water kept the uranium cool and shielded it as actinides and other fission products that had accumulated during its sojourn in the reactor decayed within the cladding, which technicians frequently checked for cracks.

Most of the rods came from experimental and submarine reactors. Fresh out of the reactor, an unshielded rod would expose a person within seconds to a lethal dose of 500 rem. After about five years in a waste pool, that would drop considerably. After forty years in a waste pool, a rod's radioactivity has declined by 99.9 percent; in about a thousand years it would approach that of the uranium ore at Mount Taylor. However, some isotopes created in reactors, including plutonium-239, would remain radioactive for much longer. Plutonium would decay by half in

24,000 years, and after 100,000 years, or four half-lives, only 6 percent of the original plutonium would remain. After ten half-lives—240,000 years—only about 0.1 percent would still exist.

"It takes plutonium one hundred thousand years to decay to the point at which you don't have to think about it," Rip said. "The international scientific consensus is that a deep geologic repository is the best place to isolate plutonium and other long-lasting nuclear waste." If the DOE can show that such a repository will be safe for ten thousand years, the idea is that it will be the same for a hundred thousand. But, geologically, it's hard to predict accurately beyond ten thousand years, which is why Congress mandated a few decades ago that the repository had to be able to isolate waste for that long. A federal appeals court judge ruled in 2003 that a high-level nuclear waste repository should be guaranteed for a hundred thousand years instead of ten thousand. In 2005, the Environmental Protection Agency (EPA), responding to another court ruling that ordered it to consider the fact that a few radionuclides would outlast that time frame, recommended that accountability extend to a span of one million years.

The estimated cost to the United States of dealing with waste from decades of nuclear activities in ways that ensure the public would receive no more than a very low dose for the next ten thousand years will probably total, in ballpark terms, in excess of $350 billion.

We stood behind a guardrail next to the pool—the water served as shielding—and watched three technicians quietly going about their work. Every action was attentive, deliberate, and dreamlike, like tai chi. Each man donned a bright yellow zippered anticontamination costume and two sets of gloves, the outer pair made of rubber. The men taped the cuffs of the gloves to the cuffs of the suit so no skin was exposed, and they taped over zippers as well. The team then stepped onto a catwalk attached to one of the overhead cranes, which were bright yellow. After a discussion between the group and a supervisor, voices echoing, the supervisor used a remote control to make the crane slide the catwalk over a particular section of the pool. Slowly the men used a tool like a long boat-hook to open a numbered hatch near the surface. They dropped a light into the water to illuminate a steel mesh basket of test-reactor fuel. Every step was executed with extreme caution and respect for the materials.

Afterward, the team walked to a special area near the exit and carefully removed the tape and the jumpsuits. The gloves came off last, so that skin never touched anything external. Everything went into a sealed bin with a radiation-warning trefoil; all items would be treated as radioactive—even though they probably were not, because there had been no direct contact with the fuel rods—and disposed of as low-level nuclear waste.

Before we left the building, we had to pass through whole-body gamma-ray detectors, undergo a frisking, and have a radiation-detector wand applied to our hands and feet. My pants passed the radionuclide test once again. I'd come close to millions of millirem without being exposed.

I'd read that the nation's quantity of nuclear waste was staggering, and early on I mentioned that to Rip. He replied that all of it could fit in the Albuquerque municipal dump with a lot of room to spare.

Greater fuel efficiency has led over recent decades to a decrease in nuclear waste generation. A 1,000-megawatt reactor today produces about twenty-five tons of spent fuel a year, and the annual national total comes to about two thousand tons. The entire American inventory of waste presently being stored at nuclear plants, after forty years of making trillions of kilowatt-hours of electricity and sparing the atmosphere billions of tons of carbon and greenhouse gases, comes to about fifty thousand metric tons. All that waste would cover an area the size of a single football field to a depth of about five yards, if the fuel assemblies were laid end to end and stacked side by side. The per capita lifetime contribution of consumers getting all their electricity from nuclear power: two pounds of waste.

By comparison, each year coal combustion in the United States alone yields one hundred million tons of ash and sludge containing toxic heavy metals; that amount will continue to expand as new plants are built and old ones increase their output. Worldwide, nuclear plants avoid a yearly average of two billion metric tons of carbon emissions. The EPA estimates that the United States annually discards about three hundred million tons of nonnuclear hazardous waste: every *day* the collective households and industries of America throw away nearly a million tons of garbage containing toxic heavy metals and dangerous chemicals, as well as plastics that never break down. That garbage will be our culture's real legacy, enduring for millions of years after all the present nuclear waste has decayed.

One day, Rip did some calculations and estimated that *all* the spent fuel generated by one of Oconee's 846-megawatt reactors in a single year—about twenty-five tons—could fit in the bed of his 350-Ford pickup. "My truck would be skwooshed flatter than a fly track by the weight, though," he said. All the spent fuel from all of America's power reactors, having generated nearly 800,000 billion kilowatt-hours per year, would fit into the beds of 960 such pickups. A year's worth of solid waste from the 500-megawatt Riverbend Coal Station alone, 16,425 tons, would be equivalent in volume to that of a building six stories high and sixty feet on a side.

In the 1960s and early 1970s, pools were just a stopover for commercial spent fuel before it went to the country's only operating reprocessing plant, in West Valley, New York. The plutonium was stored in secure bunkers and the uranium turned back into fuel pellets. The waste, small in volume, was eventually immobilized in glass and put in interim storage. The company running the plant decided that environmental regulations made it unprofitable, and it closed in 1976. Two other reprocessing plants were built in the United States but never used.

After India used plutonium from reprocessing research-reactor fuel to make an atomic bomb, in 1977 President Jimmy Carter responded with a presidential directive suspending commercial reprocessing of spent nuclear fuel in the United States. His aim was to inspire other countries to join in an effort to curb the world's plutonium stockpiles, and he believed we couldn't go to the table to negotiate with others to stop reprocessing if we continued doing it ourselves.

"Although the plutonium left over from reprocessing can be used to make bombs, it's actually more efficient to have defense reactors that are dedicated to just making pure, high-grade plutonium, which is what we used to do," Rip told me.

His job at Sandia, remember, had included counterterrorism measures in regard to nuclear waste. He thought that trying to eliminate nuclear weapons by eliminating nuclear power plants was futile. Countries could acquire bomb material through other means, such as enriching natural uranium, and, absent nuclear plants' displacement of greenhouse gases, the warming trend would accelerate. The risks to human welfare associated with the rise in global temperature were more probable and broader than those posed by hostile parties acquiring atomic bombs that might or might not be used, and might not be much different in their destructive impact from the kind of conventional bombing that the United States has carried out in Vietnam, Afghanistan, and Iraq. "Increased heat in the ocean is causing more powerful storms, and sea levels are rising. The damage done by Katrina was comparable to what a nuclear weapon would do. In the next fifty years, there's an eighty to ninety percent probability of a biological or chemical attack. I'd take a dirty bomb over just about anything else—over a chemical or biological weapon. We have a lot of people trained to handle radiological contamination and you can trace it to its source. You can't say the same about biological or chemical weapons. Those can be delivered through low-tech means." Like other counterterrorism experts, he thinks it unlikely that terrorists could succeed in initiating a nuclear explosion. Rather, it would come from a nation with a relatively sophisticated—and therefore probably detectible—bomb-producing infrastructure.

Matthew Bunn, a senior researcher at the Belfer Center for International Affairs at Harvard University who specializes in proliferation issues, says that acquisition of bomb-quality materials is so difficult that it exceeds "the plausible capabilities of terrorist groups." He thinks that "if the world community can effectively guard all of the existing stockpiles, it can prevent nuclear weapons terrorism from ever occurring: no material, no bomb." And the only way to make that material is with an enrichment or reprocessing plant. Plutonium and highly enriched uranium in a quantity sufficient to make a bomb are harder to transport than chemical and biological weapons, or than plastic explosives, which contain a large amount of energy in a relatively small package and are much easier to obtain. Bunn nevertheless calls for increased control over nuclear materials and better vigilance. "Detailed examinations by U.S. nuclear weapons experts have concluded again and again that with enough nuclear material in hand, it is plausible that a sophisticated terrorist group could at least build a crude nuclear explosive," he writes.

The United States spends $100 million a day to maintain its nuclear arsenal. That's the annual budget of the International Atomic Energy Agency (IAEA) for the protection of the whole world against the eventuality of a sophisticated terrorist group's constructing a nuclear weapon. The technology to protect nuclear stockpiles exists, and I saw sites in the United States where it is applied, but it's costly. Expertise and technology from our national laboratories could play a helpful role in the worldwide accounting of materials and their secure sequestration that is a key goal of the IAEA. Facilities in the former Soviet Union have been found wanting in regard to security, and the United States has taken limited action to address this problem. Bunn points out, "This is a big job, and a complex job, but it is a doable one. It is a matter of putting the resources and the political will behind getting the job done."

Political will can be rather like unobtainium; on the other hand, efforts to round up stockpiles in places such as the former Yugoslavia and ship them to secure locations in Russia are now under way.

"The way we manage spent nuclear fuel has had zero effect on nuclear proliferation," Rip said, "since other countries just went ahead and developed the bomb anyway—before they went to commercial nuclear power.

Despite Carter's gesture, everyone else kept on reprocessing. But here, once our fuel rods accumulate too many fission products, they're retired with ninety-eight percent of their energy unused. This should not be called waste."

Some countries are combining plutonium oxide, chemically separated from spent fuel, with uranium in order to create low-enriched (to 4.5 percent) power-reactor fuel. This mixed oxide fuel, or MOX, will also eventually be made in the United States at the DOE's Savannah River Site in South Carolina from plutonium and highly enriched uranium from bomb pits. World stockpiles of weapons-grade plutonium are estimated to total 260 metric tons. Diluted for use in conventional reactors, that tonnage could provide the equivalent of about one year's world uranium production. European countries have been burning MOX for over twenty years and thus reducing plutonium stockpiles. By 2010, worldwide plutonium production from power reactors and MOX consumption are expected to be in balance. The NRC cites "substantial world-wide experience with the use and behavior of reactor-grade plutonium, because all operating reactors contain plutonium created during the fission process." In 2006 the DOE announced a plan to establish an international partnership to ensure that plutonium is transformed into fuel that can't provide weapons material.

Antinuclear groups oppose the manufacture and use of MOX. "Given their belief system, I can understand people who want to get rid of nuclear power trying every way to do so, including going after the back end of the fuel cycle," Rip said. "But I can't imagine why *anybody* would be opposed to using up the plutonium stockpile to make electricity. It only takes about eighteen pounds of plutonium to make an atomic bomb. If you put that plutonium instead into low-enriched nuclear fuel, then it becomes useless for making a weapon."

Even if at present we were able to recycle all spent fuel and burn up all excess plutonium in reactors, some waste would still remain, along with a small quantity of nuclear refuse from medical sources and a much larger inventory generated by the Manhattan Project and the arms race. Whether or not we approve of nuclear power, the waste exists and something has to be done with it.

15 THE HUGE FACTORY

IN EARLY 1939, THE DANISH PHYSICIST Niels Bohr brought the momentous news of the discovery of nuclear fission to America. That spring, working at the Institute for Advanced Study in Princeton with the American physicist John Wheeler, Bohr correctly deduced that the component of natural uranium that fissioned with slow neutrons must be U-235. Understanding how difficult it would be to enrich uranium enough to make a bomb, he told his colleagues skeptically that such a weapon was impossible "unless you turn the United States into one huge factory." In 1943 he fled Denmark to escape the Nazi roundup of Danish Jews (his mother was Jewish) and, after arriving in America, he told the Hungarian physicist Edward Teller, who was then working at the secret laboratory at Los Alamos where the first bombs were being designed, "You have done just that."

In less than three years the United States built a secret nuclear weapons complex the size of the entire automobile industry. The assembly line stretched from state to state across the whole continent. Tens of thousands of people were given the task of rapidly constructing and running research laboratories, uranium processing and enrichment plants, reactors, metal workshops, and testing grounds. Through painstaking labor, tons of previously useless rock were transformed into a few pounds of metal that could unleash energy of a magnitude so immense that even today we have difficulty imagining it.

After World War II, the momentum of this industry continued, powered by fear of a new world war fought with an even more destructive technology. When the Soviets acquired the secret of the atomic bomb, the frenzy to expand our nuclear arsenal accelerated. Our "huge factory" eventually provided enough nuclear weapons to strike not only all the major military installations and population centers in the USSR but also Communist Party headquarters in small cities. At the height of the arms race, the United States and the USSR had seventy-eight thousand nuclear warheads. The term *overkill*, coined specifically by government think tanks to describe this lunatic state of affairs, entered the English language. Declassified documents from the 1950s indicate that, around the time that Janet Johnson and I were scooping out gravel in the arroyo for our refuge from Soviet atomic bombs, the Pentagon and National Security Council had estimated that a nuclear attack on a hundred American targets would kill or injure twenty-two million Americans.

By the early 1960s, secret federal directives contemplated preemptively destroying the Soviet Union, Eastern Europe, and China in order to end Communism forever: over thirty-two hundred nuclear weapons (some of them no doubt reposing in the hollow mountain outside my hometown) would be dropped on 1,060 targets. The United States would kill an estimated 285 million people and injure 40 million. The branches of the military, each of which was conducting its own atomic tests in Nevada and on Pacific islands, argued among themselves about whether the resulting fallout from an all-out nuclear war would be a serious problem for our side or not. The army said yes and the air force said no. Biologists spoke of all-out nuclear war producing species extinction so widespread that the living world would be reduced to insects and grass, and a small group of scientists testified before Congress about the nuclear winter that would be caused by all the dust thrown into the atmosphere by atomic blasts. (That scenario, which ultimately was not validated by the scientific community, has been replaced by the nightmare of rapid global warming, which has been heartily validated around the world by thousands of scientists from many different disciplines.) By the end of the 1980s, the two superpowers had negotiated a truce, agreeing to major reductions in their arsenals of both nuclear and conventional weapons and to a comprehensive nuclear test ban.

In 1957, when I was worrying about whether I'd be invited to a spring sock hop at my junior high, a B-36 bomber on the approach to Kirtland Air Force Base in Albuquerque accidentally dumped its atomic cargo from an altitude of seventeen hundred feet. The ten-megaton thermonuclear weapon hit empty tableland a few miles from the control tower. On impact, the bomb's trigger, made of conventional explosives, blew up, gouging out a crater in the mesa twelve feet deep and twenty-five feet wide and scattering around some small fragments of plutonium. (Unless a trigger is correctly and precisely configured around the fissile plutonium core—which in this case was being transported in a separate part of the plane—and unless a particular detonation sequence is initiated, a bomb can't explode atomically. This is why nuclear weapons that have been lost in oceans will never detonate and will over time disappear under layers of sediment.) Rapid recovery, decontamination, and a thorough radiological survey followed. No radiation escaped from the crater, and the soil and debris were isolated in a waste site. The weapon itself was completely destroyed.

In a spirit of wartime expediency, America built about thirty thousand nuclear weapons between the late 1940s and the early 1960s. During the Manhattan Project and the cold war, the military-industrial complex was careless about the garbage. It went into trenches, ponds, pits, single-walled tanks, railroad cars, and underground caissons at sites that were usually isolated. When enough rain fell on at least some temporary storage sites, the radioactive particles would get washed out. In arid locations, like Hanford and Los Alamos, most uncontained radionuclides didn't travel far because they sorbed—bonded with components of the soil.

In the early 1950s, the national laboratory in Idaho built a complex to store and bury nuclear refuse. Various other proposals for controlling radioactive waste were floating around as well and in subsequent decades the government investigated a number of them: putting the waste in a deep hole, or in a glacier, or on a remote island, or firing it into space. "We really looked into these ideas," Rip said. "Each one needed two or three pieces of unobtainium to work. In the case of shooting it into the sun, you needed unobtainium to get the nuclear waste out of earth orbit. Otherwise it would eventually wind up back on earth. But the rate of aborts of rockets was one in twenty. And we'd need *beaucoup* launches of really heavy stuff. And one out of every twenty payloads would burn up in the atmosphere or melt down on the launchpad."

The notion of dropping nuclear waste down a drill hole six miles deep was scotched because no one could be sure what the radiation and thermal energy would do to the surrounding rock or what the intense pressure and heat at that depth would do to radioactive material. Similar unknowns about deep sequestration of carbon dioxide from coal combustion abound. Disposal inside a polar ice sheet presented numerous drawbacks, including the expenses incurred by remoteness and adverse weather, not to mention instability and thawing. Islands not only tend to have a history of volcanism and earthquakes, they're also porous, and seawater seeping into their bedrock could corrode containers.

In 1955 the Atomic Energy Commission asked the National Academy of Sciences (NAS) to examine the problem of permanent disposal of all radioactive wastes—from defense, medical, and commercial sources. After two years of study, an NAS panel recommended long-term deep geological disposal in impermeable, stable geological formations: bedded salt, clay sediments, or granite tuff. But the United States felt no pressure to act. The USSR, some European countries, and the U.K., all of which by now had acquired reactors as well as medical radiological facilities and bomb factories, also began to realize that something had to be done . . . eventually.

"Everyone here and abroad in the nuclear field knew that in the future someone would have to dispose of the fission fragments that get removed from spent fuel during reprocessing," Rip said. "They didn't amount to much, so this wasn't an urgent matter. But once reprocessing stopped, and once it was clear that we really had to do something about defense-related waste, too, the national laboratories started researching the National Academy's recommendations. Among other things, I wound up looking into clays. Radionuclides just love mud. They like to bind with clay and other soils. Then they become immobilized."

"For how long?" I asked.

"At Oklo, the natural-reactor waste, which included plutonium that has now decayed away, has been geologically contained for nearly two billion years within a few meters of where it was generated. The whole ecosystem around the reactors remained intact for the one million years of their operation."

"But maybe that was a special case. Why can't we just leave the waste where it is?"

"In some instances, that's the safest thing to do. Moving it would cause more problems than it would solve. But waste is sometimes near populated areas or threatens water tables and a lot of communities want it gone. The sites are now well maintained, but in the very long run we can't guarantee that."

"Well, maybe scientists in the future would know more about what to do, so if we just left it alone . . ." I began.

"Probably better technology will appear, and that's an argument for safe interim storage while we try using things like fast neutron reactors like the IFR that can burn up waste. Some high-level residue will always be left over, and in the future the problem will be the same: How do you isolate nuclear waste from the biosphere until it decays down to harmless?"

"What a huge risk to our children, grandchildren—even our remote descendants!" I said. "They'll hate us for this. And there's just no solution."

"Right here in New Mexico, we already have the Waste Isolation Pilot Plant: one extraordinarily safe, secure nuclear waste repository in a salt bed. Sweden and Finland are planning to build repositories inside rock. And deep-ocean sediments below the seafloor are another possible location. Sure, future generations might spit on our graves. But it won't be because of nuclear. They won't forgive us for all the non-nuclear crap that will never decay, never break down. They'll hate us for burning up precious natural resources like coal and oil and putting trillions of tons of nasty solid and gaseous waste from them into the biosphere. They'll blame us for not controlling our heat-trapping greenhouse gases. The ocean will be rising rapidly and flooding coasts because we did not control *that* waste or stop burning fossil fuels even though we had a good alternative. We know a lot about nuclear waste. Thousands of studies have been done here and in other countries."

Rip maintains that nuclear waste from various sources is already being safely put away, and that long-term storage in central repositories is workable, and that even high-level nuclear waste can be stored in such a way that it poses virtually no risk to the public.

We're still quite literally paying for the reckless behavior of the government during our bomb-building spree, which, all told, has cost taxpayers nearly $6 trillion if you include the decontamination and dismantling of the "huge factory." Almost every state has affected sites. When the arms race ended, no equivalent nuclear waste-disposal race followed. The unaddressed problem was handed from one presidential administration to another, and sporadic efforts to initiate a coherent program suffered many setbacks, both technical and political. Agencies, administrations, and scientists were seldom united on any waste-disposal plan; funding was unpredictable; projects remained incomplete or were abandoned even if promising. Weapons facilities were not subject to most external regulation, and the public had little notion of what went on inside them until the 1970s, when communities near contaminated sites began to learn of radioactive waste in the water and soil. Hazel O'Leary, secretary of energy under President Clinton, made public a great deal of information, because she believed in transparency and hoped to reclaim the public trust.

In addition to government weapons facilities, power plants, and uranium mining and milling, hospitals and medical diagnostic and research laboratories have all produced or are still producing radioactive waste. Almost all of it is inaccessible to the public and almost all of it is low-level, but a small fraction remains hazardous. The nature of radioactive contamination is such that the process of cleaning it up can produce more contamination. It takes on a life of its own, as in the "Sorcerer's Apprentice" sequence in the film *Fantasia*. Imagine that you find a potentially lethal pool of radioactive sludge in your house. You don protective clothing and mop up the mess with a rag. The rag is now contaminated, so you put it in a barrel. But you may or may not have traces of sludge on your protective clothing and your shoes, so, to be cautious, at the end of the day, you add them to the barrel. Then you shower—and you leave radioactive traces in the water, and on the bar of soap, and the towel. All these go into the barrel. Eventually the whole bathroom must be torn down, because every floor tile, drywall panel, water pipe, and electric fixture has in theory become contaminated. In some cases, contamination has spread into the foundations, the surrounding soil, and the groundwater. Even the equipment that's used for decontamination is liable to become contaminated.

Bohr's huge factory has been mostly dismantled and its debris is being processed and shipped to storage sites. The cleanup of American military nuclear waste is the biggest environmental program as well as the biggest public works program in the history of the world, surpassing the cost of the Manhattan Project and the space program combined. This herculean task began in the 1980s and will take from three to seven more decades to complete. Some seven thousand buildings have to be decommissioned, and leaking storage trenches and tanks that have contaminated the soil and water around them must be remediated.

Cleanup at nearly eighty sites has already been completed, including the most contaminated ones, like a site in the town of Fernald in Ohio. There, at a factory in a rural setting, a defense contractor machined targets—that is, uranium metal fabrications that were shipped to reactors at the Hanford nuclear defense facility to be bombarded with neutrons until they turned into bomb-grade plutonium-239. Fernald's drinking water became contaminated with excess uranium—if ingested in quantity it can be about as toxic as lead—and a plume of radioactive material spread in the water table and soil. A citizens' group successfully fought for a government-mandated cleanup as well as for compensation amounting to $73 million. The DOE has built an on-site disposal facility and wastewater-purification plant; it showcases state-of-the-art technology for cleaning up nuclear materials. Eventually a lawn will replace the facility's contaminated buildings. Fernald's waste was originally designated to go to Nevada for permanent storage, but the state protested; now it's been shipped to sites in Utah and Texas run by companies under contract to the DOE. The EPA spent thirteen years and $1.4 billion to convert the site into a park with a hardwood forest, prairie, and wetland. Endangered species and other wildlife have returned.

In Colorado, Rocky Flats, where plutonium triggers for atomic bombs were once made, recently sent the last of its ninety-five thousand barrels of radioactive waste to the Waste Isolation Pilot Plant. The land is being restored to its original state with plantings of native vegetation; a portion of it will become a wildlife refuge. "There's still going to be americium and plutonium contamination in the soil there," Rip said. "Those folks were careless. But those are alpha emitters, and grass or some other groundcover is enough to stop alpha. Hardly any military spent nuclear fuel is left at Savannah, Hanford, or Idaho. It's been reprocessed—the uranium and plutonium have been removed and stockpiled. The residue in some tanks at Hanford will go into glass logs and they'll get put in deep geologic disposal."

The need to isolate high-level radioactive wastes from the biosphere has become a global issue. Ultimately, Rip feels, a solution that the public understands and finds satisfactory will have a major effect on the way nuclear power is viewed and therefore on the prospects for reducing human contribution to climate change.

16 32N164W

AS RIP COMPLETED HIS DOCTORATE in theoretical organic chemistry and chemical oceanography, some big chemical corporations tried to hire him. Disapproving of their indifference to the long-term hazards to health and the environment from the molecules that these companies create and that cannot be taken apart by nature, he refused. After being recruited by Sandia National Laboratories, he moved to New Mexico but still managed to find ways to keep going underwater: biologists needed samples collected from lakes, boats needed salvaging. And then, of course, there was his dive into that reactor pool.

By the early 1970s, when Sandia Labs began to act on the National Academy of Sciences' recommendations to research certain geological media for long-term storage of military and commercial nuclear waste, Rip led a team that investigated possible sites. One day in 1973, he heard about a sedimentologist named Charles Hollister from Woods Hole Oceanographic Institution, who claimed nuclear waste could be safely buried beneath the ocean.

Rip, in addition to researching clays and shales, had been leading a team reviewing scientific literature on Yucca Mountain. It was located in a geological basin in a remote part of the Nevada Test Site, which had been peppered by over nine hundred atomic blasts. The mountain was made of granite tuff, a foamy volcanic rock. "The choice of Yucca Mountain made sense politically, because it was already federally owned," he told me. "But it turned out to be the most geologically complex formation on the continent. Maybe in the world. But that's hindsight. Back then, we were just starting to study how to dispose of nuclear waste for ten thousand years. We had miles of data on the Yucca Mountain area that had been generated by nuclear tests. We were collecting all that information from the archives. But at the time the DOE had half a dozen other options on the table. So I wanted to hear what Hollister had to say about the seabed, even though it didn't seem to me like a good bet."

Hollister, who had developed a technique for extracting long cores from deep ocean sediments, enthusiastically described samples he had collected that revealed a vast, red-clay formation that had maintained great stability and uniformity over millions of years—far longer than the half-lives of almost all the radionuclides in nuclear waste. Made up of particles as tiny as talcum powder whose small size prevented water from circulating, the clay had low permeability and the consistency of peanut butter. It was the best clay in world for sorbing radionuclides. A pointed steel canister containing high-level nuclear waste that was dropped to the ocean floor would sink through this muck to a depth of thirty meters. The continuous rain of sediments from above would only bury the canister deeper. The location of each canister could be monitored.

One of the best potential sites was about six hundred miles north of Hawaii, where the ocean floor is a thick blanket of dirt that forms gently rolling abyssal hills over crustal rock. This piece of real estate, a marine desert situated at the center of a tectonic plate, lies in perpetual darkness under about four miles of water and covers about thirty-nine thousand square miles. The temperature stays at 2 degrees Centigrade—just above freezing. Over millions of years, skeletons of sea creatures, micrometeorites, volcanic ash, and fine grains of dust have drifted down to form a bed of viscous clay 325 feet deep. No undersea plants undulate here, no schools of shimmering fish twist and turn. The marine life in evidence in subduction zones—areas where the edge of one tectonic plate is sliding under another—is absent here. Natural background radiation is much higher than it is on dry land. Currents here are feeble: stillness reigns. This abyssal plain, vertically and laterally uniform, has remained undisturbed by any volcanic or seismic activity for thirty-five million years and is likely to stay calm for millions more. The mud is a self-healing, quicksandlike mixture that promptly repairs any breach and forms a seal around any object that happens to plummet down and sink into the bed of ooze; there it would remain for millions of years. Even individual molecules that may arrive here become immobilized by the stickiness of the clay, the enormous pressure, and the continuous hail from above. Radionuclides could not escape. The unromantic designation of this virtually extraterrestrial region on our own planet is 32N164W.

Rip joined Hollister to put together a team drawn from oceanographic institutions to explore the feasibility of the idea and to pick out potential undersea sites to study. These institutions ultimately became supporters of the project.

When Leo Gómez first heard about the project during a job interview with Rip, he said to himself, "Nuclear waste? Not in *my* ocean!" Rip welcomed that attitude. He later appointed Gómez the marine biology program manager. When Gómez spoke with deep-ocean biologists, the initial response he always got was: "Not in *my* ocean, you're not!"

"So they refused to work on the program?" I asked.

"No!" Gómez said. "You couldn't keep them away. They were committed to proving that it was a really bad idea. Rip always said to everyone, 'If you have a showstopper, bring it to us, and if you can prove it, *we'll stop the show.*' They were all really good scientists. If you're a good scientist, when new data comes in, you change your mind. The biologists in our program conducted many studies on bacteria and other marine life and concluded that even if the waste somehow escaped—really, it never could because it would just get buried deeper as sediments rained down—it would have no effect on any fish or any other life. Even if over thousands of years the canisters corroded, the radionuclides in them would bind with the clay and remain deep down in the sediments. We looked at everything. In the end we all thought it was a really good plan."

Thirteen nations were trying to find a place to put their growing accumulations of nuclear waste, and after Rip and a colleague took their plan and preliminary research to Europe, to the Nuclear Energy Agency of the Organization for Economic Cooperation and Development, the OECD established the international Seabed Working Group in 1975 to pool data, talent, and resources and to provide information that would help nations and international agencies assess the feasibility and long-term safety of the repository. Rip presided, coordinating all the contributions of the participating countries and taking responsibility for broader policy decisions. Rip usually refers to the entire effort, from the initial Sandia program to the international one, as Subseabed.

"Surely there must be some risks in putting nuclear waste in the ocean," I told Rip.

"Well, it was with Subseabed that we began doing probabilistic risk assessment. The location is remote. It's sabotage-proof. You can't get there. And, even if you can, the canisters are under almost four miles of ocean and buried one hundred yards apart one hundred feet deep in the clay. And even if you could find and pull them out and remove the spent fuel, its plutonium and uranium would have to be separated out in a reprocessing facility."

I phoned Edward L. Miles, a political scientist whose career has been in international law, environmental policy, and oceanography. Originally from Trinidad, he is now Virginia and Prentice Bloedel Professor of Marine Studies and Public Affairs in the School of Marine Affairs at the University of Washington and serves as a consultant to several international agencies and as a senior fellow of the Joint Institute for the Study of the Atmosphere and Ocean. After a somber discussion about the deleterious effects of carbon dioxide and global warming on the ocean, including species die-offs that are threatening fisheries worldwide, we turned to the Seabed Working Group. He had chaired its legal and institutional task group and advised the executive committee.

His tone brightened. "I found the idea of the dynamic duo—Rip Anderson and Charley Hollister—absolutely compelling, because I was, and am, deeply interested in the problem of nuclear proliferation. Subseabed disposal of spent nuclear fuel was the way to put the lock on the back end of the nuclear fuel cycle. You couldn't find a safer and more inaccessible place. Rip ran a group of about two hundred highly talented scientists and engineers, including quite a few prima donnas. It was mind-boggling to me how he melded the U.S. group and the international one."

In a paper on his work on environmental matters, published in *Policy Sciences*, Miles expressed his appreciation of Rip and the example he set of how to manage "very large-scale multidisciplinary, interdisciplinary programs of great complexity," and listed Rip's methods: "Always focus on the problem. See it whole at all times and know in great detail how the output of each sub-group will fit into the overall picture. Convey the big picture to participants at every meeting so that each one can see how the results cumulate and the group can identify gaps and inconsistencies that must be resolved." Another principle: "If the idea you are working on is novel and controversial, don't be afraid to create a small group of people whose job it is to invalidate the idea itself. If they fail to do so by the end of the project, you will have withstood the toughest test. If they succeed they will have saved you a lot of future pain." And Miles spoke of Rip's skills at keeping people involved: "Treat them well and gently and foster the intellectual excitement of the chase to such an extent that the very busy cannot stay away."

As we chatted, Miles reminisced. There had been frequent meetings. "We'd have three or four days of densely packed presentations. Then Rip would get up and summarize where the science was and he'd give a very clear, coherent discourse on what the problems were and what we needed to do next and how. I was interested in how he could be the maestro of such a turbulent orchestra. He told me he got the best people for each of the jobs that needed to be done. He said, 'That way, I knew that I'd get the best possible science and nobody could kill us because *we* had the guys who were doing the best possible work.' "

Many of these highly qualified representatives from oceanography, geology, biology, materials science, mathematics, physics, political science, sociology, and other disciplines were not used to talking to people outside their field, let alone sharing data with them, but Rip made sure plenty of cross-pollination went on.

"He was the first person to go about solving the nuclear waste problem in a very systems-oriented way," Miles continued. "I run a team now—an integrated natural science, social science, law team, twenty-five of us—based on how I watched Rip run us. Let me tell you, the recipe works. He was phenomenal. In my experience he was unique."

Others associated with the program told me how profoundly Rip had changed their way of thinking; they referred frequently to his leadership, democratic approach, originality, and resistance to bureaucratic pressure. He inspired energetic collaboration on a plan that they and their children, grandchildren, great-grandchildren, and residents of the distant future could live with.

"Even though we knew the waste could never escape from the sediments, we had marine biologists and microbiologists spend ten years studying what would happen just in case it did, and whether it would affect marine life or get into the food chain," Rip told me. "They couldn't find any way that it would. After doing all the research and probabilistic risk assessment, we found that subseabed disposal was the best possible option. It still is—all the others are subsets. We just needed to do some actual long-term tests to get further proof."

When the research phase was completed, the program scientists informed the DOE that they were very encouraged but that most of the work had been done only on models. "We were all convinced that we needed to put a couple of waste canisters of our design into sediments and measure the responses over twenty years, to make sure our predictions were correctly calibrated," Miles said. "We told the government that it could not initiate the program until this in-situ experiment was done. Meanwhile, the radiological assessments made by the EPA, national and international groups, and the European Commission were great. The International Atomic Energy Agency considered the subseabed option as simply another form of deep geological disposal. In everybody's final assessment, the program came out several orders of magnitude ahead of the next best option—Yucca Mountain."

No one in Washington had expected such a relatively small and quirky program to compete seriously with the nuclear power industry's choice, the Yucca Mountain project. The Nuclear Energy Institute (NEI) assumed that a land-based repository in Nevada would be more convenient and in operation sooner than one that, to ensure safety, required a twenty-year test beneath the ocean floor. However, in the mid-1980s, the NEI's public relations campaign, aimed chiefly at decision-makers, had the unintended consequence of scaring the public about nuclear waste at power plants. "The industry denies (of course) that they inflamed a fairly quiet public concern about the safety of locally stored waste," Theodore Rockwell, the nuclear engineer who for many years led Rickover's technical team and today is a gadfly to the nuclear industry, wrote me in an e-mail. "But there were large, expensive ads in *The New York Times, The Washington Post*, etc., saying that having the waste in one, well-guarded place was a lot safer than having it in seventy or so locations where it was vulnerable to terrorist attack. This, despite the fact that even such virulent antinukes as the Sierra Club's Sheldon Novick, author of *The Careless Atom*, wrote of radioactive wastes in *The Electric War: The Fight over Nuclear Power* that 'it is difficult to see in what way they are any more or less hazardous than other poisons produced by industry.' "

Congressional subcommittees had been supportive of the Seabed Working Group, thanks to persistent efforts on the part of Hollister and other advocates, including oceanographic and other academic institutes. Rip, bucking Sandia's stricture at the time on lobbying, made trips to Washington and talked informally to committee members—a defiant move that set back his career at the lab. However, all these efforts could not best those of the NEI. It had a good friend in Louisiana senator J. Bennett Johnston Jr., the powerful chairman of the Committee on Energy and Natural Resources and an opponent of the subseabed project. Johnston originated the bill that in 1987 designated Yucca Mountain as the nation's sole repository for high-level nuclear waste.

"The DOE and the government removed us as a competitor," Miles said. "The alligators got us—killed us because our science was too good! That didn't solve their problems, however. Those still exist."

The soundness of the science, the lack of showstoppers, and even the support of marine biologists, oceanographic institutions, and some receptive environmental groups, who advocated more research, were all met with indifference at the DOE. "To be killed because you're better than the alternative—that's really hard to take," Miles said. "Then DOE lied—claimed it didn't have the money for us. We were a mere fraction of what they were spending on Yucca Mountain. Without any money the seabed research couldn't continue. Without the United States in the lead, the other countries weren't going to continue."

Meanwhile, twenty years have passed since the choice of Yucca Mountain. The waste facility was supposed to open in 1998, but delays have continued to plague the project.

"As a model and as an example of leadership at work, the subseabed project still reverberates around the world," Miles told me. "I've never heard anyone say that the science was shoddy. On the contrary."

"Do you think there's any possibility of a revival of the option?" I asked.

"There would have to be a general perception that we were in really serious trouble with global warming and needed quick fixes to slow down the rate of change. So people would be asked to be willing to resuscitate the nuclear power industry. You would have to think through the whole system, including waste disposal, and think through all the options, given the very troubled experience of Yucca Mountain. I advocate nuclear energy as a transitional strategy—the span of that transition would be about a hundred years. Just to buy some time while we figure out what we can do to curb our greenhouse gas emissions. A multinational nuclear waste repository under the seabed would also reduce the risk of making atomic bombs out of reactor by-products better than a land-based solution could."

Veterans of the Seabed Working Group here and abroad still consider their work a positive preliminary step toward an international agency to set up, regulate, and monitor long-term waste disposal. At present all countries continue to use interim storage methods. A number of international supervisory agencies, such as the International Atomic Energy Agency, are experienced in meeting the challenges of forging the necessary international agreements. International law forbidding the deposit of man-made materials in deep sediments would also have to be changed. The Seabed Working Group had readily provided information about its work and its conclusions derived from modeling and risk analysis to environmental groups in order to hear their concerns and get their advice. Greenpeace eventually used selected bits of the research to pressure the representatives of the international London Dumping Convention, which meets periodically to legislate protection of the ocean, to eliminate the possibility of storage of nuclear waste under the seabed until at least 2025. John Kelly, a political scientist specializing in science and public policy who worked as the national policy adviser on the subseabed project and lobbied strenuously to keep it alive, told me, "Seabed is a classic example of an idea ahead of its time. Technically elegant and simple, but so in conflict with the culture, with the way most people see the world, that it can't be accepted until perspective shifts. That will be when political, national, and international acceptance occurs. Right now, people don't distinguish the fragile coastal areas from the deep ocean, a desert with something like one gram of carbon per square meter." He went on to note that environmental groups often object to the large-scale models used to assess the safety of nuclear waste disposal, believing the uncertainty to be too great. "But at the same time they demand huge economic international changes based on climate change models created by the same methodology and subject to similar uncertainties."

Subseabed participants believe that if the DOE had supported the program it would have gone on to demonstrate empirically to the world the inherent safety of its method. "Instead of shutting us down in 1987, if they'd let it continue at even ten million dollars a year, we'd have shown that we had all the work in place," Rip said. "We'd already done a risk assessment that was more detailed and complete than any of the risk assessments that had been done for dry land repositories. We had the transport design—the casks, the canisters, the double-hulled ship. Even in the unlikely event of a bad shipwreck, the canisters would remain shielded and contained in casks that could float. By now we'd have proven that the sediments could safely contain all the spent nuclear fuel in the world. Or Seabed could have been a backup to Yucca Mountain."

John Kelly believes that if the seabed project had survived it would have strengthened support for Yucca Mountain. "Every senator from a coastal state would be forced to take a skeptical view of seabed. Yucca Mountain would look good to them. Congress short-circuited the idea of having three sites and took away the ability to answer the question: 'Yucca Mountain is safe compared to what?' Without something to compare it to, people have difficulty making a decision. If you can't prove a need, then only zero risk is acceptable. If you can prove a legitimate need, people will accept a risk. But the Nuclear Energy Institute put out those ads about the dangerous buildup of nuclear waste at power plants. That really backfired, because that started the public worrying that the spent fuel wasn't being taken care of. And that perception has never gone away."

Les Shephard, whom Rip had recruited to be the Seabed Working Group's sedimentology chief, went on to manage Sandia's role as lead national lab at the Waste Isolation Pilot Plant. Now a vice president for energy, resources, and nonproliferation at Sandia, he thinks that Rip's work continues to have an impact on nuclear waste disposal and that one day his vision will be realized. "There has to be a move toward a multinational repository system," he said. "There are suitable subseabed sites in the North Atlantic. For each of the nuclear countries in Europe to build a separate repository on their own would be very difficult and costly. We've invested billions and billions of dollars to do that here in the United States. Those countries can't do that. Globally there will be a move toward reprocessing to reduce volumes of waste. Given the small quantity of nuclear waste remaining after reprocessing that will need to be isolated, and the many nations that will have to do that, there may be a need for future disposal in the deep-ocean sediments. I still have seen nothing that would suggest it's not feasible. It was well ahead of its time—and it's something the world should benefit from. Mohamed ElBaradei, director general of the IAEA, says that we need multinational repositories. Well, Rip Anderson was there thirty years ago. His vision has been promulgated."

"Within a few dozen years we'll be using Subseabed," Rip said to me one day when we were out by his corral. "It would be easier and cheaper than boring big, deep holes in rock. The difference between storing nuclear waste under the seabed and under the ground is perceptual. You can see the water in the ocean, but you can't see the water table under the land. Subseabed presents a political problem, not a technical one."

17 THOSE WHO SAY IT CAN'T BE DONE

AFTER MONTHS OF WRITTEN REQUESTS, security clearances, and detailed arrangements, and after a couple of hours on the road, negotiating one checkpoint after another, and after displaying ID and permission slips to one pack of heavily armed guards after another, I was finally inside a hollow mountain—Yucca Mountain, that is.

Rip, Marcia, and I sat on folding chairs in an alcove off a tunnel with a high, arched ceiling. An engineer from Los Alamos National Laboratory was pointing to pictures and diagrams. He wore a hard hat, as we all did. We also wore clear plastic safety glasses with big lenses, and sturdy boots.

Overhead, the gigantic metal tubes of a ventilation system, designed to provide fresh air and blow away natural radon, made a loud, steady whine that had a soporific effect. As the engineer was speaking I was startled by a roar familiar to me from decades of riding the New York subway: half a dozen silvery train cars rushed past.

When we began the nuclear tour, I'd heard nothing good about Yucca Mountain. The antinuclear-power argument went something like this: Nuclear power is bad, nuclear waste is bad, lasts forever, and there's no solution, and therefore Yucca Mountain as a repository is bad. Greenpeace took the position that the people who generated nuclear waste should keep it near them as a punishment. (I suppose this attitude lets off the hook Greenpeace and other consumers who use electricity, some of which comes from nuclear power, to disseminate their opinions.) Citizens' groups and politicians stated that transporting the spent fuel from nuclear plants was terribly risky, exposing schools, churches, synagogues, and pregnant women to excessive radiation; a Las Vegas newspaper editor envisioned trucks of nuclear waste overturning on the Strip and irradiating it, or terrorists shooting at the casks; and the State of Nevada argued that people living along transportation corridors could theoretically be exposed to radiation greater than background level by trucks carrying spent nuclear fuel. In 2006, a federal judge rejected the state's arguments against waste shipments, ruling that some claims were not ripe for review and others were without merit.

All opponents say that the waste would remain radioactive for possibly millions of years. Some claim that one teaspoon of spent nuclear fuel could kill half a million people; that a volcano could explode the repository, sending lethal waste all over kingdom come; that an earthquake could crack open the mountain and release its deadly contents; that floods were certain to wash out the waste and contaminate the water supply all the way to Las Vegas. Support in the technical community for a Yucca Mountain waste repository has been lukewarm. Above all, Nevadans never tire of referring to Senator J. Bennett Johnston's legislation—passed over the howls of the state's representatives at the time—as the Screw Nevada bill. The state does not want Yucca Mountain and intends to fight the federal imposition of the repository upon Nevada for as long as it takes.

After 2002, when the president and Congress gave the DOE approval to continue work on the site after obtaining assurance from the DOE that the geological formation at Yucca Mountain could safely sequester nuclear waste, project officials started putting together its licensing application. As part of a multistep process, DOE and its contractor, Bechtel SAIC, must demonstrate that the repository will meet both NRC and EPA standards—defined as a level of safety consonant with "reasonable expectations." The technology must comply with federal standards for nuclear waste protection of the kind that the NRC enforces for nuclear power plants, and the project must prove that the repository will pose no threat to public safety and the environment for millennia.

Rip is among a handful of people who have successfully gotten a long-term, deep geologic nuclear waste repository certified and open. Presently, there is only one such facility in operation on earth. The Waste Isolation Pilot Plant (WIPP) stores transuranic waste—mainly plutonium-contaminated materials from bomb-building—rather than commercial spent nuclear fuel, but the fundamental problem to be solved was the same as at Yucca Mountain: How can hazardous radionuclides be sequestered from the biosphere until they no longer pose a risk? As I mentioned earlier, as part of the Yucca Mountain Project's preparation of its licensing application, Rip had joined a panel of experts from national laboratories and academia to perform an internal peer review of the probabilistic risk assessment of the science that had been done thus far at Yucca Mountain.

I occasionally asked him how it was going. He'd explain that when any program of that magnitude was in progress there were always lapses and that rigorous attention would resolve them; this had occurred with WIPP during its journey toward licensing. "The most important thing is to be objective. We have to make sure that all the science has been done correctly, that there's quality assurance on all the data, and that the conclusions are sound. I keep an open mind as I go over the material and I identify matters that need to be cleared up."

When the licensing application is completed—a lengthy and laborious process in itself—the DOE will turn it over to the NRC, which will take about five years to study the document. If it's approved, then the next application will be for permission to operate the repository, and the third application will be to close it in three hundred years. The portals will be sealed up, the area returned to its previous environmental state, and warning markers erected to warn future visitors of the risk that lies hidden in the heart of the mountain.

Some new discovery might be important enough to terminate the project, but all the experts I talked to thought that despite the complexity of the geology, the research showed that the risk of waste escaping was extremely small. For decades the Nevada Test Site has been studied ecologically, biologically, zoologically, botanically, seismically, geologically, hydrologically, atmospherically, and radiologically. About a hundred thousand samples have been collected, and a huge amount of raw data has had to be "qualified"—jargon for, in effect, tracing the origin of every sample.

John Kelly, formerly of the seabed project, specializes in policy covering nuclear waste and the environment. He runs a consulting firm that has been contracted to deal with various aspects of the Yucca Mountain Project. "Every measurement taken has to have a pedigree, just like a purebred horse, with dates and certificates," he explained. "Qualification of data means that you have to establish a chain of custody. For example, the date a rock was sampled, when the sample was logged in, and so on."

The goals of Yucca Mountain research have been traceability and transparency. "You have to be able to track the data back to the sample and ensure that there are no errors. And the findings have to be easy to follow and clear. What the project *cannot* do is control how outsiders select certain pieces of data out of the huge context that has been created and perhaps exaggerate them or come to conclusions that the scientists find wrong or irrelevant."

The State of Nevada, along with anti-repository groups, had recently accused the project of fraud, claiming that a few scientists from the U.S. Geological Survey (USGS) who had been doing research for the project had suggested in e-mail exchanges that they'd fabricated data. The DOE had made the e-mails public in 2005. Senator Harry Reid of Nevada, then Senate minority leader, declared, "It is abundantly clear that there is no such thing as 'sound science' at Yucca Mountain." Rip disagreed, saying that most of the science is sound and that it would *all* have to be sound or NRC would not grant a license. Determining whether the science was sound was the sole purpose not only of the internal peer-review panel he was on but also of several government overseers: the DOE's Office of Civilian Radioactive Waste Management, the Nuclear Waste Technical Review Board, the NRC, and the EPA—not to mention Congress. Furthermore, the Yucca Mountain Project had already been subjected to one international peer review by the Organization of Economic Cooperation and Development's Nuclear Energy Agency in concert with the UN's International Atomic Energy Agency, and there would be more external, national, and international peer reviews in the future. "Nobody should ever even make a joke about making up data," a scientist on the project told me, "but even if the USGS researchers actually had fabricated that small bit, that would have *zero* impact on the performance and risk assessment of the repository." He considered the incident a matter of bad public relations—something the project has suffered in abundance.

A U.S. attorney investigated and dismissed charges against the researchers in 2006. Kelly told me that the brouhaha had been a misunderstanding. The scientists had been asked to qualify data—to trace sources. "On e-mail they shared their frustration, saying that maybe they would just make up the missing data. As they understood among themselves, the information certainly did exist." He explained that the scientists were busy with other matters and it was time-consuming at that moment to retrieve the data. "To them this request seemed trivial. You could go back and, with considerable effort, find all that same information outside their notebooks. But the news media didn't cover any of that."

He went on to remark that the story showed where the Yucca Mountain Project stands these days. "It's at the intersection of science with the licensing process and safety culture. In the safety culture, everything is checked and double-checked and there are forms that have to be filled out upon completion of procedures. U.S. Geological Survey people are quasi-academic scientists who've run smack into the nuclear culture, which is much more oriented toward engineering and technology than toward pure science. You have the same collision between the national laboratories and the nuclear culture. The system of data qualification is intended to prevent errors but is incapable of preventing conspiratorial efforts to falsify data. In other government programs there are procedures that are intended to prevent any one person or group from controlling the data. That's not the case with Yucca Mountain. If you step back and take a global look at how you control data, you see that Yucca Mountain is very open."

He added that the problem with Yucca Mountain was not about a few e-mails. "The real problem with the project is that the objectives have undergone revision and the conditions under which we're trying to achieve them can be mystifying. There are a few strongly antinuke people trying to stop the repository, and they're the ones with a clear objective."

In the spring of 2003, while Rip, Marcia, and I were having our breakfast before dawn in the lobby of a hotel in Las Vegas before we set out for Yucca Mountain, we watched Condoleezza Rice on television denying that the recent invasion of Iraq had anything at all to do with oil.

We then headed in our rented car northwest through the nation's fastest-growing city, where over two million people burn enough oil and other fossil fuels to create smog so bad that Las Vegas now ranks among the five worst urban areas in the country in air quality. At the farthest reaches of the city's sprawl, earthmovers gouge out craters in the desert floor and giant cranes heft slabs of prefab concrete, and practically overnight there appear, as if dropped from low earth orbit, palm-lined boulevards, artificial lakes, apartment complexes with names like "Falling Water" and malls with splashing fountains and wide plazas surrounded by chain stores: Circuit City, Borders, PetCo, Staples, and the paint not yet dry on the Starbucks and Banana Republic. Each week, five thousand new residents move here and every week a new school opens. And nearby desert serves as a staging area for still more construction equipment, road graders, and piles of building materials that will be deployed in what surely must be the nation's biggest ongoing domestic invasion of open land.

Las Vegas has lit up its neon for decades with hydroelectricity from dams on the Colorado River. However, a prolonged drought, which is predicted to worsen, is shrinking the water supply. Boat docks are now far from the edge of the water. Nevada has no nuclear plants. A big coal-fired station provides Las Vegas with some of its electricity, and plans are under way to construct big, new plants in the state. Nevada gets 86 percent of its electricity from fossil fuels, mostly from coal, and the rest from hydropower and geothermal power.

The sun rose, revealing a sagebrush plain bordered by mountains that were carved with broad alluvial fans and that progressively turned dozens of shades of blue under a soft, hazy sky. Like so much of the West, the region—basin and range country—had once been an inferno of explosions and hurtling lava. Some buried volcanoes here last erupted two and a half million years ago. "Deep down, the tectonic plate we're on is being stretched and pulled apart," Rip observed, "and so magma wells up along the fault zones and you get a weak form of volcanism—vents, ash flows, basalt, volcanoes. Not big ones like Mount Taylor, but cinder cones—some of them less than one hundred thousand years old. That's geologically fresh." Other mountains around here, like Yucca Mountain, are basically piles of volcanic ejecta, including the foamy rock called granite tuff, considered good for testing underground nuclear weapons—because it absorbs the shock wave—and good for sequestering nuclear waste.

Back in the 1970s, some Nevada politicians had made a bid to have a national nuclear waste repository at the Nevada Test Site in the hopes that eight thousand jobs would materialize in the then sparsely populated state, which had lost out on federal money after the testing of nuclear weapons stopped and thousands of jobs vanished. By the 1980s, Nevada didn't want a waste repository—new jobs were being created by a major influx of new residents. Meanwhile, the government had decided that nuclear waste from defense activities would now be stored separately from nuclear waste generated by commercial power plants. Plutonium-contaminated military waste and materials from the DOE cleanups would go to the WIPP, and Yucca Mountain would, despite the protests of Nevada's then small band of representatives, store civilian spent nuclear fuel along with a small portion of high-level military reactor waste. That decision, since relaxed somewhat, had to do with politics; military nuclear waste was not somehow fundamentally different from the kind coming from power plants. A radionuclide is a radionuclide, whether natural, man-made, civilian, or military in origin.

After reprocessing stopped in 1977 and used fuel assemblies began to accumulate in waste pools at nuclear plants, representatives for utilities began asking the government what to do about the eventuality of running out of storage space. In 1982, all nuclear U.S. utilities—actually, their customers—began paying a small tax per kilowatt-hour that would go toward a new storage facility. Congress chose Yucca Mountain for that purpose in 1987. When the DOE defaulted on its promise that the repository would be accepting spent nuclear fuel by 1998, the utilities, expecting their waste inventory to double by 2035, began suing the DOE for its failure to meet the deadline. In one lawsuit, utility companies insisted that the DOE not only stop taxing them for the repository but also reimburse them for storage costs. The utilities won, and their motive for pressing for a national repository diminished. At present, to maintain the spent fuel in temporary storage at nuclear plants costs $1 billion a year; utilities are building more on-site concrete casks for interim storage.

The Yucca Mountain repository as conceived two decades ago is no longer capacious enough for the amount of spent fuel that is now antici-pated to be generated before the repository closes; it's been authorized to store up to 70,000 metric tons, and some say that there is no technical reason why new chambers could not be excavated to hold up to 120,000 tons, maybe even 200,000. But any new additions would be likely to trig-ger lawsuits from watchdog groups and the state, causing further delays.

People who had once worked on the Yucca Mountain Project or were still involved told me that the biggest technical issues are the area's com-plex terrain, which is seismic and volcanic, and unexpected water that passes through the mountain.

After learning about the deep-ocean desert sites the seabed project had studied and how they were characterized by homogeneous layers of viscous clay and by multimillion-year-old calm, and about the imperme-able, stable, multimillion-year-old WIPP salt bed, I better understood what Rip meant by the virtues of storing nuclear waste in a geological formation that was uniform, consistent, and predictable. So why had Congress chosen Yucca Mountain?

"It was mainly political," Rip said. "It was subjectivity, not science. Yucca Mountain wasn't chosen totally on its scientific merit. It was a political compromise. The government owned the land, it was already contaminated from nuclear tests, and there was already a lot of scientific information about the land available. With Yucca Mountain you have very complicated geology that's difficult to understand. Later, when work to characterize the site began, infighting began among several national labs as to how to do the science for something that would be used for ten thousand years."

"Did the people in charge of Yucca Mountain apply the probabilistic risk assessment methodology you'd developed?"

"No. And Yucca Mountain had no single national laboratory to super-vise and vet all the science done there. Sandia had been the lead lab for Seabed and still is for WIPP. With a lead lab in charge at Yucca Mountain, the other national labs could be contracted as needed to provide inputs for performance assessment."

For about two decades, the DOE and various national labs and contractors coped with the problematic geological picture while engineering the means to adapt to it, all the while without any lead lab as an organizing force. Most of the individual issues have been studied exhaustively. The problem, according to many in the technical community, has been the drive to hew to the most pessimistic, conservative position along with the lack of understanding of how the issues are coupled together.

Politicians and Nevada activist groups say the science doesn't support having the repository there. "At this point, it's unclear what the science supports," Rip said.

"What's it like to work on a project under these conditions?" I asked.

He expressed sympathy for former colleagues from Seabed and WIPP who were now characterizing the site to see if it was safe. "They're having a tough time. The jury is still out. We work as hard as we can reviewing the analyses, but there are still a lot of questions. The companies that the government contracted to come in here tried to speed up their work as much as they could. They studied everything, but without a lead lab's guidance about what was important. And there was no prioritization of the work of the kind we'd come up with on Seabed and WIPP, using probabilistic risk assessment to model the whole picture and determine which studies were relevant and which ones could be eliminated."

"How could this situation have come about?"

"Well, the DOE people in charge of site selection didn't know just how problematic the geology at Yucca Mountain was. They could have said, 'Let's go find a simpler geology.' But the political pressure was on. In the early 1980s we were also looking at a salt bed in West Texas as well as a basalt formation at Hanford in Washington. The salt bed had a uniform, stable geology, and the Hanford basalt hadn't been characterized. A complicated political process led to their being ruled out. The single most important factor in choosing Yucca Mountain was this: The federal government already owned the land, and it had already been crapped up by atomic tests."

Had teams scoured the land and the literature to find every possible reason that the repository would fail and to apply probabilistic risk assessment in order to have a broad range of data and to reduce uncertainty? Had members of various disciplines communicated with one another? Rip shook his head.

After considerable controversy, in 2006 the project underwent a major shift, thanks to the DOE's appointment of Sandia National Laboratories as lead lab. It is presently revamping the data and calculations and establishing at Yucca Mountain the methodology that had worked so well for Seabed and WIPP. But until then the project struggled with many limitations, and on my visit I heard plenty about them.

Rip showed me a statement he'd composed, so that I could better comprehend how the project's conservative assumptions had led to far too many uncertainties. "Historically, performance assessment analyses were done in a deterministic manner," he wrote. "That means using a single value for each input in an equation or group of equations that represent some scientific process. The analysis was carried out only once to give a value to be compared to a standard or regulation. As people became more knowledgeable with deterministic analyses, they began to 'game' the system. That means that they would argue over the chosen values as being too large or too small, depending on whether they were for or against the project." He went on to say that people were gaming the system again by arguing, in essence, that "my conservative position is better than yours" and that the problem lay not in incorrect values but rather in the conservative position itself: "This subjective, conservative methodology, if followed by the industry, will move the science of performance assessment back many years—back into the dark ages." Because of the subjectivity of the conservative position, anyone trying to do an analysis can't identify or agree on the most probable case—the most normal outcome. Therefore he or she can't arrive at a margin of error for an analysis or a position regarding the safety of a facility. This ultimately leads to public apprehension. Rip wrote:

Since the conservative position is a completely subjective one, high levels of confusion occur to the reader and anyone who is trying to understand the calculations, especially those who are trying to understand how safe the repository really is. Even review groups who have experts who are leaders in the field of performance assessment and have worked in the field of probabilistic risk assessment for nuclear waste facilities for many years cannot clearly understand how safe the repository may be from the calculations using this methodology. Regulatory groups have also criticized the conservative methodology.

The fundamental reason for the huge uncertainty is that there is no way to advise the scientists on how to be consistent in the estimate of the size of a conservative position or for a program to be able to measure the degree of conservatism to be used by an individual scientist. In the Total System Performance Assessment review, it has become very obvious that the level of conservatism has and always will differ with different scientists and different scientific disciplines. Therefore, the resultant analysis using this methodology will be of questionable value for assessing how safe the repository is.

Without a strong demonstration of repository safety, the public will not accept Yucca Mountain. Now that Nevada has a far bigger population than it did twenty years ago, it has more representatives and they've acquired louder voices. Harry Reid, who rose in 2007 to the powerful position of majority leader in the Senate, has remained adamant. "The Department of Energy wants to move the most deadly, high-level nuclear waste in America to a dump at Yucca Mountain, Nevada," he's stated on his website. "Ever since I was elected to Congress, I have been fighting this project because it threatens our health and safety. Nevadans are overwhelmingly opposed to this Project, and some have been led to believe that its completion is inevitable. I disagree. We have solid scientific evidence behind us, and I believe we can defeat this ill-conceived project." In the end, it may come down to a question of whether the federal government has the right to impose upon a state a facility it does not want.

We continued driving north and west on an absolutely straight highway across sagebrush flats. Rip described the construction at Yucca Mountain as a colossal effort that would require bringing in water, electricity, roads, and concrete plants and up to two thousand workers. By comparison, the gambling industry here employs five hundred thousand people. When the Yucca Mountain facility is completed and in operation, it might require about eight hundred workers. But the final designs and concepts for it are still speculative. One leading project scientist told me, "Our job is to take care of what we have now. Our attitude: We've got waste. Let's go bury it."

On the approach to the Nevada Test Site gate we passed a tall chain-link pen where protesters have sometimes been put. Nevadans, who remember fallout from atomic tests being downplayed by the government, tend to assume that they are being lied to about safety, as are antinuclear groups around the country who occasionally turn up to demonstrate against the nuclear waste repository plan. When Yucca Mountain was officially declared by the president and Congress to be a suitable storage site, the focus of the opposition shifted from the facility itself to transportation of spent nuclear fuel, which is classified as high-level nuclear waste.

Commercial spent-fuel assemblies arriving from nuclear plants will take up about three quarters of the space in the repository, leaving the rest for high-level defense waste like DOE, research, and military reactor fuel now in pools like the one we saw in Idaho, and vitrified residues from the years of reprocessing. In the 1950s, when, as part of the Atoms for Peace initiative, the United States provided reactors to universities and labs around the world, the government promised to take back the spent fuel. Although we didn't keep that promise, these days it seems wise to retrieve foreign spent fuel.

"Bringing the waste here is the safest part of the whole story," Rip said. "There's a very low probability of any event that would cause exposure or latent cancer deaths as a result of transportation of the waste—the trucks and casks are so overbuilt, so overengineered." In 2006, the National Academy of Sciences (NAS) indeed concluded after study that risk from transport of nuclear waste was minimal. To date no radiologically related injuries or deaths have been reported as a result of nuclear waste transportation accidents, and no releases from accidents involving shipments of high-level nuclear waste have occurred.

Spent-fuel assemblies would be hauled from nuclear plants by rail or truck to Yucca Mountain using vehicles and containers that meet NRC and Department of Transportation regulations. The shell of a nuclear waste cask is fifteen times thicker than that of a gasoline tank truck; it must have three inches of stainless steel as well as thick radiation shields. Nothing can escape the double-shelled, impact-resistant steel casks, even in the worst collision. Furthermore, the transportation specialist hired by the State of Nevada to highlight problems acknowledges that these casks "are among the best containers that humans know how to make to contain hazardous materials."

They've been rammed into concrete barriers, T-boned by speeding locomotives, dropped from high altitudes onto long steel spikes, plunged into churning rivers, submerged for extended periods, burned with jet fuel for ninety minutes, and otherwise roughly handled. They've held up. Infant car seats are much more loosely regulated.

The security of the casks, which would travel to Yucca Mountain in heavily guarded caravans that would avoid Las Vegas altogether, surpasses that of other highly hazardous materials moving around by truck and rail today. Before nuclear materials are granted access, public routes must be approved and vehicles and casks must conform to strict safety requirements. Even if a cask were somehow breached, the damage would probably be small and localized. Unlike volatile chemicals, which have caused fatalities and illnesses when they are spilled in transportation accidents, radionuclides tend to be heavy and therefore to stay put rather than to disperse. In any case, for over forty years trucks and trains have been regularly hauling nuclear materials. Over three thousand shipments of spent nuclear fuel and between two and five million shipments of medical radioisotopes and radioactive sources for industrial use have been safely carried all over the country.

In 2001 a train carrying toxic waste derailed inside a Baltimore tunnel and burned for four days. (With the goal of scaring audiences more thoroughly, a feature film based on that event changed the waste to nuclear.) The State of Nevada has raised doubts about whether a cask containing nuclear waste could withstand a tunnel fire that hot. But, as with the worst-case scenario involving a burning jet in a spent-fuel pool, it is worth remembering that steel melts at 1,600 degrees Fahrenheit and the uranium fuel's protective cladding at 3,600 degrees Fahrenheit. The NRC requires transportation casks to withstand a fire of 1,475 degrees Fahrenheit for thirty minutes; it concluded after running tests that the casks wouldn't give way in a tunnel fire. In any case, as Rip and other scientists pointed out to me, such a tunnel would contain the nuclear waste if it were already containing the fire well enough to keep it burning for days.

"Can terrorists divert a truck with a cask of spent fuel and capture it?" I asked.

"It would be really difficult for them to get past the locking devices and inside the truck," Rip replied. "Global positioning satellites continuously monitor nuclear waste shipments already, so the instant one went astray, we'd know about it. The probability of a terrorist attack on a truck is very low. The consequences of such an attack would be minor—there's not much anyone could do to the truck, and even if they somehow breached all the layers of containment and got to the spent fuel, they'd die from radiation poisoning before they could do anything with the stuff. Every transport of nuclear waste today to WIPP and other sites is escorted and shipments are safe and secure. I just wish haulage of lethal chemicals got this much care and attention. You hear every year about train derailments releasing really nasty chemicals that disperse over a wide area."

At last we arrived at a collection of buildings and a security checkpoint: Mercury Gate, the portal to the Nevada Test Site. During the heyday of atomic testing, thousands of workers streamed through here every morning. Today the place seemed deserted.

A sign provided directions for tractor-trailer trucks bearing low-level radioactive waste. Thousands of cubic yards of it—including things like those gloves and contamination suits from the spent-fuel pool at the Idaho lab—have been placed in holes drilled at the Nevada Test Site and covered with soil. Almost all the residues from the nuclear fuel cycle have radiation levels so close to that of natural background that they don't pose disposal problems. Only a small fraction of the total waste generated is sufficiently hazardous for the long term to require remote isolation. How that waste is handled or sequestered depends on how long it will remain a risk as it decays and the degree of concentration of its radionuclides. The DOE has various classifications of nuclear waste.

Waste containing small amounts of radioactivity comparable to those present in household garbage—home smoke detectors, salt substitute, overripe bananas—is called exempt waste and sent to a normal garbage dump. Next comes low-level waste, which makes up 90 percent by volume of nuclear waste around the world but emits only 1 percent of the total radioactivity and contains small amounts of radionuclides that may have short or long half-lives: paper towels, rags, tools, coveralls, mildly contaminated glassware, concrete, soil. No shielding is required when cleanup workers gather, transport, compact, incinerate, and bury these materials in a shallow grave. Low-level nuclear waste interment sites operate in many countries.

Waste that emits rays and particles in greater quantity is classified as intermediate and must be shielded by lead, concrete, or water. Resins, chemical sludge, the cladding of fuel rods, and contaminated objects from reactor decommissioning fall into this category. If the radioactivity of intermediate-level waste has a short half-life, it's buried in its shielding. Long-lived waste must, by law, be isolated in a deep geological repository.

High-level waste, which contains as much as 95 percent of the total radioactivity generated in a reactor, is made up of the ash of the controlled chain reaction: the fission products and some transuranic elements—that is, those higher on the periodic table than uranium, like plutonium.

If you asked me whether I'd rather live in the nation's fastest-growing metropolis or ninety miles away, at the graveyard of quite possibly more atomic explosions than anywhere else in the solar system, I'd pick the clean air, sweep of earth and sky, and stark contrasts of light and shadow offered by the Nevada Test Site, which covers a thousand unpopulated square miles enclosed by mountain ranges. Counterterrorism units and environmental remediation teams now train here. There's so much land that they can easily avoid forbidden areas—"sacrifice zones"—that may have some residual contamination from weapons tests of long ago.

Environmental researchers take advantage of this unique outdoor laboratory. The Desert Research Institute of the University of Nevada has been conducting studies here that show that, even though deserts are as effective as forests and grasslands in reducing the increase in atmospheric carbon dioxide from fossil fuel combustion, the growth of CO_2 by 2050 will, paradoxically, be enough to trigger in plants and soil bacteria a reaction that will reduce their absorption of carbon. Instead, microorganisms under such conditions will release carbon. Definitely not good news. Other research is now suggesting that increased nitrogen from fossil fuel emissions has already had a significant impact on life in these soils.

Most of the territory through which we drove—for many miles—was arid and empty. An occasional tower or some other structure rose off in the distance. The road passed sites with radiation warning signs where numerous subsidence craters pocked the landscape, all caused by underground detonations of atomic bombs. Abandoned cables extended along the ground, a skein of neurons that splayed toward one crater or another and that had once sent back information on the underground tests to trailers crammed with instrumentation to measure every possible aspect and effect of the explosive chain reaction. In some of the swales, pale green spring grass was coming up.

On a mountain pass, we saw a collection of rusting trailers, abandoned railroad freight cars, and corrugated sheds that weapons-lab crews had used to observe test shots that took place hundreds of feet below on a huge, dry, sandy lake bed.

"What was it like?" Marcia asked Rip.

"You'd work and work setting everything up until you were really tired. Then they'd evacuate everyone and we'd go back up here to the pass and watch. The ground would wiggle. After the test everybody would come out with Geiger counters."

Rip's job at the time involved experimenting with polymer plastic components, used to protect electronics, in order to learn how they reacted to intense radiation. Over the course of fourteen underground tests, the team Rip was on would set up polymer samples in a tunnel in the side of a mesa next to a massive shutter with a peephole. The first shutter weighed ninety tons. Later, Sandians perfected a two-ton steel shutter. On the other side, at the end of a long tunnel, an atom bomb would go off, starting with its characteristic burst of X-rays. They would shoot through the peephole and strike the plastics. Then the shutter would close completely in thousandths of a second, before the contamination from the blast reached the samples. It suddenly occurred to me as Rip described this that the doors I'd seen in the side of my hollow mountain outside Albuquerque must be similar shutters that would rapidly close in the unlikely event of an accidental detonation of one of the weapons within.

We paused at Sedan Crater, a huge feature left by the underground detonation of a hydrogen bomb big enough to destroy most of greater New York City. It was strange to be in the place where atomic weapons, a source of terror to me as a child and anathema to me as a protester, had actually been exploded. It was equally astonishing to contemplate, from the crater rim, the fact that since 1945 we and other nuclear countries had restrained ourselves; no one had again used this devastating force against people. Russian warheads that, as propaganda of the era had it, were once aimed at Albuquerque were now helping make electricity for America; eventually our own warheads, once aimed at Communist nations, would be transformed into fuel for our nuclear plants as well. For so long such an accord with the Russians had seemed unachievable. But it turned out to be in the best interests of the United States and the USSR to end the arms race. This gave me hope that America and other countries might make a similar decision about a large-scale, cooperative effort to mitigate greenhouse gases and try to prevent the potential destruction of our civilization from global climate disruption.

We turned onto a two-lane road and headed toward a low mountain range in the distance, passing through another empty stretch of desert. The terrain changed to rolling hills dotted with spring flowers. We saw a squiggly road going up a rise. "That's five miles away," Rip said. "The road leads to Yucca Mountain. To get to this point from Las Vegas takes one hour and forty minutes. We're barely inside the Nevada Test Site, which is huge, and already we're really far removed. If you just put spent nuclear fuel on the ground out here with a fence around it, and some warning signs, it would be OK. The only difficulty has to do with the future. A tiny proportion of it would be dangerous for many thousands of years, so it has to be put away deep inside rock. But even if it were left out here in the desert, you and I would never get exposed. Nobody living in Las Vegas would ever get exposed. The risk is that future generations might ignore the warning markers. So the anxiety about disposal of nuclear waste is *not* about the immediate effects. It's about what happens after American society is gone."

As we passed cactus patches, Marcia spotted a roadside sign: CAUTION: DESERT TORTOISE. A research project was observing them. Wildlife abounds at the Nevada Test Site and includes mourning doves and mustangs. Here and there stood various aged, decaying structures—water towers, gas storage tanks, an old reactor building where a nuclear rocket-engine test was conducted. It's a large concrete building that has to be decontaminated.

Now we were close to the controversial mountain, a steep-sided, six-mile-long, flat-topped, undulating wedge that angles upward to a peak at one end. Of thirty-seven different legal arguments being presented by the state in its effort to stop the project, one concerns a plan to construct the waste storage area on a horizontal plane 1,250 feet under the top of the mountain rather than below the general ground level. The opposition claims that such a location fails to conform to the description of an "underground repository," but people on the project say that Yucca Mountain *does* conform to the NAS dictates of deep geologic disposal.

As we drew nearer, we saw a gaping gray mouth in the side of Yucca Mountain. At its foot were corrugated prefab sheds, a few plastic-covered Quonset huts, and a helipad. For the $8 billion spent thus far, you might expect more.

We entered one of the buildings. On a wall was a sign:

> THOSE WHO SAY IT CAN'T BE DONE NEED TO GET OUT OF THE WAY OF THOSE WHO ARE DOING IT.

Joining a small group of new workers reporting for orientation, we received the requisite safety instructions from a video about the importance of wearing hard hats, earplugs, clear plastic safety glasses, a miner's cap lamp, and a self-rescuer device that in case of fire you put over your nose to enable you to breathe oxygen. (The biggest problem in a tunnel or a mine apart from a collapse is carbon monoxide poisoning.) We were also told of the importance of wearing dosimeters. Even though there was no nuclear material of any kind within miles, we were obliged to behave as if we might find ourselves brushing up against it. According to the dictates of the ALARA mandate (As Low As Reasonably Achievable), precautions had to be followed as if the repository were already in full operation.

Conceptual Design of Yucca Mountain Disposal Plan

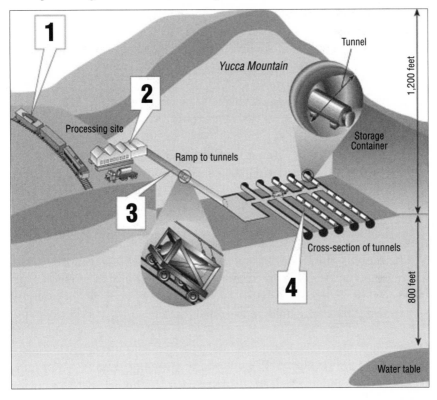

1. Canisters of waste, sealed in special casks, are shipped to the site by truck or train.
2. Shipping casks are removed, and the inner tube with the waste is placed in a steel, multilayered storage container.
3. An automated system sends storage containers underground to the tunnels.
4. Containers are stored along the tunnels, on their side.

Nuclear Regulatory
Commission

"Thanks to Three Mile Island, the NRC, which regulates this place, looks for evidence that nuclear facilities have the right safety culture," a scientist colleague of Rip's who had joined us remarked, after we'd been outfitted with our safety gear and were walking toward the tunnel. "It set out to develop and enforce a safety-conscious work environment. The goal was to go from a bunch of scientists who thought they knew everything to an attitude of 'I don't take *any*thing for granted.' The slightest lapse and everyone has to write a lot of reports. NRC wants us to look for problems and fix them. Every facility in the country has problems. It's a matter of *how* you look for them and deal with them. Yucca Mountain has a high pride of ownership. These people have been here years and years and it's their project."

Piles of pink gravel, the spoil you might see outside any mine, greeted us at the North Portal. Overhead was draped a big banner:

WELCOME TO YUCCA MOUNTAIN
A MEETING OF DESIGN, ENGINEERING, AND CRAFTSMANSHIP

Our tour guide, Bruce Reinert, an engineer from Los Alamos, wore an orange vest and his white hard hat sported an American flag decal. Before taking our group inside the tunnel, the mouth of which was plastered with sprayed concrete to keep the dust down, he ordered us to watch our step, not to wander off, and not to touch the 12,470-volt power cables that were strung along the wall next to a catwalk.

The tunnel, which was dry and did not smell dank, was mostly taken up by train tracks and lit by lamps, their glowing globes stretching off into darkness. Unconsciously reacting as if I were in a New York subway station, I tightened my hold on my purse.

People tend to assume that, geologically speaking, anything can happen—as if at random. Indeed, this has been the human perception for hundreds of thousands of years. Suddenly the ground would move. Suddenly a mountain would spurt fire. Suddenly a tidal wave would inundate the land or a river would overflow. Suddenly a storm would knock down a forest. Starting only a few centuries ago, humans began to understand that the history of the earth had been preserved in stone and that predictions about the future of a particular geological formation or about climate change could be accurately made based on observations of the record of what had happened over millions of years.

Reinert explained that at Yucca Mountain, scientists collecting data and making computer models analyze how rock was formed and how natural events and processes affect it. The models are continually refined and updated as new information comes in. For forty years at the site, thousands of core samples have been extracted from the mountain and their sources recorded in a database. The main tunnel and other side tunnels are dedicated to tests for seismic stability, heat effects, and water permeability. Pointing to an apparatus installed at the end of the alcove to test water, he asked, "How does water move through the rock? We put a drip irrigation system on top of the mountain arranged so that *all* the water would go into the mountain. Rainfall here at present is about seven inches per year, and most of that would naturally run off the mountain. With the drip system going at thirty inches per year, we tested the seepage through the rock. Water moves very slowly in this formation. Inside the mountain, we only collected from three to nine percent of the water we put in on top. Most of the water that gets inside the mountain will go around the opening made for the repository because of a natural process called a capillary barrier."

If you take an empty cardboard matchbox and try to fill it with water, you'll find that it tends to go around the outside of the box rather than into it. That's because of the capillary barrier.

"Most of the water we find inside the mountain is millions of years old," Reinert said. "The repository will be situated halfway between the top of the mountain and the water table: here the water table lies one thousand feet below." Many studies have shown that there is no way for the water below to rise up and flood the repository. Still, some in the technical community have criticized the wisdom of locating a nuclear waste repository above rather than below the water table.

In drifts off the main tunnel, experiments were simulating spent-fuel's thermal energy. How heat affects the rock over time has been a serious topic at Yucca Mountain. "We put really hot stuff, temperature-wise, in the tunnels," Reinert said. "Thanks to data we have, we can predict five hundred years of the temperature going up from the stored waste, which will be thermally hot, and then it will cool down. The rock takes a long time to cool. In regard to licensing the repository, the Nuclear Waste Technical Review Board and NRC may ask questions about the heated rock. We have the answers for them."

In one alcove we saw examples of typical kinds of nuclear waste. There was a mock-up of a fuel rod assembly inside a steel canister; nearby was a replica of a metal-clad vitrified log of high-level liquid waste the size of a giant hot water heater. In another alcove, a poster depicted the history of the land. Eleven to thirteen million years ago, an eruption occurred twenty miles away that spewed molten rock and clouds of ash rolling along at hundreds of miles per hour. The pressure and heat formed the different kinds of volcanic tuff that, after successive eruptions, became Yucca Mountain.

If the NRC grants a license, more than four years of constant mining will be required to carve out four panels—storage rooms. In them eleven thousand containers will be emplaced, once they're brought inside the mountain by train and unloaded by remote-controlled robotics. The waste in each can must be retrievable. When the repository can hold no more, the tunnels—all forty-one miles of them—will be sealed at their mouths and perhaps the main entrances to the repository will be back-filled. Estimates of when Yucca Mountain will start receiving waste range from 2017 to 2020 to never.

I felt weary thinking about how much effort and money had gone into the program already and how much more work needed to be done, and how opposition has only grown. My eye fell on that now-familiar poster, in an exhibit in the alcove, that reminded me yet again: one nuclear fuel pellet weighing 0.0007 pounds can generate the same amount of electric-ity as 1,780 pounds of coal, or 149 gallons of oil, or 157 gallons of regu-lar gas. Nevada gets about 86 percent of its electricity from fossil fuels, mostly coal. The state's carbon dioxide emissions grew by 47 percent between 1990 and 2001, the largest increase in the nation; in 2001 Nevada's per capita output of CO_2, including vehicle emissions, was 21.2 metric tons. Vermont, which gets 75 percent of its power from nuclear plants and 10 percent from dams, had per capita total carbon dioxide emissions that year of 10.6 tons.

What if all the cars and lights in Las Vegas were powered by electric-ity from nuclear plants? Right now each of those two million-plus people is contributing somewhat more than the annual national average of car-bon dioxide emissions. And what about the health costs of fossil fuels in Las Vegas? They would have to be significant. You would expect a city founded on casinos to be shrewd about odds, yet perceptions about health there seem to focus not on fossil-fuel pollution and its effects on children and the elderly today but rather on the future transport of shielded nuclear waste on routes that do not pass through Las Vegas, on shielded casks of spent fuel stored deep inside a mountain nearly one hundred miles away, and on what could theoretically befall people who might be exposed to the 0.01 percent of the spent fuel contained within the mountain that will still be radioactive in ten thousand, one hundred thousand, or—given the latest recommendation of the EPA—one mil-lion years. At present the Yucca Mountain Project is doing calculations for ten thousand years but it can extend them as far into the future as nec-essary, depending on the dictates of new regulations.

The EPA has ruled that an individual can receive no more than 4 mil-lirem a year as a result of radionuclides from Yucca Mountain that might escape into drinking water; the agency has also decreed that for the next ten thousand years, an individual next to the facility can receive no more than 15 millirem per year above natural background radiation—the equivalent of one chest X-ray. "That's what the federal regulation is about," Rip's colleague said as we emerged from the tunnel. "The moral obligation to protect the unborn of the future. In fact, the emissions will be far below that figure."

That scanty number of millirem qualifies as low-dose radiation, only a tiny fraction above natural background radiation. The Health Physics Society, whose members work in the field of radiation health protection and dose calculation, has issued a position paper decrying the present regulatory framework and its enforcement as inconsistent, inefficient, and unnecessarily expensive. Larry Foulke of the American Nuclear Society (ANS) has called for realistic estimates rather than conservative ones. In a 2004 speech to the ANS he said:

> Predicting deaths by adding up trivial individual doses over large populations or over large periods of time is scientifically indefensible. Questioning this invalid premise is not attacking an established scientific theory; it is merely challenging an administrative judgment. If the risks and costs of this administrative judgment exceed the benefits provided, it should be revoked. For example, this administrative judgment provides the basis for the EPA, DOE and NRC to set a limit of 4 millirem [in drinking water] per year from the releases from Yucca Mountain. Even though natural background radiation varies from about 80 to 800 millirem per year, with areas in which millions of people are exposed up to 8,000 millirem per year, with local doses to more than 20,000 millirem per year to people in the high-dose areas of Ramsar, Iran. I'm not asking us to repudiate the scientific findings of the advisory bodies. I am asking that we take them at their word when they say, based on their extensive reviews, that there is no scientific basis for finding that low-dose radiation is harmful and that the evidence indicates otherwise. On those grounds alone . . . it is clear that the fear of radiation built up by presuming harm where none has been shown to exist has been detrimental to the health and safety of the public and creates a dangerously fearful public attitude toward the possibility of radiological terrorism. We do not become safer by portraying the world unrealistically.

To uphold present standards of radiation protection at Yucca Mountain, the United States will spend an estimated $60 billion to comply with a regulation about low-dose radiation exposure based on an unproven assumption. The French Academy of Sciences has now concluded that at low doses and low-dose rates of ionizing radiation, programmed cell death dominates, and so the body can eliminate or control the few cells that are damaged, thereby preventing the cancer process. Low-dose and ultra-low-dose studies might be able to provide additional support for what the majority of radiation biologists and radiation protection specialists already believe to be the case. Such findings could bring about more realistic regulations. I thought of a remark my friend Chris Crawford had made before I visited Yucca Mountain: "If you misallocate health and safety spending because of hysterical concerns on the part of the public involving nuclear dangers, real people will die of otherwise preventable diseases and accidents. Suppose the excess tens of billions that will be spent on Yucca Mountain were invested in reducing coal plant emissions, or better vaccine coverage for children, or research on Alzheimer's prevention. Why worry about hypothetical residents living twenty kilometers away from Yucca Mountain ten thousand years from now when people are dying today?"

The enormous rock shield of the deep geologic repository ensures sufficient protection for anyone living at Yucca Mountain ten thousand years from now. A person making his or her home next to one of the sealed portals would, according to DOE calculations, receive an annual radiation exposure of 0.01 millirem—the amount you get by eating one banana a year. For this reason, even though the data, its traceability, and risk analysis for the repository need to be put in good order, Rip and everyone else involved believe that the probability is great that the repository will provide safe containment.

We next drove around the mountain toward the South Portal, where the borer rests. The Yucca Mucker, as it was dubbed, is a piece of equipment called a tracked alpine miner that has a white-painted rotating drum 25 feet across and 410 feet long and a face that is a mass of enormous steel incisors and canines, here somewhat worn down after munching through five miles of granite. The machine now sits in front of the tunnel exit on a broad ledge with a view of a ravine, mountains, and the plain below, awaiting a buyer with $10 million ("It's only been driven a few miles," goes the local joke) or the director of a science fiction film.

We continued upward, bouncing along a dirt road toward the crest, past some precariously balanced boulders. "They indicate the absence of big earthquakes," Rip said. He showed us the traces of old seismic faults that go through the mountain—interruptions of striations in exposed rock faces on the side of the ridge. A seismic sensor network, formerly used for underground nuclear tests and monitored by the U.S. Geological Survey, remains in place, making the Nevada Test Site one of the most studied geological sites anywhere. "We know a lot about where the faults are," he added. "The repository will be located between three faults but not on them. We have a detailed mapping of the geology." Most of the energy that is released in a seismic event occurs at the surface rather than in mountain tunnels; still, the waste containers and repository facility will be designed to withstand worst-case earthquakes.

As I would be shown later in the nuclear tour, at WIPP, 2,150 feet below the surface, an orange band runs in a straight line through the salt bed for thirty miles. "This phenomenon is due to the way the salt was evenly deposited in an incredibly stable environment that has remained stable," Rip said. "Here, along with cinder cones—which were formed when hot gas vented from under the surface—you see ash falls and thick or thin layers of lava. It's a chaotic terrain—but it can be known with precision. Still, we're left with an amount of uncertainty that you don't see in places like a salt bed."

He called our attention to the striking pink and blue striations of the nearby aptly named Calico Hills. Noting how they had buckled, he contemplated their roots deep below. "You can just look at them and see how complex the geology is and know it doesn't become less so when you go underground. I've spent a huge amount of time in the last month studying the literature on disruptive events here. There's a lot of information about how the formations react to being jolted."

As we reached the top of Yucca Mountain, five thousand feet above sea level, and got out of the van, I asked, "What's the worst that could happen to the repository?"

"There are only three ways that natural forces could transport radionuclides out of the repository," Rip replied. "One, a volcanic eruption under the mountain that could scatter the radioactive material; two, water flowing through the repository; three, corrosion of the canisters containing the waste, allowing water or a volcano to disperse it."

People on the project had been following a methodology that conservatively and subjectively postulated the presence inside the mountain of a volcanic dike—a wall created by lava from an ongoing eruption that cuts through the site of the repository. "There's absolutely no evidence of this," Rip said. "The dike is a purely imaginary one. Nevertheless, these people *assume* that a future eruption could send lava through such a dike, if it were to exist. What bothers me is that these people estimate that lava will cross fault lines, but lava or any flowing material takes the path of least resistance." He went on to describe the worst case envisioned by this very pessimistic approach: "A volcano comes up through the mountain and digests a few cans of waste and deposits ash with radioactivity of a higher concentration than it naturally would have. That's the worst consequence, and one the antis bring up, but the probability of that happening is one in a hundred million." Rip swept his arm toward the flats below, where an occasional cinder cone poked up. "There are buried volcanic dikes under the flats but not under any of the mountains. There's no record here of a volcano coming up underneath an existing mountain in the past ten million years. If anything, volcanism is growing weaker here. So you can be really, really sure that a volcanic eruption will not happen here at Yucca Mountain. But because the upper value is greater than one in a hundred million, we have to include a volcanic eruption in the mathematical anaylsis."

We had now walked to the top of the ridge. Indian paintbrush, penstemon, scarlet globe mallow, and yellow daisies poked upward among cracked, dusty slabs of granite. A clump of mop-headed Joshua trees—also referred to as yuccas—grew at one end of the crest. In the direction of Death Valley and California, the distant wall of the Sierras reared up, almost indistinguishable from blue clouds. After being inside the car and the mountain, I welcomed the huge panorama of sky, desert floor, mesas, cinder cones, and the hills banded with colors ranging from lavender to violet to red.

As we gazed around, Rip summarized what he called multiple defense layers. First, Yucca Mountain was surrounded on all sides by the vast and closely patrolled Nevada Test Site, which was bigger than Rhode Island. Below us stretched Jackass Flats, never cleaned up after atomic tests. He pointed out places where bombs had been detonated. The Nevada Test Site was adjacent to a gunnery range and a protected wildlife refuge. All this added up to over five thousand square miles. "The natural barriers at Yucca Mountain alone greatly reduce the potential for releases by a factor of about one million," he said. "The two-hundred-and-thirty-square-mile tract set aside for the Yucca Mountain Project is located in the sub-sea-level Death Valley hydrologic basin. We're in a natural bowl here. This groundwater basin doesn't drain into any river or lake and can't reach Las Vegas. But the project still has to get water permits."

Second, the waste would be sequestered inside the mountain, with a mile or two of rock on either side horizontally and about a thousand feet of rock overhead, and about a thousand feet of rock between the repository and the water table, which is, basically, a saturated zone buried by rocks and dirt.

Third, when the repository is closed, the access tunnels will be sealed up.

Fourth, inside there will be engineered barriers. All of the waste will be enclosed in thick steel canisters of corrosion-resistant alloy that will sit under drip-shields designed to fend off the corrosive plink-plink of water. "There's very little rainfall," said Rip, "and here on top of the mountain not a lot goes underground, because under a layer of surface soil there's dense rock. Most water evaporates. It can rain cats and dogs here and a half an hour later it's as dry as a bone."

Off to the south we viewed a plain bordered by hazy peaks. "There's Amargosa Valley—it's about fifteen miles away," Rip said. "WIPP's compliance boundary—where EPA assumes Fencepost Man lives—is one mile horizontal and two thousand feet vertical. At Yucca Mountain, it's about eleven miles horizontal."

Over thousands of years, could radionuclides escape from the canisters and enter the water table, which flowed downhill into the Amargosa Valley? It was in the hydrologic basin.

"If you were going to live way out here, you'd have to feed yourself," Rip said. "You'd be likely to grow your food in the Amargosa Valley, not here. In the worst-case conservative estimate, contamination could eventually escape the mountain, flow from Yucca Mountain into the water table toward Amargosa, and concentrate in the water, and through the food chain humans could then take up radionuclides."

The radiation exposure to hypothetical Amargosanos in the remote future, as calculated by worst-case, LNT estimates about the health effects of low-dose radiation, would presumably be sufficient to cause some untimely cancer deaths.

"If radioactive material could ever escape from the canisters, it would take a couple thousand years to be washed out of the mountain," one nuclear waste specialist from a national laboratory told me. "You can create a very conservative scenario in which this happens within four thousand to eight thousand years."

Thousands of additional years after that, it's remotely possible that radionuclides could make their way into crops in Amargosa. As risks of that sort are now calculated, you have to assume that the Fencepostians living in that valley would get *all* of their food and water from there and therefore consume in a single year enough radionuclides to exceed the federal regulations set forth in the twentieth century, thousands of years earlier, rules that assume—thus far, without experimental scientific basis—that there is no threshold below which radiation becomes harmless. But even with all that, the Fencepostians would be quite unlikely ever to be exposed to as much radiation as people in northeastern Washington State get today from nature.

On the visit to this desiccated tract I heard a lot about water. "There's a reason to locate facilities like this in the arid Southwest," a geologist I met there said. "You want to bury the waste deep in a dry environment." Dry soil and rock can immobilize radionuclides that might escape; water frees them to migrate.

To learn about the movement of nuclear waste from its origins over a huge time span, Yucca Mountain Project scientists examined studies done at Oklo. The fossil reactors' radionuclides, generated in that Gabon river delta, bonded with granite, sandstone, and clays. Plutonium (detectable from traces of elements associated with the now completely decayed plutonium) and other reactor products remained within ten feet of where they were created. At Yucca Mountain, 99.99 percent of the radionuclide inventory in storage would be similarly insoluble and immobile. And what about that 0.01 percent? The odds of that minuscule quantity of radionuclides leaking into the water table and migrating to the Amargosa Valley are estimated to be one in a million. "That's the only likely outcome that might be reasonably expected to affect humans in ten thousand years—and so must be included in licensing considerations," one of the project's consultants told me.

"Project scientists boxed themselves into a corner with that when they took a conservative approach to come to their conclusions," Rip remarked in the course of filling me in on the project's past efforts. "They originally guesstimated that the worst case would be tons of water flowing through the repository. You heard Reinert say they made the drip irrigation run at thirty inches a year up here at the top." Rip had just been reviewing the hydrological studies and was pleased at the thoroughness of the data. "Unlike the people writing the conclusions for the licensing application, the hydrologists didn't take the extreme of the distribution but instead used something fairly reasonable based on all of the historical climate data that they could find." Those data, which came from assessments of climate change during periods of global warming and global cooling over hundreds of thousands of years, indicated that the greatest annual rainfall would never come to thirty inches; rather, at intervals early on it would reach a maximum of 16.5 inches and then, after about two thousand years, would settle into an average of twelve inches a year for the next ten thousand years.

Water, given enough time and the right conditions, can dissolve just about anything. It's the only medium through which it is even remotely likely that natural processes could carry radioactive material out of the mountain and, over thousands of years, contaminate humans. But that would entail a film of water passing through the repository and seeping on down to the water table. The Nuclear Waste Technical Review Board, the project's occasionally stern overseer, stated in 2004 that corrosive conditions were highly unlikely to occur inside the repository. I spoke with another consultant who wished not to be identified because of the constant controversy. He had considerable expertise about conservative estimates concerning the flow of water through the repository that had been posted on a DOE public-information website. "The people who wrote those statements were talking to one another and forgetting that they would be read by a general public," he replied. "People tend to think that if something bad possibly could happen it *will* happen. That means people don't actually understand what uncertainties and probabilities are. It gets down to how to deal with uncertainties: What is the most that could be reasonably expected? It's a question of appearance."

Being a member of the public rather than an insider, I had always assumed that conservative estimates were good—cautious and sensible.

"With conservative estimates, we pretend we don't know things that we really *do* know quite well," he said. "A better way to put it: We know things, but we're uncertain about them, and that uncertainty leads us to make assumptions that are not necessarily worst-case but are a case that we could defend as being as bad as you could reasonably expect them to be—in terms of negative impact on the repository. You assume high solubility, but in another component of the model, in another process, that assumption could turn out to be unconservative and even scientifically impossible. That's where the problem of conservatisms really came in. The people on the project made some assumptions about climate change and then about the permeability of the rock in different places. There's variability within the site due to variability of time. It's really difficult to make an accurate prediction. So what the people did was to make assumptions about when we're going to have another glacial period and how long it's going to last, and what kind of rainfall will occur when it ends. They actually knew better, because they had the correct hydrological information. But when you take the conservative approach, you decide to go ahead and make what you could call 'stupid' assumptions anyway."

For years the Yucca Mountain Project relied on this methodology, which came under increasing criticism. The project's probabilistic calculations based on conservative estimates had resulted in open-ended uncertainties that could not be quantified. "The whole concept of proving the feasibility of a repository rests on minimizing the uncertainties as much as possible," Rip told me at one point. "That can't be accomplished when you use conservative assumptions for parts of the assessment." Rip and other former WIPP colleagues drafted a report about the problems that had arisen because of the methodology and the absence of a lead laboratory to supervise the science. As the memo made its way up the rungs of the DOE hierarchy, and as years passed, national peer reviews and internal ones, such as the panel of a dozen scientists that Rip was on, unanimously recommended use of realistic rather than conservative estimates. Agreement came from the IAEA and OECD's Nuclear Energy Agency; their review in 2001 had found that reliance on very conservative estimates of scientific properties had caused inconsistencies in the Yucca Mountain Project's performance assessment, which is the tool used to estimate the long-term safety of the repository. The agencies recommended a probabilistic analysis "based on a realistic rather than a conservative representation."

Finally, in 2006, the DOE named Sandia the lead national laboratory of the Yucca Mountain Project. Of all the national labs, it has the most experience with long-term nuclear waste disposal and repository certification. Andrew Orrell, a geologist who was formerly a manager at the WIPP and manager of Sandia's Nuclear and Risk Technologies Center at the Yucca Mountain Project, became head of the team coordinating all the science, including the risk assessment of the physical and environmental conditions of Yucca Mountain and the emplacement of the waste canisters. The team, to which Rip became a consultant in 2006, has been scrutinizing the conservatisms used in the project, eliminating some, and, in the NRC licensing application, flagging others as sources of concern to be examined in the future. The focus is now on putting together a realistic portrait of the repository's projected performance.

What were the reasonable expectations the repository was required to meet when it came to excess water flowing through the repository and dispersing radionuclides into the water table?

"Let me give you the *real* reasonable expectation," the consultant told me. "*Nothing* is going to happen anytime soon, that's for sure. We do know that corrosion will occur over time and waste packages will degrade; this is such a slow, long-term process that it's difficult to understand. People tackling this question are dealing with a very complex task. The EPA standard they are trying to meet, 'reasonable expectation,' has been imported into the NRC, which previously had gone by the standard of 'reasonable assurance.' This has caused tremendous uncertainty in how the whole licensing process is going to occur and the NRC has got to come to terms with what that means." Rip agreed with this assessment. He explained to me that reasonable assurance means that the process must simply meet the standard. "If you make large conservative estimates and you meet the standard, you are OK. Reasonable *expectation* means that the system must show that the system will perform as indicated by the analysis and that it will meet the standard. You can see how the two standards conflict. Reasonable assurance can use conservatisms to demonstrate that the standard is met. Reasonable expectation must show the steps in the process as well as meeting the values of the standard and in doing so must show the uncertainty for each step. Since we really don't know how any system will work ten thousand years into the future, we must rely on probabilistic analyses using mathematical models. The best estimate for the future performance of the repository has to be based on the probabilistic analyses that include all features, events, or processes that cannot be discarded as unimportant as well as the attendant tracking of uncertainty."

"Then how do you deal with the uncertainties?" I asked the consultant.

"Rip Anderson and Mel Marietta believe you can build a performance assessment that would supply the basis of reasonable expectations, and I agree. But the task is huge. The idea is to follow the lead of several of the other nuclear nations and develop a safety case that represents the most probable description of the repository performance. Then compare additional analyses to that safety case to show how any conservatisms or other perceived risks might affect the safety of the repository.

"It's important to acknowledge risks perceived by anyone and everyone and include them or show that they are unimportant. If you don't do that you will never get anywhere talking to people about risk. You have to start by asking people to describe the risks as they see them. You will be surprised at how they portray those risks. Rip understands that public comprehension is just as important a factor as any other. We've had to think in a lot of weird ways in order to include the truly improbable scenarios. To be complete, we assume that general corrosion of the canisters will occur and we assume they fail. By making such assumptions, we deal with the uncertainties in our knowledge that are difficult to estimate. This represents a cautious approach. But the people writing the documents that are posted for the public to read don't explain this clearly."

As an extra barrier, each canister will have its drip-shield—a convex, corrugated titanium canopy—to guide any potential water droplets to the rock floor. The drip-shields and canisters will probably remain bone dry, because the capillary barrier will guide any stray water around the tunnels and emplacement rooms. Rip said that because for years no lead laboratory coordinated the work and encouraged an interdisciplinary and multidisciplinary exchange of information, the engineers who designed the drip-shields didn't talk to the hydrologists. The engineers assumed that the greatest annual rainfall would be twenty inches instead of using the more realistic estimate of 16.5 inches made by the hydrologists.

"Let's look at the worst possible case," the consultant continued. "The only way, however remote the probability, that nuclear waste could ever leave the repository is if multicentury continuous heavy rainfall were to saturate the mountain so that water started to drip from the ceiling of the repository and after thousands of years corroded the shield—maybe because of a tiny pinprick in it that droplets happened to strike—to allow water eventually to reach the outermost shell of the waste canister. And suppose that the droplets by chance then happened to find a tiny flaw in the surface of a canister and to keep falling on exactly that weak spot."

To prevent any water that might make it that far from getting to the spent fuel and corroding it, the Yucca Mountain engineers have designed double-walled stainless steel storage containers with the outermost shell made of a superior corrosion-resistant alloy that is likely to last over a hundred thousand years. Furthermore, inventors have recently come up with a waterproof amorphous film—a third layer to seal the canisters. The State of Nevada, having conducted its own tests regarding canister corrosion, has doubts about the DOE plan. But Rip told me that research into human artifacts abandoned to the mercies of nature indicates that certain metals hold up very well over the centuries even when exposed to a lot of water. For example, a bronze cannon buried in wet clay off the coast of Sweden for three centuries was found barely corroded. Bronze is mostly copper, and copper can endure for thousands of years. Swedes have been looking into storing their nuclear waste in copper vessels and have also been researching a combination of steel and ceramics. Glass going back to ancient times has also been found to be quite corrosion resistant.

However, in worst-case thinking, I learned, you don't let such reassurances relax your vigilance. You continue looking for a problem, even if it means going to radical extremes. Suppose that after thousands of years, the waste package finally is breached. According to Rip and other scientists, the dripping must then go on for perhaps thousands of years more, this time through that pinhole, until the water fills what is in fact a container about eighteen feet long and seven feet in diameter. The water then must dissolve the Zircaloy cladding of the uranium pellets, then dissolve their radionuclides (most will be insoluble), and, after perhaps another thousand years, eat through the bottom of the double-shelled, film-sealed container. Possibly after thousands of years more, that contaminated liquid—maybe in droplets—has to make its way down through a thousand-foot-thick stone formation into the zone of rock and soil saturated with groundwater. This process, also highly unlikely, could take still more centuries, and meanwhile many of the radionuclides would sorb along the way with clay or rock. Water in the saturated zone would dilute the concentration of any radionuclides that happened to travel that far. Some parts of the rock in that saturated zone are drier than others. Any waterborne radionuclides carried underneath the repository would tend to bond with dry rock.

"It's very difficult to predict how long it will take for water, once it has reached the waste, to actually dissolve it, get down out of the package, and make an exit hole," the consultant said. Considering the tortuous and unlikely path of the radionuclides out of the repository, he added, with irony, "That is what's called 'reasonable expectation.' "

At the Waste Isolation Pilot Plant, which had to meet the EPA standard of "reasonable expectation," over a thousand studies had been done to track the potential path of every radionuclide out of the deep salt bed. Now, as we stood on top of Yucca Mountain, the wind riffling through our hair, I asked Rip about the differences between the two repository projects.

"The salt beds could not exist if water was present in any quantity. They've been there for two hundred and thirty million years, and will continue to be there for hundreds of thousands of years more, or millions. There's no seismic activity at WIPP. No volcanism. The geology is simple. At WIPP, we looked at everything, ran the calculations, and if some event or process had less than a one in a hundred million chance of occurring, we eliminated it. But Yucca Mountain's conservative analyses did not allow them to complete a first-class screening analysis of the features, events, and processes, or FEP." A feature might be a fracture in rock. An event might be a volcano. A process might be rainfall. "Some FEPs were screened out that should not have been, and some were screened in that should have been screened out. Interactions between FEPs were not specifically accounted for and have led to positions that were impossible."

"Were scientists on the project encouraged to find every reason Yucca Mountain would fail?" I asked.

"Yes and no," Rip replied. "The previous methodology used did not allow the scientists to assess processes that are coupled. With Seabed, we found how useful it was to couple processes in order to get various scenarios and figure out how probable they are. At Yucca Mountain, the scientists were asked to look at *all* failure modes, and many think they have done so. If you take the conservative position on many of the FEPs and on the data, it's hard to come up with *actual* showstoppers. Wild estimates cause too much uncertainty. When you don't have realism, and don't have communication and exchange of information among all the participants and the academics, then you wind up with contradictions and a lack of specifics. It's not even science anymore. So what is the public going to make of your findings? I could give you examples I came across in the review I did that defy the laws of nature. I saw a calculation that, if you take it to its logical conclusion, says that a container one hundred percent full of a solid, like metal, can somehow take in more material—say, water—without expanding the container. But you can't put ten pounds of manure in a five-pound sack. The Yucca Mountain Project was stuck with scenarios that made absolutely no sense because their basic assumptions were so extreme."

People on the project are now using the findings from the internal peer review Rip and his colleagues wrote as a guide to their new work.

After spent fuel has cooled in pools at plants for a few decades and a lot of its hottest radionuclides have decayed, and after thirty or forty more years in dry-cask storage, the fuel is cool enough that handling it is less risky. After a thousand years in any repository, the radioactivity would be reduced still further. If a future race for some reason decided to drill into the mountain and had the technology to penetrate as far as the repository, the Yucca Mountain Project assumes that the people would also have the technology to evaluate the contents and avoid them—especially because there will be images to serve as warnings to intruders. However, Yucca Mountain and its environs have no oil, gas, or mineral deposits; even if rainfall were to increase, the rock and soil are too barren for agriculture. The place is unlikely to appear inviting to citizens of the remote future.

One day when I phoned the political scientist John Kelly at his office, in the Colorado Rockies, he was working on a long document about projected radiation doses to the public from Yucca Mountain in the far future.

"Over ninety-nine percent of the waste will never get out of the repository, because it will be decayed by the time the waste package would fail," he said. "Or, if not decayed, the waste won't leak out, because it is insoluble in water and therefore can't be transported by it, or because the radionuclides sorb to the rock as water carries them out of the canister. So we're talking about the potential release of less than one percent of the nuclear waste that, under current plans, Yucca Mountain is supposed to hold."

An estimated trillion dollars, most of the public indignation, many studies, and hundreds of thousands of man-hours over decades to shape a plan for dealing with the nuclear waste turn out to be centered on some seven hundred metric tons of material, or 5,096 cubic feet. By volume, all this waste could fit into a studio apartment about twenty-five feet by twenty-five feet with a ten-foot ceiling. Some radionuclides in that waste, like technetium and neptunium, have huge half-lives and are non-sorbing, or are soluble; they will not have decayed completely after a million years, but they would be less radioactive than the uranium ore at Ambrosia Lake. Other material might be more harmful. "You wouldn't want to be carrying some of those long-lived radionuclides around in your pocket, not even after a million years," Kelly said. "So that fraction of one percent is still a reason for caution and for taking prudent safety measures."

It's extremely unlikely that any radionuclides would escape in ten thousand or one hundred thousand years. The ultimate worst-case scenario was that in one million years, *all* the waste packages would have failed, and *all* their contents would be released and transported out of the repository. Nobody connected with the project appears to believe that this scenario is the least bit realistic. But what if the virtually impossible occurs?

"If you heard that the EPA had decided that the peak dose occurring one and a half million years after closure of the repository—if all the materials remaining were released—could be as much as 200 millirem per year, would you be alarmed?" Kelly asked me.

A few years ago the prospect of a single millirem would have alarmed me. "No," I said. People who move to the Southwest from, say, Miami or Brooklyn or Harrisburg are increasing their exposure naturally by more than 200 millirem per year.

"For the first ten thousand years, the EPA is limiting the repository to an individual dose of no more than 15 additional millirem a year," Kelly said. "For the remainder of the million-year period, residents could not be exposed to more than 350 millirem per year above natural background radiation, which in Nevada is around 360 millirem per year. Where I live it's around 600 millirem." People living in Denver naturally are exposed to 700 millirem a year. "Peak dose occurs hundreds of thousands of years after closure—that is, in the unreal world created by the models. Peak dose is a function of inventory and the time interval during which the engineered barriers, particularly the cask, fail and release the inventory that exists at that time."

Like a desert mirage, the date of the ceremonial ribbon cutting at Yucca Mountain's bannered entrance keeps receding into the haze.

Before Sandia took over as lead lab, I talked to a number of people involved with the project along with members of the technical community, and often their responses would begin, "The big problem with Yucca Mountain is . . ." One former consultant to the project criticized the underlying premise that had for years guided the decision making: "They think that in order to be conservative about potential doses to the public, it's OK to violate laws of physics, mathematics, and probability. I hope Yucca Mountain will open. If DOE finally has the guts, it will do the model over and replicate the one for the Waste Isolation Pilot Plant. That one worked." Later his wish was to come true.

I heard stories about weaknesses in management, the project's missed self-imposed deadlines, budget and legal problems, hasty shortcuts, and poor accountability on the part of private contractors as well as tales about DOE-funded research that has enriched universities and contractors and permitted national labs to conduct interesting experiments without focusing on the central problem of whether the site is physically adequate.

Despite the complaints, everyone seemed to think the repository would do its job, and people pointed out that, after thirty-five years, much more is known about what might work and many sound studies have been done in spite of some irrelevant data and wrongheaded conclusions. I heard about the eagerness of Bechtel SAIC, the latest private corporation to run the project, to get its multimillion-dollar bonus by delivering a problem-free licensing application to the NRC. (Since 2006, DOE has scaled back the participation of Bechtel SAIC.) And then there are those billions of dollars in lawsuits and those formidable opponents: Senator Harry Reid; the Nevada gaming industry, with its bottomless pockets; and most of the population of Nevada—which apparently does not want the repository.

These challenges, some critics say, might inspire the NRC to delay, cancel, or reject the application, and provoke Washington to define the solution differently. Or a ruling by a federal court could put an end to the project. Meanwhile, nuclear utilities are expanding their programs to transfer waste from spent-fuel pools to dry-cask storage, and policy makers in Washington, D.C., have begun to champion other interim solutions.

"Here's the big problem with Yucca Mountain," Kelly told me. "The program began under a policy of nuclear fuel being directly disposed of instead of recycled. The overriding objective has been, for decades, to collect spent nuclear fuel that has gone once through reactors around the country and transport it to Nevada and put it in a geological repository on a fast schedule. For the licensing scenario, we assume that we transport all the material from plants to Yucca Mountain, very quickly load it in the repository over some twenty-six-to-thirty-year period, and then it's closed. As we all know now, that's not the scenario that will happen. The repository will be built in phases. Transport will not occur over the short time that had been expected. We know that the current policy of direct disposal of spent nuclear fuel is going to be revisited and we are going to have discussions about reprocessing. It's obvious now that our management of spent nuclear fuel has very little to do with whether proliferation occurs. In the future there will be advances in the development of technology, recycling of spent fuel, and fuel from that material. At this point we're in a funny situation. The policy assumption for what we are doing now is clearly out-of-date and hasn't been changed yet. So now there's no constituency for the operation of the repository as a place to put all once-through-the-reactor nuclear fuel."

If new reprocessing facilities are built, the nuclear waste issue may resurface as an important concern, tempering public support for nuclear power. Some insist on the necessity of demonstrating that the repository works, saying that the recycling question cannot be considered objectively until that occurs. But Rip says global warming will be the decisive factor.

"The success of Yucca Mountain is critical when it comes to making a choice about nuclear power," Kelly said. "Utilities all agree we have got to demonstrate this waste solution. They are very supportive of Yucca Mountain in that regard. The objective is clear: we need to get the licensing application to NRC. It's a gargantuan task. We no longer have anyone in favor of rapid disposal. Utilities do not want us to go headlong into permanent disposal, because that spent fuel is a resource we're going to need. Most utility people now believe we should be reprocessing the spent fuel, not burying it. If you put it underground, you can retrieve it later, but that's a great pain. Why the big hurry to put the waste underground? The project makes the assumption that we'll get it right the first time. If NASA decided in the 1960s that we were going to the moon and skipped Apollo missions one through nine and went directly to Apollo ten to go to the moon, would NASA have gotten it right the first time? Probably not. So why assume that these ten thousand waste packages are all going to be handled the same way?

"We need a prototypical stage, a pilot stage, and so on. We need a little bit of the repository and a little bit of spent fuel to start verifying our models while doing work on our site, building technology to emplace and monitor the waste. You wouldn't just jump to the end. It doesn't make any sense at all to approach the problem in that fashion. The NRC will force the project to develop the phases. But you don't see that in the mission of rapid wrap-up of the transportation from zero tons of spent fuel to three thousand per year from all over the country in twenty-six years. No thought has been given to beginning with just a few transportation corridors and preparing the local population for that. People need to be shown how the transport will be safe, how the police and fire departments can prepare for safety, for the correct emergency response in the unlikely event that there's an accident, and so on. The project needs to revise its plan. In reality, if we start shipping waste without doing it in small steps, there will be a repeat of what happened in Germany when spent nuclear fuel was transported. Antis from all over Europe appeared, and twenty thousand police were involved. Why not begin slowly and build confidence? Then along the way we can find out the glitches and get rid of them and we can ramp up transport slowly and carefully. Why not start with getting the first phase right? To get to the second phase, you need to maximize chances of succeeding in the initial stage. You take additional safety measures at the beginning to make it a success. Maybe it will turn out you don't need them, or you need something different. The first phase of Yucca Mountain should be to demonstrate to the world and to the people of the United States that we can solve the nuclear waste problem. The objective would be to pick up the spent fuel from the nuclear power plant, safely transport it to Yucca Mountain, and store it in the repository. Once we demonstrate that we can do this correctly, the nuclear waste question doesn't become an impediment to nuclear power and will no longer be of benefit to antis who always bring up nuclear waste as a reason not to continue with nuclear power."

The various debates could be extended for decades, perhaps until those oft-mentioned scientists of the misty future come up with a way to transmute nuclear waste into something nice.

However, until politicians, the media, and the general public stop taking as an article of faith that there is no solution to the safe disposal of spent nuclear fuel and until the nuclear industry, or Yucca Mountain, makes a concerted effort to demonstrate that this assumption is wrong, or until the DOE proposes a politically acceptable alternative, the construction of new nuclear plants will continue to meet with public resistance.

The progress, or lack of it, at Yucca Mountain is being keenly watched by other nations with nuclear waste. Meanwhile, Sweden and Finland are planning repositories that will be housed deep inside granite formations. Sweden intends to place its spent-fuel bundles in canisters surrounded by clay. Each canister will repose in its own chamber inside crystalline bedrock 1,650 feet below the surface. In contrast to Yucca Mountain, this arrangement makes for a cooler repository and a storage environment that will be nonoxidizing. Finland is investigating similarly deep subterranean disposal of spent-fuel rods in canisters with an inner shell of iron enclosed by a thick one of copper that would be placed in clay; it will test the waste package in real-world conditions. France has been exploring deep geologic disposal, possibly in clay or granite, for the unusable residues that remain after reprocessing fuel. A number of other member nations of the IAEA have also decided on interim storage with the possibility of retrieval for reprocessing. Mohamed ElBaradei, awarded the Nobel Peace Prize along with the IAEA, has called for more research and community outreach on the part of all nuclear nations to educate and reassure the public:

> The challenges we face in some ways make up a "Catch 22" situation: on the one hand, the lack of public confidence in the management and disposal of spent fuel and high-level radioactive waste hampers the effectiveness and efficiency of national efforts to construct geological repositories; on the other hand, in order to substantially increase public confidence, the nuclear community must have one or more operational geological repositories in which waste disposal technologies can be successfully demonstrated. But despite this "Catch 22," we can continue to make progress—and it is my view that, once the first country or countries have succeeded in placing a geological repository in service, the road ahead for other countries will be made much easier. In that sense, all members of the international community have a stake in the success of those national programmes that are the most advanced.

John Deutch and Ernest Moniz, cochairmen of the panel that wrote *The Future of Nuclear Power,* advocate a plan for the United States in which the federal government consolidates the spent fuel at one or more interim storage sites until a permanent facility opens. "The waste can be temporarily stored safely and securely for an extended period. Such extended temporary storage, for perhaps even as long as 100 years, should be an integral part of the disposal strategy. Among other benefits, it would take the pressure off government and industry to come up with a hasty disposal solution." The authors add that if nuclear power is tripled globally as a way to mitigate greenhouse gases in a major way, and if spent fuel is not reprocessed, enough of it and other high-level waste will be generated to fill a Yucca Mountain–size repository every three and a half years. "In the court of public opinion, that fact is a significant disincentive to the expansion of nuclear power, yet it is a problem that can and must be solved."

Given the delays and uncertain future of the Yucca Mountain repository, the House of Representatives voted in 2005 to authorize the DOE to pick interim storage sites for spent nuclear fuel and other high-level waste.

New Mexico senator Pete Domenici, a Republican who consistently gets reelected in a mostly Democratic state, has played an active role in energy policy for decades. At a hearing in 2006 in his role as chairman of the U.S. Senate Committee on Energy and Natural Resources, he announced that the United States would not store unrecycled spent nuclear fuel at Yucca Mountain. Proposing a twenty-five-year effort to develop reprocessing as well as interim storage plans that would have to be in place before the Nevada repository is opened, he remarked that it was quite obvious to him that "we are not going to be putting the spent-fuel rods in Yucca Mountain." He doesn't believe Nevadans will oppose the storage of whatever small volume of waste remains after spent fuel has been recycled several times.

We headed back down the mountain and across the Nevada Test Site, eventually pulling up at the Mercury Gate. "Look how remote we are here," Rip said. "It takes twenty-five minutes just to get from Yucca Mountain to Mercury for lunch—half an hour going fifty-five miles an hour."

We had hoped to have lunch at the test site cafeteria, but the kitchen was closed and we ended up getting food from a bank of vending machines. We sat at a table with a gregarious scientist who was also having a late lunch. He worked on the Yucca Mountain Project and was also a dedicated hiker and an environmentalist. Like others who are wary about all the controversy, he did not want to be identified.

"If we continue in an electricity-based civilization, we'll get over the belief that we can do without nuclear power," the man remarked with exasperation. "People don't put a stop to the things they need. I don't try to convince anybody to support nuclear energy. We have a job to do. This job is to comply with a federal standard for a nuclear waste repository in the state of Nevada. I encourage you if you disagree to get the standard changed. When people start talking about how nuclear power can be replaced by renewables and conservation of energy, I tell them I bet I conserve more than they do. I subscribe to *Home Power* magazine. I want to build myself a house dedicated to renewable energy and get off the grid. What people don't know is that the biggest cost to the environment isn't nuclear waste—it's lead batteries. They're thrown out in the trash. That's toxic waste that stays around forever. I drive a small, fuel-efficient car, I have only one kid, and I use compact, full-spectrum fluorescent bulbs. Incandescent bulbs are such electricity-wasters that they should be outlawed—or sold by permit only. *Everyone* should focus on conservation—that doesn't solve everything, but it buys you more time.

"Humans have a desire to find a source of energy with no impact, but every process has effluvia. I wish dams would be banned. Civilizations came and went because of the availability of firewood. Same could happen with fossil fuels. As a fuel supply runs out or gets too hard to extract, people go to the next source. Now we have a logical alternative, nuclear, but people are focused on nuclear waste. Spent nuclear fuel has its societal costs." Reining himself in, he added, "Look, we've got a lot of good science at Yucca Mountain. We've got lots of good people who live here and want to defend the project. They feel safe about it. What we're asking for is authorization to start construction. At dinner parties I don't argue with people who find out what my job is or try to convince them. I say to them when the matter comes up: take the trouble to make an informed decision. Yes, it's wrong to have SUVs, huge houses—all the things that use up energy. But you say you're not ready to give up your air-conditioning. When we decide that this problem is important enough, we'll do what's necessary—at an acceptable cost and at an acceptable risk."

18 THE GIGANTIC CRYSTAL

EARLY ONE MORNING, THE NUCLEAR TOUR set out on its final journey—to the Waste Isolation Pilot Plant (WIPP) in southeastern New Mexico. We went eastward out of Albuquerque, passing through what used to be empty land: a broad slope of sagebrush, rabbitbrush, and golden grama grass cut through by arroyos that ran down from the wall of the Sandia range. The rocket club had staged its test firings near here. Now, in a fancy delicatessen you can sniff Kenyan coffee and French Camembert, or in a big-box store buy Asian-made clothes and electronics.

I was recalling the clear dry light of my childhood here, and air so transparent you could easily follow the glinting trajectory of a homemade rocket. In fact, you could see for hundreds of miles. Now two busy four-lane streets lined with strip malls, apartment complexes, office buildings, car lots, and gas stations bustled with a continuous, noisy parade of sports utility or recreational vehicles and heavy trucks. It was just a typical inter-section in the city, and today it lay under a brown haze of dust, tailpipe emissions, and ash from a big wildfire—the drought had worsened, and on any given summer day now in New Mexico, somewhere a forest or grassland conflagration was in progress.

We passed the suburb that had replaced the ranch where I used to ride and where I first glimpsed the hollow mountain. As the bombs once stored there are being dismantled and the pits extracted and stored until they can be turned one day into fuel that will produce electricity, conta-minated materials from the process, along with other transuranic waste from the cleanup of our nationwide bomb factory, were being trucked to the WIPP.

The Chihuahuan Desert, the largest in North America, stretches from northern Mexico up into Texas, New Mexico, and Arizona. Mostly grass-land, with sagebrush, yucca, creosote bushes, agave, and the occasional juniper or piñon tree, the normally arid expanse briefly becomes lush after the August rains, with wildflowers dappling the gentle hills and flats. We headed south through ranchland, home to cattle drives and range wars and the grave of Billy the Kid, passing through the occasional cowtown, each with a dilapidated motel, some bars, and storefronts bleached by the summer sun and beaten by Blue Northers, the freezing windstorms that scream southward from the Great Plains in winter.

Heavy automobile traffic would not seem to be a problem in this sparsely populated area, but here the highway had been reinforced and widened to four lanes, and a divider had been placed in the center. Drivers have the Department of Energy (DOE) and the Department of Transportation to thank for arranging a bypass around Santa Fe to the north and improving the state roads so that they can hold up under large trucks hauling heavy casks of defense-related nuclear waste from federal sites around the United States.

We passed one of these trucks—a behemoth pulling a tractor-trailer with two big white cylinders containing transuranic waste. Luis, Marcia's youngest son, had been part of a group at Sandia that worked on the transportation plan. "They put together a computer backup system for the one that tracks WIPP trucks by satellite," she said. "They keep upgrading it."

"Hijackers wouldn't be able to divert a truck," Rip said. "They wouldn't even be able to penetrate the cab—it automatically locks the drivers inside—or the casks. You need big, specialized equipment just to open their outer and inner lids. Despite what you may see on TV shows, it's not worth it to a terrorist to steal what is basically a bunch of sludge and junk with less than one one-thousandth of the radioactivity of spent-fuel rods. There's never been a serious transportation incident involving nuclear waste. But the public is part of the equation, and if people's concerns aren't heard and clearly addressed, then you may have the best and safest design for a repository, but it's never going to open. Before WIPP opened, people needed to be assured and educated that every aspect was completely safe. The towns along the routes wanted to know what to do if by some tiny chance there was a breach. We had a demonstration truck and cask make stops along the route so that people could see them up close. Even though the chance of any bad accident or release was very, very small, we made sure local and tribal police and fire departments and hospital staffs received special training. Now, after seeing several years of safe transport, people accept that the trucks are safe."

On we sped on the superb road. The sky was deep, the air clear, and the shadows of towering white cumuli painted the hills with subtle purples, blues, and greens. Eventually we began to see cotton fields and orchards as well as livestock pens and feedlots. We breezed through Roswell, famous in certain circles for the International UFO Museum and Research Center. It's dedicated to the supposed crash of a UFO (most likely it was a weather balloon) and the government's subsequent concealment of its debris and of the bodies of space-alien pilots. We passed near a shrine commemorating the 1977 appearance of the face of Jesus on a flour tortilla a Mrs. María Rubio was making for her husband's burrito.

As we continued southward, the landscape became dryer and bleaker. The broad expanse, mostly flat, was interrupted at rare intervals by a salt pan, a sinkhole, a big cottonwood tree, a ranch gate (usually two pillars and an overhead crossbar, and a dirt road with a disappearing perspective), an occasional antelope herd around a water hole, or sheep being rounded up by a border collie. A forlorn atmosphere pervades the occasional town, which usually amounts to no more than a few adobe or brick buildings with peeling paint. You're not likely ever to find yourself in this windswept part of the state unless you're a rancher, a potash miner, or an employee of the gas and oil business or of the DOE.

As we continued south, we saw pump jacks, their heads plunging up and down like feeding dinosaurs, and refineries. Flares from natural gas burned brightly. Polished metal conduits of gas plants flashed in the sunshine. The air stank. Drilling for gas has increased in recent years. Some environmental leaders have become advocates of natural gas. However, leakage from drill sites adds methane, the worst of the greenhouse gases, to the atmosphere's burden, and natural gas in pipelines and at liquified natural gas terminals explodes from time to time, killing people. An El Paso Corporation pipeline explosion caused the deaths of twelve campers near Carlsbad in 2000. The World Nuclear Association ascribes eighty-five immediate deaths of workers and the public to natural gas per terawatt-year of electricity generated, and to nuclear power eight deaths among workers and none among the public.

Since 2000 the granting of drilling licenses for oil and gas has accelerated, with the federal government handing out thousands of them annually to companies like the El Paso Corporation. Up north, a favorite hunting ground of Rip's had been turned into a heavy-industry zone crisscrossed with roads traveled by big trucks and gashed by bulldozers and backhoes and covered with pumps, pipes, and tanks. Throughout the West on public lands and on the property of ranchers who don't own the mineral rights, the natural-gas industry has turned to coal-bed methane extraction, which has ruined grazing lands and polluted the water table, leading ranchers and hunters to make common cause with environmentalists. Without natural gas, however, we would be burning far more coal, a much dirtier hydrocarbon.

Beneath the highway we were traveling on lay one of the biggest sources of fossil fuels in the country, the Permian Basin, six hundred miles long and about one thousand miles wide. A low row of outcroppings off to the west comprises the eroded remains of the four-hundred-mile-long Capitan Reef, comparable in scale to Australia's Great Barrier Reef. Before continental drift moved the North American plate northward many millions of years ago, it was a lot closer to the equator, and the basin enclosed a shallow tropical sea that stretched from here all the way to Kansas. It's easy to picture. Just add water. When the sea evaporated, it left not only the reef, in whose varicolored calcium carbonate rock flowing water had carved out and elaborated the fairyland of Carlsbad Caverns, but also the remains of marine organisms and plants that became trapped under impermeable rock domes and turned into deep pockets of gas and oil; a hefty stratum of high-grade potash; and the two-thousand-foot-thick, contiguous Salado Formation, a vast salt bed.

Unlike the grains that fall from a shaker, this gigantic batch of sodium chloride had been transformed by immense pressure into rock covering more than a thousand square miles. The bed, 690 miles long and 260 miles wide, lies under a layer of water-impermeable, concretelike soil called caliche and on top of another stratum of impermeable rock. Salt beds are geologically unique: seismically inactive and with very little groundwater. Otherwise, the salt would dissolve. The weight of soil and rock above the bed presses the salt into rock. The Salado has stayed intact throughout the long journey of the continent; it has survived glaciations, hot epochs, heavy rains, and other climate changes over the last two hundred million years, and it has remained unaffected by the cataclysmic upheavals in the region that created the volcanoes and mountain ranges along the Rio Grande rift. The DOE chose a small piece of federal land on top of an untouched part of the Salado Formation twenty-six miles from Carlsbad for the home of the WIPP—or WIPP (in conversation, most people drop the *the*).

As we drove along, Rip told me some history. The land had been withdrawn from federal property controlled by the Bureau of Land Management. The DOE was WIPP's developer and the EPA was the regulator. The Nuclear Regulatory Commission (NRC) didn't participate. In 1981, the first shaft was drilled and tests began. Westinghouse had the DOE contract, and Sandia National Laboratories' official role was—and remains—science adviser, leading all the scientific research.

WIPP had needed to obtain certification from the EPA to open. In obedience to Congress, an EPA regulation demanded "reasonable proof" that the repository would be able to keep radioactive material out of the biosphere for the next ten thousand years, and WIPP had yet to meet that requirement when Rip began working on the project. In the mid-1980s, a coalition of antinuclear organizations took advantage of a new law that gave citizens' groups standing to bring lawsuits against projects on environmental grounds. "Some oppose WIPP because they figure it's the soft underbelly of the nuclear issue," observed Wendell D. Weart, Sandia's longtime WIPP project manager. He summed up the attitudes of the opponents this way: "If you can prevent WIPP from ever beginning disposal operations, you will prevent nuclear power and you will prevent nuclear weapons." Legal actions began slowly working their way through the courts.

In 1987, the U.S. Court of Appeals for the First District in Boston issued a decision on a challenge to WIPP, concerning EPA standards, that had been made by the Natural Resources Defense Council and other opponents of the repository. The court threw out the repository standards applicable to the WIPP project. "State legislators were saying that work on WIPP should stop because now there was no regulation," Rip said. "The political and legal obstacles were a major problem. As pressures grew, funding dwindled. The result was delays, more adversarial attacking, more time spent on defending attacks and less on research, and WIPP funding was going down fairly drastically."

While the WIPP staff maintained the shafts, tunnels, and equipment and practiced going through all the procedures that would have to be followed when the repository began operations, the DOE tried to get the project back on a timetable. "When the EPA people did an audit, they found that the project was good on quality assurance and transparency of all facets but that it lacked complete traceability of data from the point of measurement down in the mine all the way to the analysis," Rip said. "And the procedural aspects had to be squeaky clean. At that point they were not."

On visits to my parents, I followed the controversy. One independent tabloid kept asking how we could allow nuclear waste to contaminate the entire region for hundreds of thousands of years. I pictured dump trucks with glowing debris cascading out of their beds as they backed up to an open hole at Carlsbad Caverns National Park. And what if the stuff blew up? The protesters received a lot of press, and their message was clear: The consequences of all this volatile, radioactive material would be catastrophic.

Then, in 1996, a forest fire came within a mile of the waste enclosures at Los Alamos National Laboratories. And in 2000, one of the worst wildfires in the history of New Mexico swept within five hundred yards of twelve thousand waste barrels stored in negatively pressurized plastic bubble tents at the lab. The staff and firefighters fought back the blaze and saved the tents. "In Los Alamos, radioactive waste had been sitting around in barrels in bubble tents for quite a while," Rip said. "Protesters said that leaving the waste where it was in Los Alamos threatened the population and they said that transporting waste to WIPP threatened the population. And these people said that once the waste was stored at WIPP, radioactive material would contaminate the water table. It can't. And they said that the government should spend its money inventing some other way to get rid of the problem. As if we had not looked into hundreds of different ways already. These wrangles had been going on for a long time."

"But what about Carlsbad Caverns?" I asked. Their huge interconnecting vaulted chambers with stalactites and stalagmites in rainbow colors had been a favorite scenic wonder of mine since childhood. "What if the nuclear waste leaks into them?"

"Can't. They're about thirty miles west of WIPP up inside a plateau. You have to go *up* in altitude from the desert floor to get to the entrance, and the caverns only go down to a depth of 750 feet. The WIPP repository is 2,150 feet underground and there's a caliche cap preventing any radionuclides from coming up." Caliche, also called hardpan, is calcium carbonate—lime—and it cements together soils and rocks. "There's no way that the waste can travel through all the geological formations for miles and miles and get to the caverns."

"Well, I once read in the paper that if a chain reaction started in the waste, the whole place would blow up."

"Can't happen. There's just not enough fissionable material, not enough neutrons flying around. And in a relatively short time radioactive decay is going to make most of the stuff harmless."

"OK, but what if an asteroid crashed into WIPP?"

"An object that could plow that far into the crust would be so big that the entire region would be completely devastated, natural radioactive material from deep in the earth would be ejected, and WIPP would be a no-never-mind."

In 1987, after Rip and his colleagues made valiant but futile attempts to save the subseabed project, Sandia assigned him to WIPP. "He'd worked so hard to keep Seabed alive, going to bat for it in Washington," Marcia told me. "Then Sandia put him in the penalty box for that. It was awful. So he kept thinking aloud and pacing around the house, back and forth, like he does when he has a really tough problem. After a couple of days of this, he said he knew what to do: take the probabilistic risk assessment methodology from Seabed and apply it to WIPP. Nobody had thought of that. He got that tracker's gleam in his eye back."

Rip gathered a few Seabed veterans, including Leo Gómez, and they pored over the Westinghouse data, did a back-of-the-envelope performance assessment, and after a couple of hours they concluded that, if the data was accurate, WIPP was going to fail because of a brine pocket in the salt bed caused by a one-liter-per-day leak. The group wrote a paper saying that, based on the information at hand, the repository would fill with water in a few years and the salt bed would eventually wash away. Gómez noted that the projected release of radionuclides would have been "enough to cook a cow."

"When I showed DOE's WIPP manager our paper, his face got real long," Rip said. "But if you ran calculations based on that data, that flood was the outcome."

Some people associated with the project assumed that Rip and his group were troublemakers and tried to get them fired. Rip refused to back down, and he insisted that the only way to determine feasibility and break out of the legal and political impasse WIPP was in was to use probabilistic risk assessment to identify every potential pathway of every radionuclide out of the repository. "In the 1960s, an Environmental Protection Agency mathematician twenty years ahead of his time wrote a regulation that said that you had to do a probabilistic performance assessment to show the safety of geologic storage of radioactive waste in a salt bed for the next ten thousand years. Until Subseabed, that had never been done. EPA in effect said, 'Give us a probabilistic assessment and, in addition, all the likely features, events, and processes that could affect this site for the next ten thousand years.' "

Wendell Weart, in charge of the Sandia WIPP office, listened to Rip and allocated funding. Rip then set about assembling another complex, interdisciplinary, multidisciplinary, total-systems-analysis project using many of the people who had participated in the subseabed disposal program along with advisory panels comprised of sociologically oriented thinkers—archaeologists, historians, futurologists, linguists, environmentalists, and architects—to help create a warning system for distant generations. "I asked the whole bunch what they thought humans would be like in ten thousand years and how we could communicate with them." One participant, Gregory Benford, a professor of plasma physics and astrophysics at the University of California at Irvine and also an author of science fiction, was inspired to write an entertaining nonfiction book, *Deep Time: How Humanity Communicates Across Millennia*. "Our job," he writes, "was by far the furthest-out environmental impact report anyone had ever summoned forth."

"Scientists from everywhere heard about the new program and wanted to join," Rip said. "As usual, they came around with their Pet Rocks. Right away I got Mel Marietta on board. As with Seabed, we applied probabilistic risk assessment to the relevance of their projects. If the problem that they wanted to study had a chance of happening that was greater than one in a million, we took them seriously. I also wound up getting rid of a lot of unnecessary research projects that had accumulated at WIPP. That didn't make me too popular. But we saved time and money. Meanwhile, we had to answer to Sandia, DOE, EPA, and meet periodically with a panel from the National Academy of Sciences. We were also getting peer-reviewed by an outfit that was state-managed and DOE-funded, the Environmental Evaluation Group [EEG]. A scientist named Robert Neill headed it. They were the hired adversaries, brought in to assume the worst. That was fine by me. I wanted to hear about any and all problems. Bob would call me up and tell me the attack he was planning to make. It turned into a triangular sporting event: EEG would be anti, WIPP would represent the pro point of view, and the NAS panel members would be the referees."

If Rip's team could prove that neither the workings of nature nor human activities would significantly impact the repository over ten millennia, then it would be unlikely to fail even in one hundred thousand years.

"I went through Sandia and scouted out the people who thought outside the box," Rip said. "We put together a calculational machine that would not quit." Following the subseabed template, the team originated a means of identifying and examining all credible ways radionuclides could escape and estimating the contribution of each component of each pathway. "Mel Marietta kept improving on the methodology and made it just what I needed. We now had far better computer power than we'd had during subseabed and that allowed us to do a complete probabilistic analysis."

Rip's group achieved a breakthrough by applying probability analysis to *long-term* big science. "Probabilistic risk assessment had already been applied to reactor safety, and that was big science, but now we were doing it for a very long time span, and that was a first," he recalled. "It took several years for this method to be accepted, but now it's used by applied mathematicians around the world. The author of that EPA regulation was really seeing into the future." Now almost all the new EPA regulations are probabilistic, and this method is becoming the model for all performance assessments everywhere.

"We didn't stovepipe," he continued, meaning that there was plenty of interdisciplinary communication—engineers talking to hydrologists and geologists, social scientists talking to radiation biologists. "And we didn't guess or make conservative estimates. We knew very precisely from detailed geological studies what had been happening here for two hundred and fifty million years." Margaret Chu, a chemist who became Sandia's WIPP project manager, said, "There is probably no single piece of similar-sized real estate on the planet that has been more closely studied and thoroughly characterized."

Changes in the volume of rainfall over the centuries had to be included in the assessment, along with the concomitant increased flow of groundwater as well as a change in the erosion rate at the surface. The erosion turned out to be so small that it made no difference. When analyzed probabilistically, no natural features, events, or processes were found to permit the waste to escape the mine. However, what human intrusion could do to the Culebra aquifer, which lies under the desert floor and above the salt bed, did have to be taken into consideration. Drilling for oil and gas had to be included.

"The EPA said in one regulation, 'Consider drilling to go at the same rate it is today for the next ten thousand years,' " Rip explained. " 'Survey the number of drill holes in the Permian Basin and assume that number for the next ten thousand years.' We then calculated that there was a probability that twenty-four drill holes would go through the aquifer and the repository over that time period, and if they weren't adequately plugged up, water from the aquifer could flood the repository. And just the act of drilling could bring up waste. But *only* the drilling allowed the waste to come up, and the only thing to drill for would be oil, which will probably not be in use in a hundred years, let alone ten thousand. Hell, we'll be lucky if there's any oil to speak of left anywhere in *fifty* years. After the shafts are sealed, the site will be totally safe unless man fiddles with it. Even then, if some waste happened to get to the surface, it wouldn't be very dangerous." The most salient, single feature regarding safety remains the formation itself. "Those salt beds are two hundred and twenty-five million years old and will be there for another two hundred and twenty-five million years. That's over ten times the half-life of plutonium."

The completed analysis had to be approved by the EPA. "EPA people came in and lived with us until they knew minutely all the computer codes, the data of the distribution set, and they ran additional calculations to prove we'd done everything right. Those guys sat there at the terminals and told our guys what to do with data and distributions. EPA got a good sense of how robust the values were. It was like driving a Cadillac at one hundred and fifty miles per hour and seeing that it could do it. By the way, the EPA has the toughest regulations for nuclear waste disposal in the world."

After a couple of years of review, EPA accepted the results and put them out for public comment. Rip invited anti-WIPP groups to come to Sandia. "We wanted to learn from them; we wanted them to poke holes in our position. We offered them all the computer codes and use of our computers to run through the data and let us know if they found anything wrong. Most of the antis didn't take us up on the offer. One group spent a day or two with us. They never used any of the data."

"WIPP is not opening anytime soon. It will be years from now—if ever," Don Hancock, a political scientist who is the director of the nuclear waste safety program for the antinuclear advocacy group known as the Southwest Research and Information Center, announced in 1998. Hancock and others spent years leading a series of legal and regulatory challenges to block or delay WIPP's opening.

The lawsuits filed against WIPP were not on technical grounds but rather on procedural grounds. "The Environmental Evaluation Group didn't sue, and it was the only outfit that had the chops to do so and was the only one that came in and ran the models," Rip said. "DOE and EPA teamed up to defend WIPP, which is why EPA was so diligent in learning all about WIPP and running the codes. In the lawsuits, we won on the important issues because we'd anticipated every possible thing that could go wrong with the repository, we'd done the science and sensitivity analysis on each of those submodels, and we'd demonstrated to the judges that we'd addressed every potential problem and reduced uncertainties to a minimum. We generated stacks of unimpeachable data. The science was bulletproof."

In the courtroom, the DOE used the findings of Rip and his team to demonstrate that the repository was safe. Every probable scenario that could make it fail had been modeled, and every piece of data had a pedigree. The criterion held that the facility had to meet reasonable expectations of performance—not to anticipate scenarios that had only

an extremely remote possibility of occurrence or defied present understanding of geological processes. The probabilistic performance assessment, the squeaky-clean Cadillac, won the day for big science.

In March 1999, a federal judge in Washington, D.C., signed off on the
matter, and, a quarter of a century after WIPP's inception—and after
thousands of studies, tests, extensive laboratory research, computer-
modeling, and the invention of a method that fostered a whole new way
of predicting the future—workers in protective gear at sites around the
country began loading casks with radioactive waste, some of it possibly
dating back to the Manhattan Project, onto trucks that headed to WIPP,
cheered on by hundreds of Carlsbad citizens.

In Carlsbad, a pleasant town of about twenty-five thousand, with parks
that follow the gentle bends of the Pecos River, we checked into a motel
popular with engineers and cowboys. Early the next morning, as we were
standing in the lobby, a man with a graying beard and reddish hair
who wore jeans and steel-toed workman's boots entered. "Frank Hansen
knows more about salt than just about anybody," Rip said as he waved
him over. Hansen, our guide, was a member of Sandia's technical staff and
acting manager of repository investigations. He had a master's degree in
civil engineering, a Ph.D. in a branch of geophysics called tectono-
physics, and had been working on WIPP science for three decades.
(Hansen has since become part of the Sandia team at Yucca Mountain.)

After heading out of town, we passed oil tanks, pump jacks, and the
tall, dilapidated head-frames and tailings piles of potash mines, most of
them now closed. A potash refinery sent out plumes of smoke from its
stacks. Years of gray runoff from the refineries had created shallow, heavily mineralized ponds of calcium sulfate and salt that encrusted the
bushes along the banks with glittering crystals—a touch of beauty in the
midst of desolation. The water is undrinkable for humans. We saw a few
cattle cropping withered grass that must have tasted mighty bitter.

Potash, a potassium product used in fertilizer, once made this corner
of the state prosperous until Canada started exporting a superior kind,
thereby destroying the local economy. After most of the mines shut down
and the miners became desperate, the mayor of Carlsbad heard about a
search for a national nuclear repository, and city fathers lobbied government officials. Eventually the DOE, glad at last to hear from a community that welcomed nuclear waste disposal, hired the miners to excavate
the repository.

"Around here, if you don't like WIPP, you better keep it to yourself,"

Rip said. "It's brought in over a thousand jobs. Now, with the cleanup of sites winding down, shipments are going to be fewer. Carlsbad people are getting worried, because not so many people will be needed."

Politicians and businessmen in this corner of the state envision the region as a center for the study of energy, and, because WIPP has been such a success with the community, they remain hospitable to new facilities, nuclear and otherwise, that might start up here. A local branch of New Mexico State University trains scientists, engineers, and technicians to staff them. In 2006, an international consortium was granted a license to build a uranium enrichment plant in the nearby town of Eunice after the state was satisfied that the waste-disposal problem would be handled satisfactorily.

Thinking about how unhappy the state of Nevada was about Yucca Mountain, I asked, "Could commercial spent nuclear fuel and other high-level waste ever be stored at WIPP?"

"The original plan was to house the whole inventory of commercial and military spent fuel in a single national repository," Rip replied. "By the time WIPP began, the idea was that as a defense-related facility it would take spent fuel from the military branch of DOE. We also did a heck of a lot of work on salt for high-level waste. Heaters were run for a long time in salt beds to see if the thermal energy from the spent fuel would be a problem. It wasn't. Salt conducts the heat away. Technically, there's absolutely no reason why spent nuclear fuel and other high-level waste could not be stored in salt. That was what the National Academy of Sciences recommended. But first WIPP has to establish a track record of safety with military transuranic waste, because politically the idea of bringing spent fuel into the state is too much of a lightning rod. Spent nuclear fuel could go into monitored, interim storage at WIPP and be retrieved later for reprocessing."

"By the time you reprocess spent fuel half a dozen times, the volume of waste is vanishingly small," Hansen added, "more or less a thimbleful per bundle of rods, and those residues could easily go into WIPP."

"The DOE and the federal government decided that military and commercial nuclear matters had to remain separate," Rip said. "The Nuclear Regulatory Commission licenses commercial nuclear facilities and the EPA licensed WIPP. I don't even see the purpose of this arrangement. The chemical plants and oil refineries are under EPA jurisdiction and they're far more hazardous than anything nuclear. From just a technical point of view, the best place on dry land to store all nuclear waste—wherever it comes from—is at WIPP. We've proven that every way you can think of. We have traceability and transparency. Geologically and hydrologically it's the safest. There's room for it, and more panels can be mined out of the salt bed whenever we want. It's only politics and bureaucracy that stand in the way. So far, no one has made any public statements about putting commercial waste here. My gut feeling is that after the initial shock, the State of New Mexico would find it quite acceptable if paid appropriately."

We continued on through a landscape cluttered with still more pump jacks and other gas and oil industrial equipment. After a while the drilling rigs began to diminish in number, and we arrived at the four-square-mile WIPP parcel. I'd expected to encounter a vast, desert-gouging complex, and so at first I mistook the modest collection of buildings, shafts, and mounds of spoil for just another potash mine.

We were immediately immersed in the now-familiar world of plastic-picture-ID-radiation-badge-wearing personnel, security checkpoints, acronyms, pamphlets and posters written in the passive voice ("steel-toed boots must be worn in designated areas"), and prefab buildings. We were served glasses of juice and handed hard hats, protective glasses, and other accessories. Next came the obligatory safety video, perhaps scripted by an overprotective grandma. It admonished us to watch our step; to be careful of the elevator door because it could swing out and pinch us; in case of fire in the mine to evacuate along a prescribed route and use the self-rescuer we'd each been given. Open-toed or canvas shoes were forbidden. We were warned about slipping on ice in winter, about tripping; about ladders, forklifts, and trucks; about off-limits radiation areas, labs, confined spaces, and places where active mining was going on. In areas marked with the radiation trefoil, we were not to eat or chew gum. No smoking anywhere. Claustrophobes were advised that the tour might include tight spots. We were taught how to adjust the miner's lamps we wore on our heads to the smallest spotlight and were warned to keep the lamps on at all times but to be careful about shining them inadvertently into the eyes of others. This meant that during conversations I had to remember to keep my head cocked. And then there was the alarm system: a bell meant a fire or fire drill; the evacuation alarm imitated the yelping of the red-alert siren on *Star Trek*. We had to stay with our escort at all times.

Once we had all our equipment strapped on and before we took the elevator down into the mine, we were each given a numbered brass token so that our bodies could be identified in case we perished in an underground fire. The old-timer who handed out the discs said sweetly, "Y'all be careful now, you hear?"

Thus fortified against every possible misfortune, we went outside. Hansen led us past tall rectangular towers, each housing a mine shaft.

Behind the salt-handling shaft were neatly scalloped mounds, each of a different color: white, lavender, and pink—just like those expensive sea salts that gourmet stores sell. We entered a tall, corrugated-steel structure and passed through two air locks. Hansen opened a massive tornadoproof door and ushered us to the personnel elevator, which he described as the best conveyance of its kind anywhere. The ride was lurch-free, with the cables evenly humming.

The operator asked if we wanted to experience total darkness, and briefly turned off the light. Far above our heads hovered a tiny shard of brightness—the top of the shaft. We plunged into the black void, down and down at the rate of about four hundred feet a minute and backward through geological time. It took us over a minute to drop through layers of dune sand and alluvium. We next passed that stratum of cementlike caliche. Then we passed through limestone formed from skeletons of marine organisms, reddish-brown sandstone, mudstone, and siltstone. The rock bounced the noise of the cables back at us until about a thousand feet down, when the racket abruptly became muted, absorbed by the crystalline structure of extremely dry, nearly pure rock salt. We'd entered the Salado formation and kept going until we reached its center. After about six minutes, the journey concluded.

About half a mile underground, we emerged from the cage into a hushed, warm, dimly lit, shadowless, high-ceilinged room: a man-made cavern. A fresh salt breeze blew, thanks to gigantic fans. Hansen assured us that the repository was well ventilated and the air was filtered before it was cycled back to the surface. "The site is set up to start clean and stay clean," Hansen said. He added that some of the electricity to run all this equipment came from wind turbines; the DOE has mandated that its facilities get 7.5 percent of their energy from renewables.

The air was so much drier than the desert that our lips and hands became parched. Despite safety glasses, our eyes were gently stung by fine salt dust. Adding to the uncanny impression that we were near the ocean was the taste of salt on our lips. In fact, a quarter of a billion years ago this had been the tidal flats of the Permian Sea, which had lapped here for so long that layers upon layers of salt had been deposited. They were eventually compressed into rock into which tunnels and chambers had now been carved by gigantic drill bits studded with monster teeth that left deep, sparkling, concentric grooves. Our footfalls on the scoured floor were silent; the sounds from a group of men and equipment in the middle distance were muffled and echoless, despite the volume of air and the hard, glassy appearance of the walls. Vanishing perspectives and the large scale of the chambers toyed with the senses. Here there were no straight lines: the plasticity of the salt caused walls to bulge, floors to hump, ceilings to bow. The fluorescent lighting cast no shadows. The circular patterns along walls and ceilings fooled the eye into seeing sculpted columns, vaults, niches, bas-reliefs—as in an ancient temple. That level band of rose-orange that Rip had described ran along the wall at shoulder height (it continued for thirty miles). This non-Euclidean labyrinth, something out of a myth, seemed endless but actually covered a little less than a square mile.

When I got my bearings, I noticed a sign:

WELCOME TO THE WIPP UNDERGROUND.
YOU HAVE JUST ENTERED AN ENVIRONMENT COMMITTED TO SAFETY.

We walked toward a high-ceilinged tunnel with mining machines parked along corridors, ready to do more work when needed. They could never be brought back to the surface again, because they'd promptly rust. Ever-present salt dust, making rainbow halos around the light fixtures, corrodes everything the instant the slightest amount of humidity comes into contact with it. But the environment here is so very dry that the corrosion only occurs when equipment is taken back to the surface. DOE cameras remain here, too, because aboveground their salt-dusted innards would attract moisture and disintegrate, even in the arid desert. I'd been warned not to wear a watch for that reason, but I'd forgotten and thought it would be safe in a zippered pocket. Some months later it stopped working and the repairman who replaced the rusted gears asked me if I'd worn it in the ocean.

Salt does not permit right angles to last long. The miners keep scraping the floors flat and planing the walls, but inevitably all such efforts are obliterated, and surfaces bulge and bow. Like the red clay of the midplate seabed, salt under pressure becomes plastic. And it never sleeps. From the point of view of the salt bed, the mine is a hole that has to be filled in. "The earth tries to heal itself, growing back together again the way a wound in the body does," Rip said.

Hansen pointed upward. "Salt creeps." We saw mesh covering the ceiling and a system of bolts and metal plates that automatically adjust to the continuous pressure exerted by the salt as it seeks to fill any empty space it encounters. At a predictable rate, the floors of the tunnels are always growing upward, the ceiling downward, and the walls closing in. WIPP opponents have claimed that this tendency indicates just how "unstable" the mine is. But it's this very characteristic that secures the

The Waste Isolation Pilot Plant

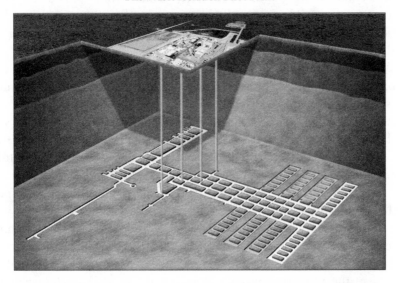

U.S. Department of Energy
Waste Isolation Pilot Plant
Below the surface operations, the colored bands show the different strata, the palest one being the Salado formation, the location of the underground repository, with its corridors and panels (large chambers containing emplacement rooms). The four vertical lines indicate the shafts. To the right is the waste-disposal area, and to the far left, the chambers designated for research projects.

waste. The warmer the temperature, the more quickly salt creeps, filling in fissures and smoothing out surfaces, making it an ideal medium for isolating nuclear waste, which, stored in barrels and placed in chambers in the salt bed, will eventually be enclosed by the salt, entombed in this virtually impermeable medium where it will remain for millions of years.

"The creep deformation of salt is extremely well understood," Hansen told us. Since 1974, he'd been studying the conditions here and his research was documented in over forty technical publications.

"An average rock weighs approximately 250 pounds per cubic foot," Rip said. "For every foot deeper you go you add on 250 more pounds. At a depth of 2,150 feet the weight is 537,000 pounds on a one-foot square. That pressure deforms salt at the rate of three inches per year."

We climbed into an electric cart and rode through a central gallery that displayed plaques with the names of supportive senators and governors of states eager to get rid of their military transuranic refuse. From time to time, a large, specialized, noisy vehicle would pass us carrying heavy equipment. The tunnels had street signs—S-550 was one—and big air-lock doors that blocked our route. Hansen would stop and reach up and pull a rope to open and close them.

As we rode along, Rip and Hansen described how Wendell Weart, the project manager, had scientists conduct a test of permeability in the tunnels, mainly to refute persistent charges by Don Hancock and environmental groups that there was so much brine in the repository that the salt bed would melt, and radioactive waste would be washed into the water table and ultimately into the Pecos River. An impossibility, Rip had told me, chiefly because of the impermeable rock strata above and below the bed that prevent any seepage into or out of the repository. "The anti-WIPP coalition was using that old brine scenario and assuming it was the likeliest outcome when in fact it was the result of a typical deterministic calculation done with worst-case values. Some droplets on the ceiling in a research section of the mine had been *conservatively* assumed to represent the amount of water throughout the mine. Which is why we wrote the report saying if that assumed scenario were true, the repository would fail."

"So what happened to the unrepresentative brine pocket that started the whole thing?" I asked.

"It dried up."

The plasticity of the salt bed relies on a tiny amount of brine. It helps the broken pieces to heal; it restores bonds. Instead of flowing, the water in the formation is trapped in the salt crystals in tiny bubbles that Rip used his miner's lamp to illuminate in a chip of rock salt. This liquid, being millions of years old, is an elixir for climatologists who use it to gauge the oxygen content of earth's ancient atmosphere. Dinosaurs were bathing in that water when our mammalian ancestors were insignificant bits of furry flesh cowering in burrows or clinging to treetops.

Because a half mile of soil and rock kept out most cosmic radiation and because the natural level of uranium in the salt bed is minuscule, background radiation here is extremely low. For this reason, Gómez and many of his radiation biologist colleagues around the world believe this location would be optimum for ultra-low-dose research. The subterranean laboratory would contain microorganisms and small mammals, and a surface section would house control groups. By observing how organisms fare in an ultra-low-dose environment as compared to a normal one in which exposure to natural background radiation is vastly greater, scientists, as I've already mentioned, hope to test which is more accurate: the linear nonthreshold hypothesis or the threshold hypothesis. Does low-level radiation in fact cause malign health effects even in the smallest doses? Or below a certain threshold of exposure is radiation insignificant, or even healthful? To get accurate data, natural isotopes such as potassium-40 and radioactive trace elements normally present in the feed would have to be eliminated, and all the water would have to be distilled for the same reason. Even though nuclear waste might be stored a few tunnels away, the lab would remain free of contamination because of the natural shielding provided by the salt and because of special ventilation and air locks. WIPP officials have heartily endorsed the project, in part because polls and focus groups here and abroad indicate that people don't mind a nuclear waste disposal site if research and experimentation are done there: what might be a negative perception becomes positive if learning is attached.

When WIPP is completed, there will be eight panels, subdivided into seven emplacement rooms, each 33 feet wide and 13 feet high. We visited one chamber whose entrance was covered with a canvas curtain. Inside, about ten thousand drums were stacked in a horseshoe pattern rising almost to the mesh-covered ceiling, which was beginning to descend. The floor had recently been scraped down forty inches but, when the room was full, would be allowed to resume growing to fill the empty space. "Mother Earth is going to come together here," Rip observed.

The barrels looked oddly small. White packs of magnesium oxide surrounded them to neutralize gases that might be released by the biological decay of any carbon-containing material inside. Once a panel is filled, it will be sealed off and the salt will be left to do its work, ultimately enfolding and compacting the drums to about a third their present size. The waste will thus be immobilized, very difficult to reach, and isolated from the biosphere, probably for millions of years.

Just what was in those barrels?

Back on that chilly day when we were touring the Idaho National Laboratory (INL), we'd stopped at the Radioactive Waste Management Complex, a major cleanup operation, and Jeff Hahn, who did public relations for the contractor in charge, BNFL—British Nuclear Fuels Limited—had given us a tour. He looked to be in his late twenties; he'd previously worked for the DOE in the alternative energy field on wind and solar power, fuel cells, and even the hydrogen car prototype. Although enthusiastic about renewables, he also approved of nuclear power plants, saying with a shrug, "All energy is nuclear."

Next to a barracks at the site were some boxes with Plexiglas windows that let us peer at examples of waste headed for WIPP: tools, plastics, wood, glassware, discarded coveralls, unidentifiable pieces of equipment, and sludge that had been contaminated with transuranics—plutonium, americium, curium, and the like. Materials that also contain toxic chemicals must be classified as "mixed waste." Most transuranic waste is "contact-handled," that is, the radiation it emits doesn't require heavy lead shielding or remote manipulators.

Starting in the 1950s, bomb-production refuse began arriving at the Idaho lab and was interred in railroad cars covered with tarps, plywood, and soil. Recent concerns on the part of environmental groups about the aquifer have led the State of Idaho to decree that all nuclear waste had to be removed from INL in a timely fashion. In the 1990s, the DOE put a temporary building—at seven acres, the biggest single-span structure in Idaho—over the burial site and began excavating the containers. The day we had visited INL, people in protective gear on top of a berm were using various tools to get at a buried box, laboring as carefully as archaeologists I'd once seen on a New Mexico dig. Under bright lights, a huge vacuum cleaner roared as it sucked dirt away. Forklifts and cranes moved boxes and drums. The project went on around the clock: sixty-five thousand cubic meters of Idaho waste had to be characterized and shipped to WIPP by 2015.

Had people been exposed to any radiation here? My dosimeter registered none, and Hahn told us that we were not going to receive any more than we'd have gotten in our hotel in Idaho Falls.

The resurrected containers were taken to another big temporary building to be warmed up after their sojourn in the cold ground, inspected, and X-rayed to determine whether the contents included any liquids, such as gas, oil, or water. Operators viewed video displays of the containers and employed remote manipulators or glove boxes to search out paint cans, gas cylinders, and other contents prohibited at WIPP by the New Mexico Environment Department.

A lot of nuclear waste turns out to be sludge: cement sludge; parfait sludge, which is soft material layered with cement; and soft sludge, with a consistency of peanut butter. Each drum of sludge glided along a conveyor belt to a chamber with a ventilation and filtration system where a

hole was drilled and gases tested. Workers assayed debris for fissile contamination in a room with thick walls to prevent the delicate sensing
machines from picking up natural background radiation. Core samples of
waste went to a chemistry laboratory for analysis before winding up back
in drums bound for WIPP. Material from opened drums was transferred
to new ones. At the end of the line, each one was given a bar code so that
it would have complete traceability from here in Idaho all the way to a
room in the WIPP salt bed.

All this exquisite attention has a price. BNFL earned about $5,500 per
cubic meter of waste it processed. In order to make a profit, BNFL ran
the operation around the clock, seven days a week. It takes three person-
days per drum. To characterize a single drum of transuranic waste so as to
meet all regulatory requirements costs about $9,000.

A Sandia scientist in the Carlsbad office later told me about giving a
talk to an Albuquerque Chamber of Commerce group about WIPP.
He'd expected to receive the usual denunciations about how risky it was
and was surprised when the businesspeople instead complained about
taxpayers' money being irrationally lavished on so much overengineering
and redundancy. "Billions of dollars could be saved if characterization
were simplified," he said. "The problem is all the politics and paperwork.
Most countries don't even spend the time." Many people in the nuclear
world told me that because of fundamental misperceptions on the part
of the public about low-level radiation, the process of cleaning up transuranic and other contact-handled waste required many steps. "All that
effort then has the effect of confirming the worst fears in the public
mind—that those contaminated coveralls must be terribly deadly," one
person said. "The elaborate fortifications at nuclear plants encourage a
similar point of view about nuclear power generation: that it must be an
immeasurably lethal threat if so much concrete has been poured and so
many barriers thrown up."

At INL, in another high-ceilinged, barnlike building that was open
on one side so that trucks could enter, we watched as a young woman
operating a crane picked up a package of seven barrels bound together
with shiny clear plastic wrap and lowered it into a white cylindrical cask
eight feet wide and ten feet long called a TRUPACT-II.

As part of the education program for the citizenry along the routes
to WIPP, the DOE had a traveling exhibit displaying a model
TRUPACT-II with a window in its side. I was able to see it at WIPP. I
don't believe that anyone who takes a look at a TRUPACT-II up close
could remain concerned about it ever breaking apart and leaking. The
cask has multiple layers of protection: a protective stainless steel skin

about half an inch thick, ten inches of shock-absorbing material and padding, an outer containment vessel, and an inner containment vessel. The cask weighs over six tons empty and nearly ten tons when filled.

Using remote manipulators, the crane operator placed a small inner lid on the cask and then a big one. It would be quite a feat to remove the lids without appropriate equipment. Then the cask was vacuum-sealed and loaded on a truck that would ultimately head for WIPP. But because of the nationwide orange security alert thanks to the American invasion of Iraq that week, all shipping of nuclear waste had been halted.

As is the case with casks transporting spent nuclear fuel, the TRUPACT-II has had to pass the same very demanding tests before being certified by the NRC. Drivers are specially trained and must maintain impeccable driving records, even off duty. Transport, always monitored by satellite, takes place at night, when there's the least traffic. "To the best of man's ability, we've done everything we can to prevent accidents," a WIPP official told me.

At WIPP, in the room with the ten thousand barrels, we saw some deep horizontal boreholes a few feet in diameter that reminded me of the catacombs of Rome; they will entomb steel canisters containing "remote-handled waste"—high-level waste, contaminated by cesium and strontium, that is very hot both radioactively and thermally. This heavily shielded material will be robotically moved and inserted. In 2007, after WIPP had demonstrated an unblemished safety record, shipments of this hotter waste began. Present regulations, born of political compromise, permit only 2 percent of all the waste here to be of the remote-handled kind. Most of the high-level military waste is supposed to go to Yucca Mountain, but everyone I met at WIPP believed that, from a scientific and engineering point of view, WIPP could take all of it.

"WIPP is a better place for it anyway," Rip said. "WIPP could safely hold all the nuclear waste in the world. Six million cubic feet—585,000 thousand barrels—is the limit by regulation here. In practice, the mine, or another mine next door, could take millions. But it's better to have a small success now than go for the whole enchilada. We have to show everyone how safe it is."

"There's overkill on safety here," Hansen said. "Fourfold redundancy. The EPA demands a rigorous process and the New Mexico Environment Department is even tougher. If a barrel were accidentally pierced it would shut down the repository for months. Since the program started over twenty years ago, only one person has died as a result of being here, and he took a fall in a shaft. Right now, you're safer than you would be in any other place in the world."

Two outside watchdogs have concurred. Robert H. Neill, head of the EEG group, appointed by the state to oversee WIPP when Rip was working on the project, stated after seven years of successful operation that "we have not had any major spills, problems, transportation accidents." No one in New Mexico had been more vociferous about the site than Don Hancock of the Southwest Research and Information Center. Even he now agrees that WIPP has been operating much better than he had anticipated.

Rip pressed his miner's lamp against a tunnel wall, illuminating a world I had never suspected. When I gazed at that soft, diffuse light passing into the interior depths of the rock, I realized that we had penetrated into the heart of a vast crystal.

Ears popping, we rode the elevator half a mile back up from the repository to the surface, where we were given a tour of the waste-handling facility, a cavernous structure like an airplane hangar.

When the TRUPACT-IIs arrive, they're sniffed by detectors, swabbed down, and then opened. Of all the thousands of shipments made thus far, the exterior of only one cask had been found slightly contaminated—and it was sent back to Idaho for correction. We passed negatively pressurized bays set up for receiving the casks, unpacking the cargo, and ushering it into a huge elevator for the trip down to the salt bed. Following a cordoned-off path that kept us well away from any waste packages, we walked among catwalks, cranes, forklifts, massive devices for removing the cask lids, and enough other gleaming machinery to furnish the sets of many science fiction films. It was the cleanest place I'd ever seen—and at this point I'd visited a lot of spotless facilities. There were continuous air monitors everywhere; all the air was passed through the finest of filters.

After going through a series of air locks—steel doors a couple of feet thick filled with lead shot—we entered the remote-handling facility, where shielded waste would one day arrive via a separate entrance to the building. Resembling the hot cells we'd seen at Argonne-West, the place had three-foot-thick steel and concrete walls, and thick panes of glass with mineral oil sandwiched between them to prevent distortion. Technicians outside the room will use manipulators to weld and repair contaminated equipment and otherwise deal with highly radioactive material.

Over the years Rip has made presentations at various national and international conferences on nuclear waste, probabilistic risk assessment, and computer simulation of geological scenarios. "WIPP has become a model for the international community," he told me. "Russia and Europe look to WIPP, since deep geological disposal is an international issue. We're transferring the technology to other countries. But their people say that a repository like ours is too expensive. Almost $2 billion have been spent on WIPP. To operate and maintain it costs $180 million a year. That's why at some point nations are going to have to get together and share a repository."

After the plutonium-contaminated defense sites are cleaned up, the next material to go into WIPP will be components from the dismantled bombs that were in the hollow mountain and elsewhere during my childhood. According to present plans, by 2034 the repository will be filled, and all aboveground traces of this temple of technology will be destroyed along with the rest of the buildings, the roads, and the parking lots, and the debris will be bulldozed into the shafts, which will then be backfilled. Hansen led the team that determined the best method for sealing them up: layers of concrete, salt, and bentonite, which is, basically, clay cat box litter, a sorber that will trap any water and that will immobilize any escaping radionuclides—and more salt, and more concrete. In a century, as the salt bed heals the fracture that constitutes the repository, the tunnels and rooms will be obliterated.

The land, restored to its previous condition with desert plantings, will be actively guarded for one hundred years. Warning markers will be erected and information about the site preserved in several major languages, and some minor ones, including Navajo, both there and in archival repositories around the world.

Part 6 **BORROWING FROM OUR CHILDREN**

We do not inherit the earth from our ancestors.
We borrow it from our children.

— AUTHOR UNKNOWN

We regard our native land as a power which acts of itself, and relieves us each of exertion. While with them I thought only about doing the Indians good. But back among my fellow countrymen, I had to be on my guard not to do them positive harm. If one lives where all suffer and starve, one acts on one's own impulse to help. But where plenty abounds, we surrender our generosity, believing that our country replaces us each and several. This is not so, and indeed a delusion. On the contrary the power of maintaining life in others lives within each of us and from us does it recede when unused. It is a concentrated power. If you are not acquainted with it, your Majesty can have no inkling of what it is like, what it portends, or the ways in which it slips from one.

— CABEZA DE VACA'S
LETTER TO THE KING OF SPAIN,
FROM THE INTERLINEAR
VERSION BY HANIEL LONG

19　THE IRON CHAMBER

ABOUT A DECADE AFTER THE FLIGHT into Albuquerque over that mysterious mountain, my questions about its secrets ultimately led me to a small room walled with iron a foot thick that shielded the interior from natural background radiation. Wearing a hospital gown, I lay on a table. Under it, and just above me, the most sensitive detectors in the world silently searched my body for internal radioactive contamination.

The chamber, licensed by the National Institutes of Health, was located in the Chihuahuan Desert at New Mexico State University's Carlsbad Environmental Monitoring and Research Center. The center analyzes natural and man-made radiation in the local soil, air, water, plants, animals, and—thanks to its Lie Down and Be Counted community outreach program—humans. Before the procedure, I'd filled out a questionnaire that asked, among other things, whether I'd been anyplace where I have might have been exposed to excess radiation. I'd grown up near a nuclear weapons complex, where research reactors still operated and nuclear waste from the Manhattan Project and the cold war remained in temporary storage. My mother had been intensively irradiated to treat cancer when I was a teenager, and over the years I'd had a couple of routine computerized tomography (CT) scans, dozens of mammograms, and so many dental X-rays I'd lost count. And then there was the fallout from atomic bomb tests, and the Chernobyl reactor's plume that had spread around the globe. And, by the way, I'd also visited a uranium refinery; experimental reactors at national laboratories; the graveyard of the Three Mile Island II reactor; a nuclear submarine; nuclear power plants; a coal-fired plant; the Nevada Test Site; diagnostic medical radiation clinics; and nuclear waste facilities.

David Schoep, who administered the scan and who had the interesting title of *In Vivo* Bioassay Manager, had looked over the form and asked, "Are you a nuclear worker?"

"No, just curious," I replied. No point in telling him about the hundreds of conversations with Rip and other scientists, mathematicians, and engineers; stacks of peer-reviewed documents; security checks by burly guards holding big guns; excursions across sagebrush steppes, deserts, and mountain ranges and inside a mountain and down to the underworld; glimpses of yellow rock crystals, a butterfly garden, a fish-friendly dock, a shimmering Dresden-blue spent-fuel pool, people in a slow, contemplative dance remotely manipulating objects in hot cells; and enough flame-red FiestaWare to stock a large china cabinet.

The scan took half an hour. I could have listened to music or kept the lights on, but I preferred darkness and quiet. I contemplated the possibly extensive radioactivity I might be harboring and what the findings would mean to me. What would I do if Dave told me that I'd accumulated a problematic internal dose? Despite my potentially radionuclide-loving microfiber pants, so handy for travel, I'd never managed to set off any of the previous radiation detectors—and by now I'd encountered over a dozen of them. If it turned out that I'd picked up internal contamination, usually not detectable with external monitors, would I revert to my original assumptions? Would I decide that the nuclear world was intolerably dangerous, that I'd been lied to, that all reactors should be banned?

I'd come to realize that in the controversy about nuclear power and the disposal of nuclear waste, radiation is the most misunderstood and misapplied of topics. As the novelist George Eliot described a controversy in *Middlemarch*, "Everybody liked better to conjecture how the thing was, than simply to know it; for conjecture soon became more confident than knowledge, and had a more liberal allowance for the incompatible."

Any dose I might have received had to be low. The chance of a relatively few extra rays and particles depositing enough energy in my DNA to break chemical bonds, thus leading to cancer, would be extremely small. The probability of cells in such a cancer undergoing genetic changes sufficient to cause it to spread was even smaller. In any case, as an American fortunate enough to have health care and therefore a much greater probability of living longer than if I had been born in, say, Haiti or Sudan, my lifetime chances of developing a malignancy were around one in three or four—since in prosperous countries more people survive long enough for cancers to arise. Statistically the greatest risk I'd faced during the entire tour was from all the automobile travel. Frequent jet flights and journeys through the high-altitude uranium-rich West, however, had exposed me to greater radioactivity than if I'd remained at sea level in my wooden house, which rests on low-radon sandy loam.

But any discovery of extra radionuclides in my tissues would not, I realized, alter the cumulative impact of the discoveries I'd made during this surprising adventure. My conclusions about the importance of nuclear power and its risk relative to other forms of large-scale energy generation would stand. Respected professors, radiation biologists, epidemiologists, and other researchers who have seen firsthand the worst that radiation can do all told me that they favored nuclear energy and would not be averse to living near nuclear plants. I'd seen that, for the public, uranium is cleaner and safer throughout its shielded journey from cradle to grave than our other big baseload electricity resource, fossil fuel.

The Environmental Protection Agency calculated in the 1990s that during a period of ten thousand years the nuclear waste stored at the Waste Isolation Pilot Plant (WIPP) would probably cause no more than a thousand deaths and, more likely, about a hundred deaths. The EPA calculations came from conservative estimates based on the linear non-threshold hypothesis. It's likelier that far fewer, if any, deaths caused by the toxicity of the waste will occur, because the naturally impermeable repository, once sealed, will be so inaccessible and remote. To be prudent, agencies follow standards of radiation protection derived from that hypothesis, which equates zero risk with zero radiation, moderate risk with moderate exposure, and high risk with high exposure. Predictions of cancer deaths in large populations from low-dose exposure (below 10,000 millirem) are derived from this assumption. But, as I've said, such collective-dose estimates are considered by many in the field to be very unrealistic. Populations experiencing higher levels of natural background radiation in places like Denver and Albuquerque do not have higher rates of cancer. But average exposure to Americans from natural background radiation has now been surpassed by diagnostic medical radiation. A study by the National Council on Radiation Protection released in 2007 reported that the per-capita dose of ionizing radiation from clinical imaging in the United States increased almost 600 percent from 1980 to 2006. Dr. Fred A. Mettler Jr., principal investigator for the study, called the finding a "sentinel event." Recent low-dose studies have indicated that below a certain threshold, not yet determined with exactitude, the body responds to radiation by destroying and eliminating damaged cells, thereby helping to *prevent* the initiation of cancer.

The National Academy of Sciences panel on the Biological Effects of Ionizing Radiation VII (BEIR VII), while continuing to endorse the nonthreshold linear hypothesis, has called for further study of low-dose radiation. There is wide agreement on how high doses affect the body, but any exposure we receive is probably going to be low-dose unless we are in a cancer-treatment program or undergo medical imaging. Discoveries by the Department of Energy's existing low-dose radiation laboratory, as well as by the projected ultra-low-dose laboratory at WIPP, may change public perception about radiation. This research may have far-reaching implications not only for those of us who are alive now and making decisions about the future of our energy usage but also for our children and grandchildren and their offspring as they deal with the consequences of our reckless experiment with global temperature disruption.

Although the EPA prohibits nuclear facilities from exposing the public to more than 15 millirem per year, they actually emit far less radiation than would cause even that minimal exposure. To meet that rather arbitrary standard, we spend billions of dollars to remediate sites to a far lower level than natural background radiation. Meanwhile, vacationers who fly from New York to the West and ski in the Rockies are increasing their exposure by hundreds of millirem. And the U.S. Capitol's natural construction materials exceed local background radiation dose rates and emit 550 percent of the typical dose rate allowed around nuclear plants and about thirteen thousand times more than the average individual dose rate from nuclear power production worldwide. Our elected representatives work in a place that has a much higher dose rate than will be permitted at the Yucca Mountain high-level nuclear waste facility.

And what about low-dose exposure from those hypothetical dirty bombs? Should New Yorkers permanently abandon midtown or downtown Manhattan if they're receiving 25 or 50 or even 100 millirem a year more than they were before a radiological dispersal device went off? If they moved to New Mexico they'd be increasing their exposure by 200 millirem. Over 80 percent of scientists polled think that below an exposure of 100 millirem per year, radiation is unlikely to cause negative health effects. Should we spend a projected $60 billion to fortify Yucca Mountain with many redundant barriers on the assumption that they'll protect a hypothetical human in the very remote future from less radiation than Mother Nature showers upon people living today in northeastern Washington State?

I'd now seen many elaborate ways to protect humans for the next ten thousand years from low-dose radiation exposure. Estimates about the cost of this protection range from $350 billion to $1 trillion. Concerning these enormous sums, nuclear-power pioneer Ted Rockwell points out that the billions thus spent—in his view, unnecessarily—"don't disappear down a rat-hole. They go into some rat's pocket. That's why the expenditure is tolerated and encouraged—by some rat's lobbyist."

Some cancers may occur because of low-dose radiation but would be hard to detect among those caused in the population by, among other things, spontaneous, random mutations. Most scientists on both sides of the threshold debate agree that the possible increase in malignancies following irradiation is always minuscule. We already expose ourselves to other weak environmental carcinogens—from fossil-fuel combustion, petrochemicals, and the chemical industry, for example—that are not regulated at all or loosely so. Even strict proponents of the linear nonthreshold hypothesis say that it can be misapplied. Continued experimentation and better results using newly available technology could provide a more accurate basis for regulation, as the BEIR-VII report has suggested.

I'd signed up for the special high-tech scan while attending the international Ultra-Low-Level Radiation Effects Summit in Carlsbad. It had been arranged by Leo Gómez in his role as principal scientist at ORION International Technologies. For years Gómez had been trying to get funding to explore the concept. A DOE grant led to the conference in January 2006. His cochairs were David Brenner, who is a professor at Columbia University and an advocate of the linear nonthreshold hypothesis, and Otto Raabe, who is professor emeritus of the University of California at Davis and an advocate of the nonlinear threshold hypothesis. Twenty-six radiation biologists and health physicists from universities, from radiation-effects research agencies here and abroad, from the U.S. EPA, and from U.S. national laboratories met and discussed their findings and their wishes, hopes, and doubts.

The scientists toured WIPP and marveled at the unusual environment. They all agreed that the negligible background radiation in the subterranean salt bed made an ideal location for an ultra-low-dose lab, and the majority felt that experiments there could contribute significantly to clearing up mysteries presently veiled by the overwhelming presence of natural radiation.

In a report about the summit, Gómez, Brenner, and Raabe wrote:

> The facility will allow radiation standards to be safely and scientifically established with full scientific consensus. The economic benefits of relaxing current safety standards that are based on an unproven, conservative linear nonthreshold model would far outweigh the cost of this facility. The benefit of replacing current perceptions in the minds of the American public with data collected in an ultra-low radiation environment will allow progress especially towards increasing the use and efficiency of nuclear energy and reducing our reliance on fossil fuels, with all of their accompanying international and ecological problems.

Experiments at such a lab at WIPP may have results that could affect us as well as our children on a personal level. The report concluded:

> Cancer is a fact of life. DNA damage is one of the causes of cancer. Natural mechanisms exist that either repair the damaged DNA or signal the cell to die (apoptosis); however, in carcinogenesis some mechanisms fail. Research conducted in an environment where most confounding factors can be controlled . . . should contribute immensely to the understanding of carcinogenesis, with a potential to contribute information to preventing and curing cancer.

As I lay on the table in the dark, my thoughts soon traveled from rays and particles outward, into the world of gross matter, shining with the radiation of the nearest star: the desert, the continent, the oceans, the entire shimmering blue Earth as seen from space, her translucent ice caps glowing like white lanterns.

When I'd asked Rip about the hollow mountain full of bombs, I didn't expect to hear about melting polar ice, rising sea levels, acidification of the ocean, thawing of the tundra and the concomitant release of more greenhouse gases, more frequent and more violent hurricanes, disruptions to agriculture, environmental refugees, more wars fought over dwindling resources, and the potential destruction of our present way of life, all because of how we get most of our energy. Almost everyone I met on the nuclear tour was preoccupied with the risks from our reliance on fossil fuels. The United Nations Intergovernmental Panel on Climate Change (IPCC) has been issuing peer-reviewed reports based on the work of hundreds of scientists. In May 2007, the IPCC's draft report on mitigation of greenhouse gas effects stated that only a concerted action requiring a huge monetary investment and a profound change in policy can avert a climate collapse, and that intervention must be accomplished by 2020. According to the IPCC, the massive deployment of several technologies will be essential: hybrid cars, bio-fuels, capture and storage of carbon emissions from fossil fuel plants, and nuclear power. The world's biggest contributors to the catastrophe, the United States and China, see the conclusions of the draft report as a threat to prosperity. Meanwhile, the UN expects a large increase in human fatalities, especially in poor countries, as micro-climates disappear and whole ecological systems give way. Would the civilization that has been elaborated for hundreds of generations withstand the greatest challenge it had ever faced? Ultimately, the whole nuclear tour had led me to that question.

———

The Environmental Monitoring and Research Center trains first responders in how to deal with radiological emergencies, and it also watches over the environment, WIPP employees, and other residents of the Carlsbad area. So far no one has come in with a dose traceable to the repository operations or the transport of nuclear waste. Schoep told me that a local man who makes periodic visits to Chernobyl always gets scanned when he comes home. No contamination has ever been detected in his body.

My scan picked up cosmic radiation as it passed through my body and found, in my lungs, a slight spike in lead-210, a radon daughter that Schoep considered either a false positive or from natural radon in desert dust stirred up by the morning's wind. Thorium-232 and thorium-238 were slightly above average, probably also from that dust. The most significant emitter was potassium-40, occurring naturally in the tissue of all humans, and my level was average. I was negative for cesium-137, which is produced in reactors.

From my descent into salt caverns where thousands of barrels of nuclear waste were being stored, from a walk to the rim of the biggest hydrogen-bomb crater at the Nevada Test Site, and from visits to experimental reactors, nuclear plants, a nuclear submarine, nuclear medicine departments, hot cells, a spent-fuel pool, a nuclear waste processing center, and a working nuclear waste repository, I'd come away unscathed.

FOR OVER THIRTY YEARS WE'VE BEEN exhorted to "Think globally, act locally," in the words of René Dubos, a microbiologist, environmentalist, and author who introduced the slogan at a UN conference on the human environment.

While Rip was thinking—and acting—globally about environmental health and nuclear safety, he and Marcia were working hard locally. They attended meetings, talked to officials, rallied citizens, hired lawyers, testified at hearings, and did research, all with the goal of fostering cooperation to protect the watershed, the "air shed," and open land. One fall day after the conclusion of the nuclear tour, I drove through the South Valley—the air redolent with the aromas of cut alfalfa and roasting green chiles—to a ten-acre field owned by Bernalillo County that Rip and Marcia had seen on their walks and had, after about a year of lengthy negotiations with county officials, turned into a wildlife sanctuary with the help of other dedicated volunteers. The previous spring, after spending his own money to hire a farmer to disk the soil, Rip had harrowed it himself, using a makeshift rig attached to his truck, and then spent eighteen hours cleaning out an irrigation ditch and watering the field, assisted by fellow activists wielding shovels. He bought seed—corn, millet, and barley— that the group planted. By the summer, golden eagles, cattle egrets, hawks, doves, geese, ducks, raccoons, and skunks were paying calls. Today, as we sat at a little picnic table under the shade of cottonwoods, flocks of sandhill cranes migrating south along the Rio Grande flyway descended, filling the field with their odd burbling.

This was one of the first citizen-powered environmental projects of its kind in Albuquerque. Volunteers have planted hundreds of native trees and show up to help with irrigation and other chores.

"This land can never be developed now," Marcia said.

"These ten open acres will keep recharging the aquifer," Rip added. "It's all about the water. Without water, we die."

They were already engaged in their next campaign, saving four hundred acres from developers and using them to establish an agricultural campus in the South Valley to help preserve resources and teach environmental stewardship while supporting local farming traditions going back centuries.

I had returned to Albuquerque to go camping with Rip, Marcia, their family, the Critchfields, and other assorted friends in a patch of wilderness on the Chama River that Rip has owned for several decades and that he refers to with some reverence as The Land. I rode with him in his old pickup of many colors.

Leaving town, we got stuck in traffic on an approach to Interstate 25. Rip patiently advanced a yard or two at a time. It was a hot morning. The truck heater was stuck in the On position. I kept the window down. The air was smoggy and thick with fumes from the eighteen-wheelers hemming us in. Tasting diesel in the back of my throat, I imagined what it would be like if all these vehicles ran instead on cheap electricity from nuclear plants, and I reflected on how we import most of our oil from countries with undemocratic, oppressive regimes, how many of our taxpayer dollars go toward a military presence to make sure the oil gets extracted and tanker routes get protected, and how we use that oil to make almost everything we grow, use, wear, eat, and discard. The U.S. consumes about 21 million barrels a day. Just my portion, my number of barrels over my lifetime, would be significant, even though I consider myself only a modest consumer. How could we ever relinquish or even cut back on fossil fuels when our entire way of life depends so intimately upon them? I wanted to visit New Mexico; flying here emits a pound of carbon to the atmosphere per person per mile flown. I wanted to go camping; just this one truck was exhaling at least a pound of carbon for each mile we drove. My only solace was the thought that if the whole group were to stay in Albuquerque we'd be driving around burning gasoline and using electricity mostly generated by fossil fuels; at least at The Land, for four days our only pollution would be from a nightly campfire.

I looked around me in despair. The clean, clear future that as a child I'd taken for granted—assuming, of course, that we kept the Soviets at bay—had been used up: the open mesa and arroyos where I'd played were paved over, densely populated, crammed with energy-devouring, carbon- and particulate-emitting vehicles and houses and buildings, and littered with the junk of contemporary, fast-buck American culture. Yes, I could still find beauty here—the rugged face of the Sandia mountains, the enclaves of traditional adobe houses, the bosque and the fields along the river and preserves in the foothills where wild cholla cactus blooms—

but they're mostly either tourist attractions or they're shrinking in size. For every Rip and Marcia there seemed to be two dozen developers. For every activist coalition there were a hundred corporations with head-quarters in New Jersey, Texas, or Germany solely intent on making a short-term profit. A new suburb planned for south of Albuquerque will house 120,000 families. Where will the water come from? That problem has been put off for another generation to solve. When dust comes out of the tap the developers will be long gone. The sprawl seems destined to expand and require more and more water and electricity and burn more and more fossil fuels and destroy more and more resources and emit more and more greenhouse gases. And all over the country and the world there are countless Albuquerques where once there had been villages or farms or desert or forest or open shoreline.

Our world does not have to continue in this way. We do not have to pollute the earth in order to have modern civilization. Conservation, energy-efficient architecture, transportation, and agriculture, stricter pollution regulations, and an intelligent, long-term national energy plan could all make a big difference locally and globally for humanity and the environment. We still have enough resources to create a variety of realis-tic solutions tailored to local needs, and innovations have already brought improvements. But we need to do far more, and we must act quickly. Many people now speak of a mobilization of money and brains to develop a broad spectrum of emission-free energy resources, a nonparti-san, science-based project comparable in scope and intensity to the Man-hattan Project. In order to protect the planet and civilization, we will also need an extraordinary degree of international cooperation—nature does not recognize any political party or country.

As the IPCC has indicated, any pragmatic plan must include more nuclear power. The American Wind Energy Association hopes that by 2020 wind farms will be supplying as much as 6 percent of our electricity. The Energy Information Administration estimates that the figure will be closer to 0.5 percent. Regardless of whose predictions are accurate, the DOE projects electrical demand to grow by half by 2025. Even with increased conservation of energy, the need for baseload electricity will still have to be met by either fossil-fuel or nuclear power. The Nuclear Regulatory Commission (NRC) has made licensing of nuclear plants simpler through new regulations and through advance approval of the location and design. This should prevent long delays and cost overruns. Some estimates indicate that a plant of standardized, streamlined design, with many more built-in passive safety features, and therefore fewer pumps, valves, and other components, could be built in five years, as is already the case in France. The price per plant comes to about $3 billion—the cost of maintaining the U.S. presence in Iraq for one week. Reactors could make hydrogen for fuel cells as well as electricity while burning up waste residues. Although meeting baseload demand means that new nuclear plants are likely to be large, designs now come in different sizes: smaller reactors can supply electricity to local consumers or feed supplementary power to the grid during peak demand. Toshiba is now offering to provide and maintain a nuclear power plant about the size of a spruce tree; it features an underground, replaceable, sealed reactor core that can electrify a remote village—say, a small settlement way off the grid in Alaska, where diesel-fired generators are the norm—and supply all of a town's heat for thirty years.

An investment firm, Kohlberg Kravis Roberts, turned to environmental groups for advice when it bought out the biggest utility in Texas, TXU, in early 2007. TXU had planned to add eleven new coal-fired plants, but now it will build only three and instead will build two 1,700-megawatt nuclear plants with advanced reactors. To some in the financial world this move, which appears to be associated with concern about climate change, indicates the beginning of a nuclear renaissance. Fifteen companies or consortia are preparing reactor licensing applications for as many as thirty-three reactors of advanced design to be built in the next ten to twenty years, according to the Nuclear Energy Institute. Some of these will replace the fleet's oldest reactors. For nuclear power to maintain its 20 percent share of the American electricity pie chart, the industry will have to add thirty-five new reactors by 2030. There will almost certainly be future boondoggles and attempts at cover-ups on the part of industry, because, like other major business enterprises, energy corporations have a history of such maneuvers. On the other hand, continuing technological improvements and the high standards originated by Admiral Rickover for the industry and reinforced after the failure at Three Mile Island have helped create a culture of safety. The NRC will continue to exert control over nuclear utilities; it has evaluated the fleet's ability to deal effectively with terrorist attacks and has pledged to scrutinize new plant designs accordingly. The agency's and the industry's best interests lie in maintaining a strong safety record and encouraging nuclear utilities to continue to police one another.

As politicians of both parties are beginning to understand, along with some members of the environmental lobby, America must make decisions soon about nuclear power. As the biggest single producer of greenhouse gases, the nation has an ethical obligation to the rest of humanity to take action and to make long-range plans to replace the oldest operating plants, which will be retired in a few decades. To prevent the future rise of carbon dioxide emissions, by 2100 the world will require ten times the number of reactors we have today. Although reactor technology was forged in the United States, the country's leadership in that role has weakened. France, Japan, and other countries have added innovations and are continuing to build new nuclear plants, whereas until very recently the only large-scale additions in the United States have been fossil-fuel plants.

Economic growth usually means burning more coal. But fossil fuels should be considered precious natural resources to be husbanded for use in products such as pharmaceuticals. As more countries develop, they will be inclined to use fossil fuels unless the industrialized nations serve as exemplars in greenhouse gas mitigation and also help supply alternatives. Given the present trend, with Germany, China, and others building large numbers of coal-fired plants, we can expect their waste to continue to be pumped into the environment and the toll of preventable death and disease to rise. Emissions-control technology has been a low priority, and applications of it worldwide have been few. In the United States, carbon waste goes unregulated and large-volume sequestration of CO_2 remains just a plan, with drawbacks, such as leakage, that have yet to be addressed. Voluntary efforts on the part of American industry to control greenhouse gases have been rare. New American coal-fired plants— around 154—that have been proposed for the near term will cost $136 billion. They'll be constructed along traditional lines and will still be burning coal fifty years or more from now . . . if our coal supply lasts that long. (Again, should we decide to use coal to make synthetic automobile fuel, our reserves will be depleted more rapidly.) The DOE foresees over 300 new coal-fired plants in operation by 2030. Plans exist for a small, government-sponsored experimental plant that would turn coal into cleaner-burning gas and produce commercial chemicals while capturing its own carbon, but if that project goes forward—environmental groups are protesting it—that prototype won't be making electricity until 2015. Fossil fuels are expected to account for almost all new electricity generation on the planet and, given present regulation, will continue to emit more greenhouse gases. A new coal-fired plant is commissioned every week, mostly in China.

Some say we should just stop using so much electricity, and certainly we can be thriftier. Some say, as we have seen, that we should replace fossil fuels with wind and solar power—and renewables can indeed play a role. But until technology for storing energy from them advances, we'll have to keep getting most of our electricity from fossil-fuel and nuclear plants. To alleviate human suffering, to help people live longer, healthier lives, electricity is essential and today fossil fuels are usually what developing countries employ to make it.

Around 140 new nuclear plants are in the works in other countries. But the United States, without enough nuclear energy to make sufficient electricity, hydrogen, or synthetic fuels cleanly, is likely to remain a major contributor of greenhouse gases and continue to become bogged down in distant wars over dwindling oil and gas supplies, perhaps as a result sustaining more attacks on American soil. Because of general ignorance and misinformation about rays and particles and the realities about the risks and benefits of large-scale forms of energy generation, we're in danger of making poor choices and blindly accepting energy policies that harm the planet and darken prospects for our children and grandchildren.

Two billion people subsist without electricity. Their life expectancy is forty-three years on average. I once visited a Caribbean island where a diesel generator lit up a few bulbs and ran pumps and other equipment a day or two every week—and that made life a lot better than it was for the residents of a neighboring island who had no electricity at all. In impoverished areas remote from the grid, solar panels and wind turbines could make a real difference in survival rates. For more urbanized areas, nuclear power could provide clean energy reliably and so could coal plants engineered to contain all their waste. But as I sat in Rip's truck inhaling that sickening petrochemical exhaust and I reflected on my own blinkered notions prior to the nuclear tour and the difficulty of keeping in mind a broad vision for the coming decades, I became gloomy.

At present, the developed countries with the lowest carbon emissions per capita are the ones that rely mostly upon nuclear and hydroelectric power. A roundup of recent independent studies by the World Nuclear Association concludes that in most major countries, "new nuclear power plants offer the most economical way to generate baseload electricity." The reality is that all energy production is subsidized in the United States, renewables included. In 2004, the estimate for worldwide subsidies for nuclear power, which produced 2,260 billion kilowatt-hours, came to $16 billion, or an average of 0.6 cents per kilowatt-hour.

By mid-century, the world's need for energy is expected to increase by 160 percent. A simultaneous expansion of global nuclear capacity to around 1,350 reactors would cut the predicted increase in carbon emissions by a quarter. Additional, smaller reductions could be obtained by renewables, conservation, and cleaner fossil-fuel technology. *The Future of Nuclear Power*, the 2003 MIT-Harvard study that called for a tripling in the number of American nuclear plants, recommends an increase in government support for their construction as a means of not only delivering emission-free electricity and but also of achieving a reduction in the number of coal-fired plants. The authors propose a tax on carbon emissions to support nuclear expansion. Control of carbon output through emissions trading or taxation would raise the cost of electricity from fossil-fuel plants considerably. Nuclear power, once established, is not intrinsically more expensive than other means of electricity generation. France sells cheap electricity to other countries from its nuclear plants. Over the long run, uranium is and will continue to be inexpensive. We have enough of it to last indefinitely, as well as the technology to keep recycling uranium fuel and to burn useless residues in reactors.

Some argue that until present stores of spent nuclear fuel are safely put away, new nuclear plants must not be built. But I saw that nuclear waste is being safely stored right now and that there are some workable solutions for its long-term sequestration. The world's inventory of spent nuclear fuel would be tiny if all the enriched uranium in the pellets were fully exploited through existing technology. Residues not burned in reactors of advanced design could be secured in deep geologic repositories. The Waste Isolation Pilot Plant has demonstrated to the world that nuclear waste can be securely transported and isolated in a virtually inaccessible location. Many nations lack suitable geological formations or the wherewithal to construct such a repository, but international cooperation, along with new deep-drilling technology that has considerably advanced our ability to gain access to deep-ocean sediments, could one day lead to permanent disposal of waste with no future uses in sites where it would be naturally shielded and immobilized for millions of years. The Seabed Working Group has shown that the world's scientific and technological communities and existing international organizations can be effectively mobilized, given the right leadership and an environment of scientific objectivity. The group succeeded in putting together a cooperative way to control nuclear waste, and the methodology could be applied to all aspects of the nuclear fuel cycle.

After the accident at Three Mile Island, the NRC required every American nuclear plant to improve training of operators and to institute upgrades, multiple backup systems, redundancies, and fail-safe controls. Since September 11, 2001, security at all plants has been considerably enhanced.

Uranium for most nuclear power plants is not sufficiently enriched to make a fission bomb. To do that a rogue state would have to build a reprocessing plant or a uranium enrichment facility—both are detectable. For physical reasons, plutonium from spent fuel is highly inadequate for a national military program and, as Harold McFarlane of Idaho National Laboratory (INL) has said, "such plutonium would pose substantial technical challenges for a subnational organization without access to extensive national resources."

Turning weapons stockpiles into low-enriched, proliferation-resistant fuel reduces the chance of these materials being diverted into weapons. The International Atomic Energy Agency vigilantly patrols nuclear facilities around the world to prevent weapons production while fostering the long-range goals of controlling technology and resources and supporting the spread of responsibly operated nuclear power. It makes no sense for countries to spend their treasure making reactor fuel when they can lease it from a central source in a secure nation that would recycle it. This arrangement already exists informally and ought to be regulated and strengthened by international agreement

Often people in the nuclear world told me that nuclear power did not have to be perfect—it just had to be cleaner, safer, and more efficient than fossil-fuel generated power. And, of course, it is so. Some spoke about how nuclear plants, while displacing greenhouse gas production and meeting electricity needs, could buy some time to make technological advances that would bring superior methods of producing clean energy. Others say fission as we now employ it is in fact that longed-for superior method. Even if we use all of our ingenuity and willpower to apply a broad spectrum of solutions, even if we can reduce our greenhouse gas emissions by a large factor, it will still take the earth centuries to recover. But if we do nothing we'll be collaborating in a suicidal act of selfishness that could well result in the destruction of the only home we have.

Earlier in the summer, I'd gone up to The Land with Rip and Marcia to help water about fifty little mountain mahogany bushes and other native trees and shrubs they'd acquired from the Forest Service and planted in the spring on a rocky slope. As a pair of bald eagles floated over the Chama River and its limestone cliffs, Rip hauled water uphill in big plastic spackle buckets, and from them we filled saucepans to douse the sprigs. Most were pitifully dry, many with just a shriveled leaf or two. Some were bare. "They may be alive—water them anyway," said Rip.

Now, as he and I sat stalled in traffic on our way to The Land on that hot morning, I recalled that optimistic command and asked Rip what he thought should be done.

He replied that the human race, or at least some important part of it, had to understand how grave the situation was and identify the best, most practical way to reach the goal. "The only way we can make it is if we all pull together," he said. "We don't really have any other choice." He went on to say that we had to keep in mind the long-term consequences of our actions, and the way events and processes interact on a large scale over many years. "When we make choices, we tend to have a knee-jerk reaction to the scenario du jour. Then the next day there's a new scenario, a new reaction, and decisions and policies are made that contradict yesterday's decisions and policies. Without a big vision, we get nowhere. We wind up with a big mess. No water, no energy, no habitat."

"How do you break that pattern?"

"You take the politics out of energy, out of long-term planning. Other countries have done that with greenhouse gas emissions, nuclear waste disposal, and energy resources. France and Finland educated their children early on about nuclear power so that when they grew up and started making decisions, they understood the big picture. It took France just twenty years to switch over to nuclear. We have to think of future generations and the long term. You can't keep changing your policies and your budget allocations every few years the way the United States has been doing. Politicians hardly ever take the long view. We have to work at the grassroots level to change their thinking."

He thought that nationally and internationally a lot could be done. "As more data come in and risk-modeling about global climate disruption becomes more accurate, as uncertainties are reduced, priorities are becoming clearer to everyone. And people are already directly experiencing the effects. You can't say a particular hurricane or drought is because of global warming, but the overall trends, the increase in anomalies, are becoming evident and, compared to a few years ago, the public is hearing much more about the problem and how dire it really is. When you look at the big picture, there is no way you can leave out nuclear power. It always comes out ahead of all the other solutions. Problems can be solved and must be. We've got to have international control over the fuel cycle and weapons technology. We've got to do everything right. I'm optimistic that we'll make the right choices."

―――――

On the drive back down to Albuquerque after the camping trip, the late-afternoon sky was clear and endless; in a grass valley, deer stepped out into the open. The world unstintingly offered up its beauty and suggested enormous possibilities.

"I love this time of day—the long shadows, the golden light," Rip said.

I thought of his decades of effort in environmental health and safety and his battles to protect the water, air, and land where he lived. "You know, Rip, you really are a Green," I said.

"Most environmental activists are terribly right in the long run and terribly wrong in the short run," he said. "But almost no one is thinking about what is—in the long run—a matter of life or death."

"Why is it so difficult to keep in mind something as vast as global climate change? To deeply sense something on that scale?"

After a silence he replied, "I believe it's because most of the people who make policy live in cities or suburbs and they think differently from rural people. Or politicians are gentleman ranchers who relax on the weekend on land that someone else looks after. They've lost touch with Mother Nature and don't think they really need her to survive. Nature to them is just a form of amusement. They deal in abstractions, in the problem of the day, not the problem of the decade or of the century. The price of neglecting the future can be really high on a real working ranch or farm. If you don't take care of Mother Nature, she won't take care of you. You learn to respect that and to work with her."

Rip had been imprinted from his earliest childhood with concerns that hunter-gatherers and farmers have shared for thousands of years. "How much grain do I have to save in the fall in order to have enough seed grain for the spring planting? When I was a little guy, we had a small area in the granary where we put all the seeds for the next year. We protected them from mice, and never touched them, never fed those seeds to the animals. We sold all the potatoes in the cellar except a small amount that we kept aside, cut up in the spring, and planted as early as we possibly could, making the best estimate that the crop would not come out of the ground until the last frost had passed. It was natural to think ahead, and everybody worked on the future. You had to keep in mind that your livestock might get a disease, or bad weather would ruin your crops. That's just the way nature operates, and you have to remember that or you do not survive. You learn to prepare, to think ahead, to make sacrifices today so that your grandkids will have something after you're gone.

"In a city, everything is available all the time. Meat comes in plastic wrap and nobody thinks about the steer it came from. Death and birth are going on all the time, and you witness that when you're living on the land. But in cities and suburbs, that's all hidden or ignored. It's as if nobody dies and so you can pretend to be immortal. So if there's a problem, you believe that you have all the time in the world to deal with it and that you can be very particular about exactly how you deal with it. There's another factor: every man for himself. Most Americans live far away from where oil is drilled or coal is mined and they never see what a messy, dirty business it is. They ignore what makes a city or suburb possible in the first place and feel independent and separate and self-involved. They don't see themselves as a part of larger systems or of nature—nature is going on somewhere else and is of little or no importance. Even ranchers now think like this—I see how they overgraze and don't worry about the consequences. The idea is to make a profit today and forget about the future and about the welfare of others. People want goods and services and energy but they don't want to think about the costs to the environment. This is why I'm so scared of global warming. Almost nothing is being done."

"So what do we do?"

"Nuclear power has its drawbacks, but the ratio of benefit to risk is the best I have been able to find. I've looked and looked, and thought far ahead and, given all that I know, I believe that if we're going to make it, nuclear will have to be our main resource. The choice gets down to this: What kind of world do we want for ourselves, for our kids, for our grandkids? We make sure they get vaccinations and bike helmets, but what about the world they'll have to survive in? The experts at the WIPP conferences about the future believed that civilization would still be around in ten thousand years. Everyone emphasized that the decisions we make in the present could have a tremendous impact. If we don't make the right choices, if energy becomes scarce or really expensive, people will give up their big houses and their lawns, and they'll look at nuclear power differently. But there is something we can do now."

Rip told me the story about a man who has been warned that a flood is coming. The man says, "It's all right—God will save me." As the flood approaches, the police come to his house and tell him to evacuate. He chooses to stay, saying, "God will save me." When the water reaches halfway up his house, rescuers come by with a rowboat, and he refuses it, saying, "God will save me." Finally he's on the roof of his house and his feet are getting wet. A helicopter comes to take him to safety, but he waves it off, saying, "God will save me." Finally the rising water inundates the entire house and he drowns. When he gets to heaven and meets God, he demands an explanation. "Why didn't You save me?" God replies, "I sent someone to tell you to evacuate. I sent you a rowboat. I sent you a helicopter. You turned down every chance I gave you."

Rip shrugged. "One day God could say to us: I gave you the brainiest men and women in human history to come up with an understanding of the atom and its nucleus. I gave you enough uranium and thorium to last you for thousands of years. I gave you an understanding of how when uranium decays it releases energy. You didn't need to invent anything else. You had everything you needed to provide energy for yourselves and your descendants without harming the environment. What else did you *want*?"

21 THE POWER WITHIN

A human being is a part of the whole, called by us "Universe," a part limited in time and space. He experiences himself, his thoughts and feelings as something separated from the rest—a kind of optical delusion of his consciousness. This delusion is a kind of prison for us, restricting us to our personal desires and to affection for a few persons nearest to us. Our task must be to free ourselves from this prison.

—ALBERT EINSTEIN

WHEN I FLEW OUT OF ALBUQUERQUE, the plane made a wide turn. I took in the sweep of the Rio Grande Valley, from the Jémez Mountains to the north, where Los Alamos lies hidden, to Mount Taylor, a gigantic broken cone anchoring the western center of the panorama, to the river and its bosque, to the sprawling city, and, toward the south, the military reservation and its hollow mountain, which may or may not still be storing bombs and which I now know was called Site Able. I've heard that deep inside there's an emergency presidential relocation center that had been built as a refuge for President Eisenhower in case of nuclear attack. I'd asked Rip a couple of times to find out if we could pay a visit there, and he said he would try. But he really didn't care to talk about the place, and ultimately I concluded that it remained heavily restricted.

Above the plane there curved the dome of the sky, clear today, with tall cumulus clouds. Beyond them, I pictured the thin glowing shell of our atmosphere, the interplanetary void, and the unimaginable emptiness of interstellar space.

I thought about the power of assumptions, and how they can form a false backdrop. Years earlier, after I'd returned to Albuquerque for a visit, I finally got around to taking the tramway that had been built after I moved to New York and that was popular with tourists who wanted to be carried from the foothills, about a mile above sea level, up to almost eleven thousand feet at the top of Sandia Crest. As the car silently swooped up in an arc and we glided over arroyos, boulders, escarpments, and ravines, I discovered that my fixed image of the mountain face that formed the eastern boundary of my childhood had in fact been little more than a shallow bas-relief—a notion of a mountain range as simple as a child's drawing. For the first time I became acutely aware of the depths and complexities that lay behind the Sandias' muscular facade. That upthrust of billion-year-old stone that had loomed over my early days concealed deep canyons, precipices, and vertical granite slabs, and behind them, still more canyons, mica-flecked plinths and pinnacles, a silver thread of a waterfall, caves, aspen glens, and pockets of hardwood forest turning red near the crest. I felt the presence of that ancient range and of the ongoing force of nature, doing all it could to bring us into being and all it could to take us out again, and I sensed my own tiny and temporary spark of consciousness within the panorama.

How amazing it was to find that something so completely familiar turned out in reality to be so very different from what I had assumed all my life.

The power to save our world does not lie in rocks, rivers, wind, or sunshine. It lies in each of us.

NOTES

Part 1 **ORIGINS**
1 **SURVIVAL**

8 **For decades he had been watching an alarming trend**: Michael P. Farrell, "Energy and Global Climate Change: Why ORNL?":

> In the late 1950s, global climate change was an unknown threat to the world's environment and social systems. Except for a few ORNL researchers who had just completed their first briefing to the U.S. Atomic Energy Commission (AEC) on the need to understand the global carbon cycle, the connection between rising carbon dioxide concentrations and potential changes in global climate was not common knowledge, nor were the consequences of climate change understood. It would not be until almost 15 years later—the mid-1970s—that a comprehensive Department of Energy (DOE) research program was established to study the effects of increased atmospheric carbon dioxide concentrations on the world's climate—the first global climate change program. http://www.ornl.gov/info/ornlreview/rev28_2/text/egc.htm.

8 **Bill McKibben**: Bill McKibben, *The End of Nature* (New York: Anchor Tenth Anniversary Edition, 1997).

9 **Annual per capita coal waste**: National Research Council (U.S.), *Managing Coal Combustion Residues in Mines*, (Washington, D.C.: National Academies Press, 2006), http://www.catf.us/projects/power_sector/power_plant_waste/news/NAS_Coal_Ash_Full_Report.pdf.

9 **U.S. greenhouse gas emissions**: U.S. Energy Information Administration, "Emissions of Greenhouse Gases in the United States 2005," http://tonto.eia.doe.gov/FTPROOT/environment/057305.pdf; U.S. Environmental Protection Agency, "Global Warming—Emissions," http://yosemite.epa.gov/oar/globalwarming.nsf/content/emissionsindividual.html.

10 **World energy resources**: Energy Information Administration (EIA) Annual, 2004, www.eia.doe.gov/iea.

11 **Electricity demand increases**: Energy Information Administration, "International Energy Outlook 2006," http://www.eia.doe.gov/oiaf/ieo/electricity.html.

16 **Sierra Club "crisis report"**: Modesto Irrigation District, "Greening of Paradise Valley," http://www.mid.org/about/100-years/chpt_19.htm. See also Thomas Raymond Wellock, *Critical Masses: Opposition to Nuclear Power in California, 1958–1978* (Madison: University of Wisconsin Press, 1998).

16 **Renewables require space**: John M. Ryskamp, "The Need for Nuclear Power,"

IEEE Power Engineering Society Meeting, April 28, 2003, http://nuclear.in1.gov/docs/papers-presentations/nuclear_need_2-26-03.pdf.

17 **Health consequences of coal-fired plants:** Abt Associates, "Dirty Air, Dirty Power," Clean Air Task Force, June 2004, http://www.catf.us/publications/view/24; "SAMI Key Conclusions and Talking Points," Clean Air Task Force, August 22, 2002, http://www.catf.us/resources/other/CATF_SAMI_detailed_talking_points.pdf; American Lung Association, "Particle Pollution Fact Sheet," http://www.lungusa.org/site/pp.asp?c=dvLUK9O0E&b=50324. Keith Bradsher and David Barboza, "Pollution from Chinese Coal Casts a Long Shadow," *The New York Times,* June 11, 2006, http://www.nytimes.com/2006/06/11/business/worldbusiness/11chinacoal.html?ex=1307678400en=e9ac1f6255a24fd8ei=5088partner=rssnytemc=rss.

17 **China coal combustion:** Keith Bradsher and David Barboza, "Pollution from Chinese Coal Casts Shadow Around Globe," *The New York Times,* June 11, 2006, http://select.nytimes.com/search/restricted/article?res=F40C13F739550C728DDDAF0894DE404482.

17 **Survey favors nuclear power:** Bernard L. Cohen, *The Nuclear Energy Option: An Alternative for the 90s* (New York: Plenum Press, 1990), pp. 42–43.

17 **Carbon emissions from nuclear power:** Mohamed ElBaradei, "Nuclear Power: Preparing for the Future," Statements of the Director General, March 21, 2005, International Atomic Energy Agency, http://www.iaea.org/NewsCenter/Statements/2005/ebsp2005n004.html.

18 **Internal plutonium doses:** William Moss and Roger Eckhardt, "On the Front Lines: Plutonium Workers Past and Present Share Their Experiences," in *Radiation Protection and the Human Radiation Experiments* (Los Alamos Science, number 23), ed. Necia Grant Cooper (Los Alamos, N.Mex.: Los Alamos Laboratory, 1995), pp. 42–150.

19 **"Carbon dioxide poses a dilemma":** Alvin Weinberg, *The First Nuclear Era—The Life and Times of a Technological Fixer,* pp. 236–237 (Springer, 1997).

20 **"Nuclear power, designed well.":** Al Gore, "Remarks by the VP at Chornobyl Museum Kiev," Clinton Foundation, July 23, 1998, http://www.clintonfoundation.org/legacy/072398-remarks-by-the-vp-at-chornobyl-museum-kiev.htm.

2 ALWAYS LOOK AT THE WHOLE

21 **greenhouse gases:** U.S. Environmental Protection Agency, "Global Warming—Emissions," http://yosemite.epa.gov/oar/globalwarming.nsf/content/emissionsindividual.html.

3 AMBROSIA LAKE

31 **Sources of electricity by state:** Energy Information Administration (U.S. Department of Energy), "State Energy Consumption, Price, and Expenditure Estimates (SEDS)," http://www.eia.doe.gov/emeu/states/_seds.html.

34 **Uranium promotions:** Michael A. Amundson, *Yellowcake Towns: Uranium Mining Communities in the American West* (Boulder: University Press of Colorado, 2002), pp. 84, 95.

35 **Hazards of natural uranium, depleted uranium:** Agency for Toxic Substances and Disease Registry (ATSDR), "ToxFAQs™ for Uranium," September 1999, http://www.atsdr.cdc.gov/tfacts150.html.

35 **depleted uranium:** World Health Organization, "Depleted Uranium" (Fact Sheet No. 257, revised January 2003), http://www.who.int/mediacentre/factsheets/fs257/en/.

38–39 **Manhattan Project acquires uranium:** Richard Rhodes, e-mail to author.

39–40 **Uranium discovery, booms:** Amundson, *Yellowcake Towns*, pp. 23–24, 105–106, 139–140.

41 **Radon exposure in small mines:** Author interview with Charles Key, M.D., pathologist and epidemiologist, medical director, New Mexico Tumor Registry, Cancer Research Institute at the University of New Mexico; participant in studies of health effects of radon on uranium miners.

42 **Remediation of Navajo mine sites:** Author interview with Charles Key, M.D.

42 **Percentage of Native American miners:** e-mail to author from University of New Mexico Bureau of Business and Economic Research.

43 **cancer incidence rates:** Janet J. Kelly, Anne P. Lanier, Steven Alberts, and Charles L. Wiggins, "Differences in Cancer Incidence among Indians in Alaska and New Mexico and U.S. Whites, 1993–2002," *Cancer Epidemiology Biomarkers & Prevention* 15 (August 2006): 1515–1519.

43 **Environmental conditions on Navajo lands:** Judy Pasternak, "Blighted Homeland" (four-part series), *Los Angeles Times*, November 19–22, 2006.

44 **"the exposures of the cases.":** Kiyohiko Mabuchi, Charles E. Land, and Suminori Akiba, "Radiation, Smoking, and Lung Cancer: A Binational Study Provides New Insights into the Effects of Smoking and Radiation Exposure on Different Histological Types of Lung Cancer," *RERF Update* 3, no. 4 (1991): 7–8, http://www.rerf.or.jp/eigo/rerfupda/death/radiat.htm.

46 **Boice study finds no increase in cancer:** John D. Boice Jr., Michael Mumma, Sarah Schweitzer, and William J. Blot, "Cancer Mortality in a Texas County with Prior Uranium Mining and Milling Activities, 1950–2001," *Journal of Radiological Protection* 23 (September 2003), 247–262, http://www.iop.org/EJ/abstract/0952-4746/23/3/302.

47 **Ambrosia Lake:** Amundson, *Yellowcake Towns*, pp. 84–85.

53 **Rising uranium prices:** Susan Moran and Anne Raup, "A Rush for Uranium; Mines in the West Reopen as Ore Prices Reach Highs of the 1970s," *The New York Times*, March 28, 2007, http://select.nytimes.com/search/restricted/article?res=F30A1EFB38540C7B8EDDAA0894DF404482.

58 **Ample uranium supply:** John M. Deutch and Ernest J. Moniz, "The Nuclear Option," *Scientific American*, September 2006, http://www.sciam.com/article.cfm?chanID=sa006&colID=1&articleID=0000137A-C4BF-14E5-84BF83414B7F0000.

59 **"I have been a committed environmentalist":** Hugh Montefiore, "Why the Planet Needs Nuclear Energy," *The Tablet*, October 23, 2004, http://www.thetablet.co.uk/articles/1963/

60 **Debunked studies:** Uwe R. Fritsche, "Comparing Greenhouse-Gas Emissions and Abatement Costs of Nuclear and Alternative Energy Options from a Life-Cycle Perspective," Institute for Applied Ecology (Öko-Institut, 1997); Jan Willem Storm van Leeuwen and Philip Smith, "Nuclear Power—the Energy Balance," http://www.stormsmith.nl/, 2003; World Nuclear Association, "Energy Analysis of Power Systems," http://www.world-nuclear.org/info/inf11.html, March 2006.

60 **Nuclear power scarcely cleaner than coal:** Friends of the Earth, "Nuclear Power, Climate Change, and the Energy Review," June 2006, http://www.foe.co.uk/resource/briefings/nuclear_power.pdf, pp. 7–8.

60 **Exhaustion of uranium supply:** World Nuclear Organization, "Supply of Uranium," March 2007, http://www.world-nuclear.org/info/inf75.html.

60 **"Nuclear power emits virtually no greenhouse gases":** Mohamed ElBaradei, "Nuclear Power: Preparing for the Future" (Statements of the Director General, March 21, 2005), International Atomic Energy Agency, http://www.iaea.org/News Center/Statements/2005/ebsp2005n004.html.

60 **100-watt lightbulb/comparative fuel figures:** Nuclear Energy Institute, "Nuclear Facts," Up Front, http://www.nei.org/doc.asp?catnum=2&catid=106.

61 **"A bundle of enriched-uranium fuel rods":** Peter W. Huber and Mark P. Mills, "Why the U.S. Needs More Nuclear Power," *City Journal*, Winter 2005, http://www.city-journal.org/html/15_1_nuclear_power.html.

Part 2 THE INVISIBLE STORM
4 MOTHER NATURE AND FENCEPOST MAN

69–70 **U.S. food irradiation:** John Henkel, "Irradiation: A Safe Measure for Safer Food," *FDA Consumer*, May–June 1998, http://www.findarticles.com/p/articles/mi_m1370/is_n3_v32/ai_20625879; Robert V. Tauxe, "Food Safety and Irradiation: Protecting the Public from Foodborne Infections," *Emerging Infectious Diseases* 7, no. 3 (June 2001, Suppl.), http://www.cdc.gov/ncidod/eid/vol7no3_supp/tauxe.htm.

70 **U.S. medical sterilization:** Division of Bacterial and Mycotic Diseases, "Food Irradiation," Centers for Disease Control and Prevention, Department of Human and Health Services, http://www.cdc.gov/ncidod/dbmd/diseaseinfo/foodirradiation.htm #whatis.

70 **"360 millirem per year":** Idaho Department of Environmental Quality, "INL Oversight: Guide to Radiation Doses and Limits," http://www.deq.idaho.gov/inl _oversight/radiation/radiation_guide.cfm. See this website for general information on common sources of radiation exposure. It also has a dose calculator so you can figure out how much radiation you're getting.

72 **Permissible dose limits:** Idaho Department of Environmental Quality, "INL Oversight."

75 **"Then we have coal combustion":** Lawrence Berkeley National Laboratory, "Natural Sources of Radioactivity," http://www.lbl.gov/LBL-Programs/tritium/natural-dosage.html.

75 **coal-fired plants concentrate:** Alex Gabbard, "Coal Combustion: Nuclear Resource or Danger," Oak Ridge National Laboratory, http://www.ornl.gov/info/ornlreview/rev26-34/text/colmain.html.

76 **"On average, a third to a quarter":** American Cancer Society, "Lifetime Probability of Developing or Dying from Cancer," http://www.cancer.org/docroot/CRI/content/CRI_2_6x_Lifetime_Probability_of_Developing_or_Dying_From_Cancer .asp?sitearea=. The lifetime probability of developing cancer in American men is one in two for all cancers, excluding basal and squamous cell skin cancers and in situ cancers except urinary bladder. The rate for American women is one in three with the same exclusions. The worldwide rates may be less, since the life expectancy in the United States is greater than in many other parts of the world.

77 **Radionuclides in common foods:** Susan M. Stacy, *Proving the Principle: A History of the Idaho National Engineering and Environmental Laboratory, 1949–1999* (Idaho Falls, Idaho: Idaho Operations Office of the Department of Energy, 2006), http://www.inl.gov/proving-the-principle/chapter_07.pdf, p. 60. See also Idaho State Uni-

versity's Radiation Information Network, "Radioactivity in Nature," http://www
.physics.isu.edu/radinf/natural.htm.

5 UNDARK

83 **Follow-up studies of radium-dial painters:** R. E. Rowland, "Radium Dial
Painters: What Happened to Them?" R. E. Rowland, http://www.rerowland.com/
dial_painters.htm; also see: R. E. Rowland, *Radium in Humans: A Review of U.S.
Studies* (Argonne, Ill.: Argonne National Laboratory, 1995).

83 **last surviving "radium girl":** *The Waterbury Republican-American,* "Last 'Radium
Girl' Marks 100th Birthday," reprinted in *Journal of Nuclear Medicine* 47, no. 7 (July
2006), p. 44, http://jnm.snmjournals.org/cgi/reprint/47/7/38N.pdf.

87 **Public knowledge of science:** Cornelia Dean, "Scientific Savvy? In U.S., Not
Much," *The New York Times,* August 30, 2005.

89 **headlines in the *New York Post*:** Marcin Rotkiewicz in collaboration with Henryk
Suchar and Ryszard Kamiński, "Chernobyl: The Biggest Bluff of the Twentieth
Century," Polish weekly *WPROST,* no. 2 (January 14, 2001), http://www.wonuc
.org/xfiles/chern_03.html.

89–90 **Bhopal:** World Almanac Books, *The World Almanac and Book of Facts, 2007* (New
York: Press Pub. Co.), p. 312.

90 **Unfolding of Chernobyl reactor failure:** Richard Rhodes, *Arsenals of Folly: The
Making of the Nuclear Arms Race* (New York: Knopf, 2007), chapter 1.

90–91 **Effects of Chernobyl explosions, fire:** International Atomic Energy Association,
The Chernobyl Forum Report, http://www.iaea.org/Publications/Booklets/Chernobyl/
chernobyl.pdf.

93 **"It is important that public misperceptions":** Neil Wald quoted in Rosalie
Bertell, "Avoidable Tragedy Post-Chernobyl," *Journal of Humanitarian Medicine* 2,
no. 3 (2002): 21–28, reprinted by International Institute of Concern for Public
Health (IICPH), http://www.iicph.org/docs/chernobyl.htm.

98 **Studies of liquidators:** Richard Stone, "Nuclear Radiation: Return to the Inferno:
Chornobyl After 20 Years," *Science* 312, no. 5771 (April 14, 2006): 180–182.

98 **estimated deaths:** Table 16.4, Chernobyl Forum Report.

99 **"There is no scientific evidence":** UNSCEAR secretariat, "The Chernobyl
Accident: UNSCEAR's Assessments of the Radiation Effects," UNSCEAR
(United Nations Scientific Committee on the Effects of Atomic Radiation), updated
April 14, 2007, http://www.unscear.org/unscear/en/chernobyl.html.

99 **Better management could have prevented psychic trauma:** American Psychological Association, "One Year After Katrina," http://www.apa.org/releases/katrina
06.html.

100 **"People have developed a paralyzing fatalism":** Elizabeth Rosenthal, "Experts
Find Reduced Effects of Chernobyl," *The New York Times,* September 6, 2005.

101 **1996 report:** J. C. Nénot, A. I. Stavrov, E. Sokolowski, and P. J. Waight, Back-
ground Paper 8, "The Chernobyl Accident: The Consequences in Perspective," *One
Decade After Chernobyl: Summing up the Consequences of the Accident,* sponsored by the
International Atomic Energy Agency (IAEA), European Commission (EC), and
World Health Organization (WHO) in cooperation with the United Nations
Department of Humanitarian Affairs (UNDHA), United Nations Educational,
Scientific and Cultural Organization (UNESCO), United Nations Environment
Programme (UNEP), United Nations Scientific Committee on the Effects of
Atomic Radiation (UNSCEAR), Food and Agriculture Organization (FAO) of the

United Nations, and the Nuclear Energy Agency (NEA) of the Organization for Economic Cooperation and Development (OECD) (April 1996), http://www .iaea.org/worldatom/Programmes/Safety/Chernobyl/paper8.html.

103 **Range of dose rates:** There were 135,000 evacuees from the thirty-kilometer zone around Chernobyl. They received an average dose of 100 millirem above natural background radiation, which is the same dose limit set by the U.S. Department of Energy for public exposure and less of an increase than evacuees would have received if they had left Ukraine, which has low natural background radiation, for Colorado. Excess cancer deaths over a lifetime might afflict slightly over one tenth of a percent of the evacuees. The 270,000 residents of "strict control zones" received an average dose of 500 millirem and the potential deaths as a result come to a little over 0.5 percent. Their exposure was equal to that received naturally by Coloradans. UNSCEAR reported that large parts of the countries now known as Belarus, the Russian Federation, and Ukraine were contaminated, chiefly by cesium-137 and iodine-131. The seven million people living in these areas received an average dose of 70 millirem. Under normal circumstances, about 800,000—slightly less than 10 percent—would die of cancer over a lifetime. The conservative estimate of predicted excess cancers as a result of Chernobyl: 0.05 percent.

103 *Chernobyl Heart:* For a comparison of the facts as presented in *Chernobyl Heart* with those in the report of the Chernobyl Forum, see Paul Lorenzini, "Hearts and Minds," Nuclear Engineering International, Feature/Viewpoint, February 16, 2006, http://www.neimagazine.com/story.asp?sectionCode=188&storyCode=2034355.

105 **"While these figures":** International Agency for Research on Cancer, "The Cancer Burden from Chernobyl in Europe," Press Release no. 168, April 20, 2006, http://www.iarc.fr/ENG/Press_Releases/pr168a.html. See also: F. M. Mettler, "Causation and Attribution of Radiation Effects."

106 **Nuclear capacity in Ukraine:** Yuriy Leonidovych Kovryzhkin, "Nuclear Power in Ukraine: Past, Present, Future," World Nuclear Association, Annual Symposium 2004, http://www.world-nuclear.org/sym/2004/kovryzhkin.htm.

106 **Cancer rates in proximity to nuclear facilities:** National Cancer Institute, "No Excess Mortality Risk Found in Counties with Nuclear Facilities," http://www .cancer.gov/cancertopics/factsheet/Risk/nuclear-facilities. (Study published in the *Journal of the American Medical Association* in 1991.)

107 **"The overall average thyroid dose":** Richard D. Klausner, "Estimated Exposures and Thyroid Doses Received by the American People From Iodine-131 in Fallout Following the Nevada Atmospheric Nuclear Bomb Tests," National Cancer Institute, http://www.cancer.gov/legis/testimony/i-131.html.

108–109 **cancer-cluster rumors:** Harry Otway and Jon Johnson, "A History of the Working Group to Address Los Alamos Community Health Concerns: A Case Study of Community Involvement and Risk Concerns," Office of Scientific and Technical Information, www.osti.gov/bridge/servlets/purl/751963-rNM6nM/webviewable/751963.pdf.

109 **no environmental carcinogens in Los Alamos:** Ibid.

110 **"The claims of dire health consequences":** W. G. Sutcliffe, R. H. Condit, W. G. Mansfield, D. S. Myers, D. W. Layton, and P. W. Murphy, "A Perspective on the Dangers of Plutonium," Lawrence Livermore National Laboratory, April 14, 1995, http://www.11nl.gov/csts/publications/sutcliffe/.

110 **Human plutonium exposures:** George L. Voelz, James N. P. Lawrence, and Emily

R. Johnson, "Fifty Years of Plutonium Exposure to the Manhattan Project Pluto-nium Workers: An Update," *Health Physics* 73, no. 4 (October 1997): 611–619. In 1995, Los Alamos encouraged some of those men to tell their stories for publication in Los Alamos Science's *Radiation Protection and the Human Radiation Experiments*, edited by Necia Grant Cooper.

6 INTO THE STRANGE CITY

112 **Survey favors nuclear power:** Bernard L. Cohen, *The Nuclear Energy Option: An Alternative for the 90s* (New York: Plenum, 1990), pp. 42–43.

112 **Survey finds benefits outweigh risks:** Carol L. Silva and Hank C. Jenkins-Smith, "Precaution and Scientific Judgment: Scientists' Environmental Policy Beliefs in the US and EU," The Bush School of Government and Public Service, Texas A&M University, 2002, http://bush.tamu.edu/research/working_papers/csilva/APPAM04.pdf.

116 **Survey of 1,737 scientists about radiation thresholds:** Carol L. Silva, Hank C. Jenkins-Smith, and Richard P. Barke, "The Politics of Caution: The Bases of Scientists' Precautionary Choices," The Bush School of Government and Public Service, Texas A&M University, April 2005, http://bush.tamu.edu/research/working_papers/csilva/PoliticsofCaution.pdf.

117 **Setting low-dose radiation safety standards:** Ibid.

117–118 **Threshold dose model proponents:** Ibid.

118 **Exposure to high natural background radiation:** M. Ghiassi-nejad, S. M. J. Mortazavi, J. R. Cameron, A. Niroomand-rad, and P. A. Karam, "Very High Background Radiation Areas of Ramsar, Iran: Preliminary Biological Studies," *Health Physics* 82, no. 1 (January 2002): 87–93. See also Health Physics Society on effects of low-dose radiation, http://www.hps.org/publicinformation/ate/q1254.html.

118 **Health effects of very high natural background radiation:** Chandrasekara Dissanayake, "Of Stones and Health: Medical Geology in Sri Lanka," *Science* 309, no. 5736 (August 5, 2005): 883–885, http://www.sciencemag.org/cgi/content/full/309/5736/883.

120 **Radiation effects based on extrapolations:** Greta Joy Dicus, "Radiation Protection Standards: Past, Present and Future," U.S. Nuclear Regulatory Commission, 2001, http://www.nrc.gov/reading-rm/doc-collections/commission/speeches/2001/s01-003.html.

127 **Report of Steven M. Becker:** Steven M. Becker, "Addressing the Psychosocial and Communication Challenges Posed by Radiological/Nuclear Terrorism: Key Developments Since NRCP Report No. 138," *Health Physics* 89, no. 5 (November 2005): 521–530, http://www.healthphysics.com/pt/re/healthphys/abstract.00004032-200 51100000013.htm;jsessionid=FQQKXFQd30KxszbY4byGwyCqRkJ55tFvM2 FQywVJMphjgq5bpnfp!736553971!-949856145!8091!-1.

129 **"At the levels created by":** U.S. Nuclear Regulatory Commission, "Fact Sheet on Dirty Bombs," http://www.nrc.gov/reading-rm/doc-collections/fact-sheets/dirty-bombs.html.

129 **"The 9/11 Commission found that preparedness":** Jamie Gorelick, in an e-mail to author, 2006.

130 **"No reproducible evidence exists":** Rosalyn Yalow, "Concerns with Low Level Ionizing Radiation," *Mayo Clinic Proceedings*, 69: 436–440.

Part 3 **THE HIDDEN WORLD**

133 **"Out beyond ideas":** Jalaluddin Rumi, translated by Coleman Barks, *Essential Rumi* (San Francisco: HarperSanFrancisco, 1995).

7 RISK AND CONSEQUENCE

137 **"the way we regard uncertainty":** John F. Ross, "Pascal's Legacy," *EMBO reports* 5 (Suppl. 1, 2004): S7–S10, European Molecular Biology Organization, http://www .nature.com/embor/journal/v5/n1s/full/7400229.html.

137 **Report on catastrophic reactor failure:** Presidential Commission on Catastrophic Nuclear Accidents, *Report to the Congress from the Presidential Commission on Cata- strophic Nuclear Accidents,* Washington, D.C., August 1990, http://www.state.nv.us/ nucwaste/news/rpccna/pcrcna01.htm. Also see Norman C. Rasmussen's descrip- tion: http://www.state.nv.us/nucwaste/news/rpccna/pcrcna12.htm. Also: Bernard L. Cohen, *The Nuclear Energy Option,* pp. 89–93.

138 **1991 NRC study:** Nuclear Regulatory Commission, "Severe Accident Risks: An Assessment for Five U.S. Nuclear Power Plants (NUREG-1150, Vol. 1)," Wash- ington, D.C., December 1990, http://www.nrc.gov/reading-rm/doc-collections/ nuregs/staff/sr1150/v1/.

140–141 **twelve thousand cumulative reactor-years:** Uranium Information Centre, "Safety of Nuclear Power Reactors," Nuclear Issues Briefing Paper 14, January 2007, http://www.uic.com.au/nip14.htm.

142 **NRC disclaimer:** Wikipedia, "NRC disclaimer of CRAC-II and NUREG-1150, http://en.wikipedia.org/wiki/NUREG-1150#NRC_disclaimer_of_CRAC-II_and _NUREG-1150. An e-mail to author from Scott Burnell, Public Affairs Officer, Nuclear Regulatory Commission, confirmed that this is the official position of the NRC (February 22, 2007).

8 GOING TO EXTREMES

149 **Covert weapons technology in India and other nations:** e-mail from Richard Rhodes to author, June 14, 2007.

151 **"it is curious and promising":** Richard Rhodes, "The Genie is out of the bot- tle," *The Guardian,* August 6, 2002, http://www.guardian.co.uk/nuclear/article/ 0,,770601,00.html.

151 **Global Nuclear Energy Partnership:** U.S. Department of Energy, "Global Nuclear Energy Partnership," http://www.gnep.energy.gov/gnepProgram.html.

152 **Nuclear fuel bank:** William J. Broad, "$50 Million Offer Aims at Curbing Efforts to Make Nuclear Fuel," *The New York Times,* September 20, 2006, http://select.nytimes. com/search/restricted/article?res=F10F1FF73B550C738EDDA00894DE404482.

152 **hub-and-spoke arrangement:** Harold Feiveson, "The Search for Proliferation- Resistant Nuclear Power," *The Journal of the Federation of American Scientists* (Sep- tember/October 2001, Volume 54, Number 5), http://www.fas.org/faspir/2001/ v54n5/nuclear.htm.

153 **energy equivalent:** Megatons to Megawatts: "Recycling Nuclear Warheads into Electricity," USEC http://www.usec.com/v2001_02/HTML/search.htm.

153 **Nuclear weapons converted to fuel:** United States Enrichment Corporation, "Megatons to Megawatts," http://www.usec.com/v2001_02/HTML/Megatons _chronology.asp.

153 **International consortium to enrich uranium:** Matthew L. Wald, "Uranium Enrichment Project Gets License," *The New York Times*, June 24, 2006, http://select .nytimes.com/search/restricted/article?res=FA0B1FFD3D540C778EDDAF0894D E404482.

154 **Alarming discovery in France:** James Lovelock, *The Ages of Gaia* (New York: Norton, 1988), pp. 122–124.

154 **natural reactors:** Alex P. Meshik, "The Workings of an Ancient Nuclear Reactor," *Scientific American*, November 2005, pp. 82–91, http://www.sciam.com/article .cfm?chanID=sa006&colID=1&articleID=00078840-5C1A-1359-9B5C83414B7 F0119.

156 **Oklo fossil reactors:** Office of Civilian Radioactive Waste Management "Oklo: Natural Nuclear Reactors," U.S. Department of Energy, http://www.ocrwm.doe .gov/factsheets/doeymp0010.shtml.

157 **discovery and application of nuclear energy:** Richard Rhodes, *The Making of the Atomic Bomb* (New York: Simon & Schuster, 1986).

163 **"Making electricity had little in common":** Susan M. Stacy, *Proving the Principle: A History of the Idaho National Engineering and Environmental Laboratory, 1949–1999* (Idaho Falls, Idaho: Idaho Operations Office of the Department of Energy, 2000), p. 26.

166 **many other considerations:** Ibid., pp. 45–49.

167 **"It appeared that boiling-water reactors":** Ibid., p. 130.

168–169 **BORAX development, AEC activities and findings:** Ibid., pp. 132–134.

177 **Fire at Browns Ferry:** U.S. Nuclear Regulatory Commission, "Backgrounder on Nuclear Power Plant Fire Protection," July 2006, http://www.nrc.gov/reading-rm/ doc-collections/fact-sheets/fire-protection-bg.html.

177 **Congressional decision to stop funding:** Charles Till, interview, *Frontline.*

9 TINY BEADS

182 **"One thing after another":** Susan M. Stacy, *Proving the Principle: A History of the Idaho National Engineering and Environmental Laboratory, 1949–1999* (Idaho Falls, Idaho: Idaho Operations Office of the Department of Energy, 2006), p. 224.

182 **Hydrogen bubble:** J. Samuel Walker, *Three Mile Island: A Nuclear Crisis in Historical Perspective* (Berkeley: University of California Press, 2004), pp. 151–155, 186–189.

183 **"What shook the public":** Victor Gilinsky quoted in ibid., p. 241.

183 **News bulletins were alarming:** Peter M. Sandman, "Tell It Like It Is: 7 Lessons from TMI," International Atomic Energy Agency, http://www.iaea.org/Publications/ Magazines/Bulletin/Bull472/htmls/tmi.html.

184 **"On an average, there were 120 entries":** Bernard Cohen, *The Nuclear Energy Option: An Alternative for the 90s* (New York: Plenum Press), pp. 58–59.

185–186 **Multiple studies investigated health issues:** Walker, *Three Mile Island*, pp. 234–237.

186 **multiple studies:** The Hatch-Susser Study (Columbia University, 1990), National Cancer Institute Study (1990), and Pennsylvania Department of Health Studies (1981, 1984, 1993). Additional independent studies include:

186 **"For accident emissions":** Maureen C. Hatch, Jan Beyea, Jeri W. Nieves, and Mervyn Susser, "Cancer near the Three Mile Island Nuclear Plant: Radiation Emissions," *American Journal of Epidemiology* 132, no. 3 (September 1990): 397–412, http://www.ncbi.nlm.nih.gov/entrez/query.fcgi?cmd=Retrieve&db=PubMed&list _uids=2389745&dopt=Abstract.

Population Exposure and Health Impact of the Accident at the Three Mile Island Nuclear Station (1979), by the Ad Hoc Population Dose Assessment Group (technical staff members from the NRC, the U.S. Department of Health and Human Services, and the Environmental Protection Agency). Conclusion: no immediate health effects. Latent or long-term effects, if any, would be minimal.

Report of the President's Commission on the Accident at Three Mile Island, 1979. Commissioned by President Carter and chaired by John G. Kemeny, then president of Dartmouth College. Conclusion on health effects: no detectable cancers or genetically related instances of ill health from the accident expected. Most significant health effect: mental stress.

Three Mile Island: A Report to the Commissioners and to the Public, 1980. Commissioned by the NRC and conducted by the Washington, D.C., law firm of Rogovin, Stern & Huge. Conclusion: health effects on the population as a whole, if they existed at all, would be nonmeasurable and nondetectable.

Report to the Nuclear Regulatory Commission from the Staff Panel on the Commission's Determination of an Extraordinary Nuclear Occurrence, 1980. Published by the NRC and based on work by representatives of the NRC, Environmental Protection Agency, U.S. Department of Health and Human Services, the former Pennsylvania Department of Environmental Resources, and U.S. Department of Energy. It confirmed the population dose estimates of the Ad Hoc Population Dose Assessment Group, the report of the President's Commission, and the report commissioned by Metropolitan Edison.

Investigations of Reported Plant and Animal Health Effects in the Three Mile Island Area, 1980. Published by the NRC and based on the findings of investigators of the NRC, Pennsylvania Department of Agriculture, U.S. Environmental Protection Agency, and Argonne National Laboratory. Conclusion: "It appears that none of the reported plant and animal health effects (reviewed in the report) can be directly attributed to the operation of or the accident" at TMI.

Follow-up Studies on Biological and Health Effects Resulting from the Three Mile Island Nuclear Power Plant Accident of March 28, 1979. Conducted by the Committee on Federal Research into the Biological Effects of Ionizing Radiation and published by the National Institutes of Health. The research subcommittee was made up of representatives from the National Institutes of Health; Food and Drug Administration; Alcohol, Drug Abuse and Mental Health Administration; Communicable Disease Center; Environmental Protection Agency; Nuclear Regulatory Commission; Department of Energy, and Department of Defense. Conclusion: the accident would produce no detectable health effects.

Report of the Governor's Commission on Three Mile Island, 1980. Commissioned by Gov. Richard Thornburgh. Conclusion: agreed with the findings of the President's Commission that health effects would be negligible and found that the mental stress from the accident would be transient for the general population.

Impact of TMI Nuclear Accident Upon Pregnancy Outcome, Congenital Hypothyroidism and Mortality, 1981. Conducted by the Pennsylvania Department of Health. Conclusion: pregnant women exposed to accident releases showed no measurable differences for prematurity, congenital abnormalities, neonatal deaths, or any other factors examined. Follow-up reports are issued at five-year intervals. Conclusions: no increase in infant hypothyroidism as a result of exposure from radioactive iodine. The finding was supported by an independent Hypothyroidism Investigation Committee organized by the Health Department.

Cancer Mortality and Morbidity around TMI, 1985. Study conducted by the Pennsylvania Department of Health and is being followed up by the department. Conclusion: no increased cancer risks to residents near TMI.

Assessment of Off-Site Radiation Doses from the Three Mile Island Unit 2 Accident, 1979—commissioned by Metropolitan Edison and conducted by Pickard, Lowe and Garrick, Inc., a Washington, D.C., consulting firm. Conclusions: generally consistent with those studies that concluded that radiation releases from the accident were too small to cause detectable health effects.

(List paraphrased by author from the American Nuclear Society, "Health studies find no cancer link to TMI," http://www.ans.org/pi/matters/tmi/healthstudies.html.)

187 **mental stress:** Ibid., Hatch.

188 **persistent assumptions:** Walker, *Three Mile Island*, pp. 204, 226–227, 242–243.

188 **presidential commission:** Ibid., pp. 210–212.

189 **$6 billion:** Dan Fagin, "Lights Out at Shoreham," Long Island: Our Story, *Newsday*, 2007, http://www.newsday.com/community/guide/lihistory/ny-history-hs9shore,0,563942.story.

189 **Utility charges:** New York State Office of the Comptroller, press release, December 12, 2005, http://www.osc.state.ny.us/press/releases/dec05/121205.htm.

189 **Long-term costs of Shoreham:** Matthew C. Cordaro, "Powerful Regrets," *The New York Times*, February 11, 2007.

190 **2005 random survey:** Bisconti Research, Inc., "Questions for EPZ Survey: FINAL August 2005," Nuclear Energy Institute, http://www.nei.org/documents/Survey_Plant_Neighbors_10-12-05.pdf; NEI, "Nuclear Power Plant Neighbors Accept Potential for New Reactor Near Them by Margin of 3 to 1," October 12, 2005, http://www.nei.org/index.asp?catnum=4&catid=851.

190 **Rickover, Deutch, and Three Mile Island:** Francis Duncan, *Rickover and the Nuclear Navy: The Discipline of Technology* (Naval Institute Press, 1990), p. 273.

Part 4 **THE KINGDOM OF ELECTRICITY**
10 **MAN'S SMUDGE**

193 **Duke statistics:** e-mail to author from Tom P. Shiel, Corporate Communications, External Relations, Duke Energy Carolinas.

193 **Having already reduced emissions:** Thomas C. Williams, Duke Energy Carolinas, e-mail message to author, 2006.

194 **Clean Smokestacks law:** North Carolina Department of Environment and Natural Resources, Division of Air Quality, "Key Facts about the Clean Smokestacks Act," http://daq.state.nc.us/news/leg/stackfacts.shtml; "Commission Adopts Rules for Curbing Mercury Emissions," http://daq.state.nc.us/news/pr/2006/hg_rule_11092006.shtml.

195 **14,300 train cars:** Union of Concerned Scientists, Clean Energy, "How Coal

Works" backgrounder, http://www.ucsusa.org/clean_energy/fossil_fuels/offmen
-how-coal-works.html.

195 **Most important commodity:** nationalatlas.gov, "Overview of U.S. Freight Railroads," http://www.nationalatlas.gov/articles/transportation/a_freightrr.html?one.

195 **154 mining fatalities:** U.S. Department of Labor, Mine Safety and Health Administration, "Coal Daily Fatality Report—Year End 2006" http://www.msha.gov/
stats/charts/coal2006yearend.asp.

195 **Environmental depredations:** Energy Information Administration, "Country
Analysis Briefs—United States—Coal," http://www.eia.doe.gov/emeu/cabs/Usa/
Coal.html.

196 **Methylmercury sources, health impacts:** Committee on the Toxicological
Effects of Methylmercury, Board on Environmental Studies and Toxicology,
National Research Council, *Toxicological Effects of Methylmercury,* http://books.nap
.edu/catalog/9899.html#toc, National Academies Press, 2000; U.S. Environmental Protection Agency, "Mercury," http://www.epa.gov/mercury/effects.htm, also
http://www.epa.gov/mercury/.

196–197 **Low-level radiation exposures:** U.S. Environmental Protection Agency, "Calculate Your Radiation Dose," http://www.epa.gov/radiation/students/calculate.html.

197 **Uranium, thorium produced from coal-fired plants:** Rip Anderson, personal
communication to author. His calculations use figures from the EPA.

197 **When uranium thought to be rare:** Alex Gabbard, "Coal Combustion: Nuclear
Resource or Danger," Oak Ridge National Laboratory, http://www.ornl.gov/info/
ornlreview/rev26-34/text/colmain.html.

198 **EPA exemption of fossil-fuel waste:** United States Environmental Protection
Agency, "Special Wastes: Fossil Fuel Combustion Waste": http://www.epa.gov/
epaoswer/other/fossil/index.html.

198 **Carbon dioxide emissions:** Simon Romero, "2 Industry Leaders Bet on Coal but
Split on Cleaner Approach," *The New York Times,* May 28, 2006, http://select
.nytimes.com/search/restricted/article?res=F10A11F9355A0C7B8EDDAC0894D
E404482; Energy Information Administration, U.S. Carbon Dioxide Emissions
from Energy Sources, "Carbon Dioxide Emissions by Fossil Fuel Type," June 2006,
based on data from *Monthly Energy Review,* May 2006, http://www.api.org/ehs/
climate/economics/upload/EIA_CO2_2005_Flash_Est_2006_6_28.pdf.

199 **Four tons a year:** Peter W. Huber and Mark P. Mills, "Why the U.S. Needs More
Nuclear Power," *City Journal,* Winter 2005, http://www.city-journal.org/html/15_1
_nuclear_power.html.

199 **Coal production, consumption:** U.S. Energy Information Administration,
"Annual Energy Review–Coal," http://www.eia.doe.gov/emeu/aer/coal.html.

199 **twenty-four thousand premature deaths:** Clean Air Task Force, "Publication:
Dirty Air, Dirty Power: Mortality and Health Due to Air Pollution from Power Plants,"
http://www.catf.us/publications/view/24.

199 **Better coal technology:** Romero, "2 Industry Leaders Bet on Coal."

200 **600 million metric tons:** World Nuclear Association, "Nuclear Energy: Meeting
the Kyoto Targets," http://72.14.253.104/search?q=cache:4fWLnCwK7LcJ:www
.world-nuclear.org/pdf/meeting_kyoto_targets.pdf. See also Nuclear Energy Institute, Up Front, "Quantifying Nuclear Energy's Environmental Benefits," http://
www.nei.org/index.asp?catnum=2&catid=43.

200 **a variety of solutions:** Robert H. Socolow and Stephen W. Pacala, "A Plan to
Keep Carbon in Check," *Scientific American* 295, no. 3 (September 2006): 50–57.

200 **154 new coal plants:** Bret Schulte, "A Texas Mess over Coal," *U.S. News & World*

Reports, posted November 26, 2006, http://www.usnews.com/usnews/news/articles/061126/4coal.htm.

200 **Gregory H. Boyce:** Ibid.

201 **Ad campaign for clean coal:** Brian R. Hook, "Peabody Energy: Fueling the Future," *St. Louis Commerce Magazine,* March 2006, http://www.stlcommerce magazine.com/archives/march2006/cover.html.

201 **"Climate change is real":** James E. Rogers, chief executive of Duke Energy, quoted by Steve Lohr, "The Cost of an Overheated Planet," *The New York Times,* December 12, 2006, http://select.nytimes.com/search/restricted/article?res=9807 E5DF1531F931A25751C1A9609C8B63.

201 **"The way to go is nuclear":** John McPhee, "Coal Train," *The New Yorker,* October 10, 2005.

11 FROM ARROWHEADS TO ATOMS

208 **Consequences of closing Indian Point nuclear plant:** National Research Council (U.S.), National Academy of Sciences Board on Energy and Environmental Systems, *Alternatives to the Indian Point Energy Center for Meeting New York Electric Power Needs* (Washington, D.C.: National Academies Press), 2006, http://www.nap.edu/catalog.php?record_id=11666.

209 **Uranium yields:** Nuclear Energy Institute, "Nuclear Facts," Up Front, http://www.nei.org/doc.asp?catnum=2&catid=106.

209 **Production cost of electricity:** Nuclear Energy Institute, "Nuclear Statistics," http://www.nei.org/doc.asp?catnum=2&catid=106.

211 **Plant operating capacity:** Nuclear Energy Institute, "Nuclear Facts," Up Front, http://www.nei.org/doc.asp?catnum=2&catid=106.

212 **"The 2003 M.I.T. study estimated":** John M. Deutch and Ernest J. Moniz, "The Nuclear Option," *Scientific American* 295, no. 3 (September 2006): 77–83.

212 **Increase in nuclear capacity:** Ibid.

212 **"All these reductions in the cost":** Ibid.

212 **"Emissions trading is":** United Kingdom Department for Environment, Food, and Rural Affairs, "Emissions Trading Schemes," http://www.defra.gov.uk/Environment/climatechange/trading.

214 **"Insurance pools have paid":** Nuclear Energy Institute, "Price-Anderson Act Provides Effective Nuclear Insurance at No Cost to the Public," Up Front, June 2006, http://www.nei.org/index.asp?catnum=3&catid=595.

12 BARRIERS

219 **"Mortality rates of these workers":** Mailman School of Public Health, Columbia University, "U.S. Nuclear Power Workers Show No Unexpected Radiation-related Cancer According to Mailman School of Public Health Study," November 9, 2004, http://www.mailman.hs.columbia.edu/news/radiation-nuclear-pow.html; Geoffrey R. Howe, Lydia B. Zablotska, Jack J. Fix, John Egel, and Jeff Buchanan, "Analysis of the Mortality Experience Amongst U.S. Nuclear Power Industry Workers After Chronic Low-Dose Exposure to Ionizing Radiation," *Radiation Research* 162, no. 5 (November 2004): 517–526, http://www.bioone.org/perlserv/?request=get-document& doi=10.1667%2FRR3258.

219 **"The hydraulic barrier"**: Sara Leach, "ERDC Increases Security at All Facilities," U.S. Army Corps of Engineers, Engineering Research and Development Center, August 2003, http://www.hq.usace.army.mil/cepa/pubs/aug03/story16.htm.

225 **"They are credited"**: Edward McGaffigan Jr., "Remarks," speech, 14th Annual Regulatory Information Conference, Washington, D.C., March 6, 2002, http://www.nrc.gov/reading-rm/doc-collections/commission/speeches/2002/02-007.html.

226 **Safety in chemical industry**: U.S. Chemical Safety and Hazard Investigation Board, "CSB Current Investigations," http://www.chemsafety.gov/index.cfm?folder=current_investigations&page=index.

226 **"to build a capacity to police"**: Eben Raplan, "Flynn: Homeland Security Report Card," Council on Foreign Relations, October 25, 2006, http://www.cfr.org/publication/11814/. See also Stephen E. Flynn, *The Edge of Disaster: Rebuilding a Resilient Nation* (New York: Random House, 2007).

226 **"This vulnerability has been raised"**: Senator Susan Collins quoted by Eric Lipton, "Debate over Security for Chemical Plants Focuses on How Strict to Make Rules," *The New York Times*, September 21, 2006, http://select.nytimes.com/search/restricted/article?res=F30F1FFD3A550C728EDDA00894DE404482.

227 **Silent Vector**: Nuclear Energy Institute, "Security Effectiveness: Independent Studies and Drills," Up Front, http://www.nei.org/index.asp?catnum=2&catid=279.

228 **"a large passenger aircraft"**: James Muckerheide, quoted by Matthew Wald, "Threats and Responses: Reactor Vulnerability; Experts Say Nuclear Plants Can Withstand Jet Crash," *The New York Times*, September 20, 2002, http://select.nytimes.com/search/restricted/article?res=FA091EFB3A540C738EDDA00894DA404482.

230 **"Starting about thirty years ago"**: Foulke, "The Need for Realism."

230 **Rebuttal**: Douglas M. Chapin and others, "Nuclear Power Plants and Their Fuel as Terrorist Targets," *Science* 297, no. 5589 (September 20, 2002): 1997–1999, http://www.radscihealth.org/rsh/docs/Chapin20Sep02Science-PolicyForum.pdf.

231 **"outrageous public statements"**: Summary of Chapin and Others, "Nuclear Plants and Their Fuel as Terrorist Targets," *Science* 297, no. 5589 (September 20, 2003), http://www.sciencemag.org/cgi/content/summary/297/5589/1997.

231 **Risk of terrorism**: Frank N. von Hippel and others, "Revisiting Nuclear Plant Safety," *Science* 299, no. 5604 (January 10, 2003), http://www.sciencemag.org/cgi/content/citation/299/5604/2016?ct=.

231 **hypothetical worst-case scenarios**: Robert Alvarez and others, "Reducing the Hazards from Stored Spent Power-Reactor Fuel in the United States," *Science and Global Security* 11, no. 1 (2003): 1–51, http://www.princeton.edu/~globsec/publications/pdf/11_1Alvarez.pdf.

231 ***Science and Global Security***: Princeton University, "The Program on Science & Global Security," http://www.princeton.edu/~globsec/.

232 **"Natural forces"**: Herschel Specter, "Nuclear Risk and Reality," *The New York Times*, May 20, 2003, http://select.nytimes.com/search/restricted/article?res=F30B13F6385A0C738EDDAC0894DB404482.

232 **"Even if the [spent-fuel] pools"**: Herschel Specter, "The Low-Risk Pool," October 17, 2004, http://select.nytimes.com/search/restricted/article?res=F50E1FFB3E5E0C748DDDA90994DC404482.

232 **Failure to address events realistically**: U.S. Nuclear Regulatory Commission Staff, "Fact Sheet on NCR Review of Paper on Reducing the Hazards from Stored Spent Fuel," U.S. Nuclear Regulatory Commission, August 2003, http://www.nrc.gov/reading-rm/doc-collections/fact-sheets/reducing-hazards-spent-fuel.html

233 **Princeton authors defend findings:** Robert Alvarez and others, "Response by the Authors to the NRC Review of 'Reducing the Hazards from Stored Spent Power-Reactor Fuel in the United States,' " *Science and Global Security* 11 (2003): 213–223, Princeton University, http://www.princeton.edu/~globsec/publications/pdf/SGS _213-223_response.pdf.

235 **"These studies assessed the capabilities":** U.S. Nuclear Regulatory Commission, "Frequently Asked Questions About Security Assessments at Nuclear Power Plants," February 15, 2007, http://www.nrc.gov/security/faq-security-assess-nuc -pwr-plants.html.

235 **NRC ruling on plant defenses:** U.S. Nuclear Regulatory Commission, "NRC Approves Final Rule Amending Security Requirements," News Release, January 29, 2007, http://www.nrc.gov/reading-rm/doc-collections/news/2007/07-012.html.

236 **Emergency reactor features:** Matthew Wald, "U.S. Takes Step to Address Airline Attacks on Reactors," *The New York Times*, April 25, 2007, http://www.nytimes.com/ 2007/04/25/us/25nuke.html?_r=1&oref=slogin.

237 **Rickover and the lost screw:** Theodore Rockwell, *The Rickover Effect: How One Man Made a Difference* (Annapolis, Md.: Naval Institute Press, 1992), p.3.

13 UNOBTAINIUM

243 **Hydrogen research at Idaho National Laboratory:** "Producing Hydrogen with Nuclear Energy," http://www.inl.gov/docs/factsheets/producing_hydrogen_with _nuclear_energy.pdf.

243 **Cost of Middle East peacekeeping:** John Dizard, "Bush, Iraq and the Hydrogen Economy," *Financial Times*, January 31, 2005, http://www.energybulletin.net/4189 .html.

244 **Need for more reactors:** Dan Gallagher, "Groups Pan Using Reactors for Clean Hydrogen Fuel," *Casper* [WY] *Star-Tribune*, April 24, 2004, NewsMine, http:// newsmine.org/archive/nature-health/environment/pollution/hydrogen-fuel-cells -props-up-coal-industry.txt.

244 **"Hence, we need to generate":** Paul M. Grant, "Hydrogen Lifts Off—With a Heavy Load," *Nature* 424, no. 6945 (July 10, 2003): 129–130, http://www.nature .com/nature/journal/v424/n6945/full/424129a.html [subscription].

247 **Gaia concept:** James Lovelock, *The Ages of Gaia*, 2nd ed. (Oxford: Oxford University Press, 2000), p. 30.

249 **Christie Brinkley:** Lloyd Grove, "Brinkley's Nuclear Cloud" *The Washington Post*, January 23, 2002, p. C3, NucNews.net, http://nucnews.net/nucnews/2002nn/ 0201nn/020123nn.htm#700.

249 **Solar panel toxic waste:** V. M. Fthenakis, "Overview of Potential Hazards," chapter VII-2, in *Practical Handbook of Photovoltaics: Fundamentals and Applications*, ed. T. Markvart and L. Castaner (New York: Elsevier, 2003), http://www.pv.bnl.gov/art _170.pdf.

249 **Disposal of solar panels:** Richard Rhodes and Denis Beller, "The Need for Nuclear Power," *Foreign Affairs*, January/February 2000, http://www.foreignaffairs .org/2000010faessay4/richard-rhodes-denis-beller/the-need-for-nuclear-power.html.

251 **Wind power highly unreliable:** David Dixon, "Wind Generation's Performance During the July 2006 California Heat Storm," *Energy Pulse*, September 8, 2006, http://www.energypulse.net/centers/article/article_display.cfm?a_id=1332.

251 **Utility wants to build "wind park":** ENSR International, "Essential Fish Habitat Assessment for the Long Island Offshore Wind Park Within New York State

Waters and Federal Waters of the Outer Continental Reef," May 16, 2005, FPLEnergy (Florida Power and Light), http://www.fplenergy.com/projects/pdf/liwind/04c.pdf.

252 **Endangered species:** U.S. Army Corps of Engineers, *Public Notice*, June 9, 2005, http://www.savejonesbeach.org/full-public-notice-lioswp.pdf.

253 **"ridgelines industrialized":** Eleanor Tillinghast, "Many Costs, Few Benefits of Wind Power," *Berkshire Eagle*, March 4, 2006, Industrial Wind Action Group, http://www.windaction.org/articles/1864.

254 **comparative study of impact on ecosystems:** John M. Ryskamp, "The Need for Nuclear Power," presentation, IEEE (Institute of Electrical and Electronics Engineers) Power Engineering Society Meeting, April 28, 2003, http://nuclear.in1.gov/docs/papers-presentations/nuclear_need_2-26-03.pdf.

254 **Department of Energy budget:** "The Office of Energy Efficiency and Renewable Energy ($771 million) budget includes considerable funding increases for hydrogen technology, fuel cell technology, vehicle technology, biomass, solar, and wind research programs. The Office of Fossil Energy ($444 million) supports the Coal Research Initiative and other power generation/stationary fuel cell research programs. The Office of Nuclear Energy, Science and Technology ($392 million) includes $250 million for the Global Nuclear Energy Partnership (GNEP); and also supports Generation VI, Nuclear Power 2010, and the Nuclear Hydrogen Initiative." U.S. Department of Energy, "Department of Energy Requests $23.6 Billion for FY 2007," press release, February 6, 2006, http://energy.gov/news/3150.htm.

255 **"Wind power has been promoted":** ABS Energy Research, "Wind Power Report 2006," Industrial Wind Action Group, http://www.windaction.org/documents/4446.

255 **German energy outlook:** Bernard Benoit, "Berlin Open to Liberalising Energy Market," *Financial Times*, January 12, 2007, http://www.ft.com/cms/s/507d626a-a1e1-11db-8bc1-0000779e2340.html.

255 **Renewables and growth in fossil-fuel use:** Paul Lorenzini, "A Second Look at Nuclear Power," *Issues in Science and Technology*, vol. 295, no. 3, March 22, 2005, http://www.issues.org/21.3/lorenzini.html.

256 **Research and development funding declined:** Daniel M. Kammen, "The Rise of Renewable Energy," *Scientific American* (September 2006): 84–93.

258 **tritium leaks:** U.S. Nuclear Regulatory Commission, "Fact Sheet on Tritium, Radiation Protection Limits, and Drinking Water Standards," July 2006, http://www.nrc.gov/reading-rm/doc-collections/fact-sheets/tritium-radiation-fs.html; "Groundwater Contamination (Tritium) at Nuclear Plants," http://www.nrc.gov/reactors/operating/ops-experience/grndwtr-contam-tritium.html.

258 **Bruno Comby:** Bruno Comby's speeches, articles, and interviews can be found at Environmentalists for Nuclear Energy (EFN) at http://www.ecolo.org/.

258 **"Nuclear power can play a significant role":** William C. Sailor and others, "Nuclear Power: A Nuclear Solution to Climate Change?" *Science* 288, no. 5469 (May 19, 2000): 1177–1178, http://search.ebscohost.com/login.aspx?direct=true&db=f5h&AN=3176192&site=ehost-live; http://www.sciencemag.org/cgi/content/full/288/5469/1177 [subscription].

259 **"If we had 50 years":** James Lovelock, "Nuclear Power Is the Only Green Solution," *The Independent*, May 24, 2004, http://comment.independent.co.uk/commentators/article61727.ece.

259 **"Opposition to nuclear energy":** Ibid.

260 **"Renewable energy might"**: James Lovelock, "The Selfish Greens," remarks at the Adam Smith Institute, March 15, 2004, World Nuclear Association, http://www .world-nuclear.org/reference/pdf/lovelock.pdf.

Part 5 CLOSING THE CIRCLE

263 **"An outstanding advantage"**: James Lovelock, "Nuclear Energy: The Safe Choice for Now," http://www.ecolo.org/lovelock/nuclear-safe-choice-05.htm.

14 TEN THOUSAND YEARS

265 **"as a nation"**: Dale Klein, interview by Jeff Nesmith and George Lobsenz, *The Newsmakers*, C-Span, October 18, 2006, http://www.c-span.org/newsmakers/klein .htm.
268 **Waste repository life span**: Steve Tetreault, "Yucca Radiation Limits Unveiled," *Las Vegas Review-Journal*, August 10, 2005, http://www.reviewjournal.com/lvrj _home/2005/Aug-10-Wed-2005/news/27026244.html.
268 **Cost of storing nuclear waste**: Leo S. Gómez, David J. Brenner, and Otto G. Raabe, co-chairs, "Executive Summary," *Ultra-Low-Level Radiation Effects Summit: Final Report*, Carlsbad, New Mexico, January 15–16, 2006, p. 1, http://www.orionint .com/ullre/report-2006.pdf.
269 **Greater efficiency decreases nuclear waste**: Nuclear Energy Institute, "High-Level," Up Front, http://www.nei.org/doc.asp?catnum=2&catid=62.
269 **Coal combustion waste increase**: U.S. Environmental Protection Agency, "Electricity from Coal," http://www.epa.gov/cleanrgy/coal.htm#fn3.
269 **Nuclear plants reduce carbon emissions**: Nuclear Energy Institute, "Nuclear Waste Disposal—High Level."
271 **"if the world community can effectively guard"**: Matthew Bunn, "Preventing Nuclear Terrorism," September 24, 2002, Belter Center for Science and Public Affairs, John F. Kennedy School of Government, Harvard University.
272 **Mixed oxide fuels (MOX)**: World Nuclear Association, "Military Warheads as a Source of Nuclear Fuel," Information and Issues Briefs, October 2005, http://213 .130.40.135/info/inf13.htm.

15 THE HUGE FACTORY

273 **"unless you turn"**: Richard Rhodes, *The Making of the Atomic Bomb* (New York: Simon & Schuster, 1986), p. 500.
277 **bomb-building spree**: For a detailed account of enormous public deception and fearmongering that characterized the arms race, see Richard Rhodes's history, *Arsenals of Folly: The Making of the Nuclear Arms Race* (New York: Knopf, 2007).

16 32N164W

283 **"Always focus on the problem"**: Edward L. Miles, "Personal Reflections on an Unfinished Journey Through Global Environmental Problems of Long Timescale," *Policy Sciences* 31, no. 1 (February 1998): 1–33, http://search.ebscohost.com/login .aspx?direct=true&db=afh&AN=626976&site=ehost-live.
285 **"The industry denies"**: Theodore Rockwell, e-mail message to author, 2005. Rockwell has given permission to quote from his correspondence.

17 THOSE WHO SAY IT CAN'T BE DONE

289 **Objections to transporting nuclear waste:** *60 Minutes,* "Yucca Mountain," July 25, 2004, CBS, http://www.cbsnews.com/stories/2003/10/23/60minutes/main 579696.shtml; Robert J. Halstead, "Radiation Exposures from Spent Nuclear Fuel and High-Level Nuclear Waste Transportation to a Geologic Repository or Interim Storage Facility in Nevada," State of Nevada/Office of the Governor/Agency for Nuclear Projects/Nuclear Waste Project Office, http://www.state.nv.us/nucwaste/ trans/radexp.htm.

289–290 **Federal ruling:** Associated Press, "Nevada Loses Decision on Atomic Waste," *The New York Times,* August 9, 2006, http://www.nytimes.com/2006/08/09/wash- ington/09brfs-005.html?ex=1171083600&en=bd9f1a0d303fc09b&ei=5070.

292 **"It is abundantly clear":** Harry Reid, "Press Release of Senator Reid: Reid State- ment on Falsification of Yucca Mountain Documentation," March 16, 2005, http:// reid.senate.gov/newsroom/record.cfm?id=233737.

293 **Nevada energy sources:** Energy Information Administration, "Nevada Quick Facts," last update April 12, 2007, http://tonto.eia.doe.gov/state/state_energy_ profiles.cfm?sid=NV.

299 **Casks "best containers":** *60 Minutes,* "Yucca Mountain," July 25, 2004, CBS, http://www.cbsnews.com/stories/2003/10/23/60minutes/main579696.shtml.

299 **Transport of nuclear materials:** American Nuclear Society, "Keeping Nuclear Transportation Safe," http://www.ans.org/pi/np/transport/.

302 **Effects of elevated carbon dioxide levels:** U.S. Department of Energy, Nevada Site Office, "Research at the Nevada Test Site: The Nevada Desert FACE Facility (NDFF)," http://www.nv.doe.gov/library/factsheets/DOENV_1116.pdf.

302 **massive shutter:** Bill Murphy, "Boomtown: Time-traveling around Mercury, Nev., and the Nevada Test Site with Tom Hunter and Dan Bozman," *Sandia Lab News,* p. 7, December 10, 2004, Sandia National Laboratories, http://www.sandia.gov/ LabNews/LN12-10-04/labnews12-10-04.pdf.

308 **carbon dioxide emissions:** Mongabay, "State Carbon Emissions Data for the United States, 1990–2001," June 21, 2006, http://news.mongabay.com/2006/0621 -co2.html.

309 **position paper:** Health Physics Society, "Compatibility in Radiation-Safety Regu- lations," March 2001, http://www.hps.org/documents/regulations.pdf.

309 **"Predicting deaths":** Larry Foulke, "The Need for Realism," speech, 10th Inter- national Conference on Radiation Shielding, Madeira, Portugal, May 13, 2004, http://www.itn.pt/ICRS-RPS/foulke.htm.

310 **Radiation exposure from repository:** U.S. Department of Energy Office of Civil- ian Radioactive Waste Management, "Facts About Radiation," Factsheet, January 2005, http://www.ocrwm.doe.gov/factsheets/doeymp0403.shtml.

312 **Waste sequestered:** U.S. Department of Energy Office of Civilian Radioactive Waste Management, *Annual Report to Congress OCRWM FY 2000,* September 2001, http://www.ocrwm.doe.gov/info_library/program_docs/annualreports/00ar/ 00ar.htm.

314 **Long-term movement of products of nuclear decay:** George A. Cowan, "A Nat- ural Fission Reactor," *Scientific American* 235, no. 36 (July 1976): 36; also Robert W. Holloway, DOE Savannah River Laboratory, "Plutonium Fission in the Oklo Nat- ural Reactor," U.S. Department of Energy, Savannah River Site, http://sti.srs.gov/ fulltext/dpms8075/dpms8075.pdf.

316 **Probabilistic analysis recommended:** OECO Nuclear Energy Agency and the

International Atomic Energy Agency, *An International Peer Review of the Yucca Mountain Project* TSPA-SR: *Total System Performance Assessment for the Site Recommendation*, (OECO, 2002), p. 65, http://www.nea.fr/html/rwm/reports/2002/nea3682 -yucca.pdf.

316 **Yucca Mountain administrative changes:** Steve Tetreault, "DOE Takes Steps to Reorganize Yucca," *Pahrump Valley Times*, January 20, 2006, http://www.pahrump-valleytimes.com/2006/01/20/news/yucca.html.

325 **"The challenges we face":** Mohamed ElBaradei, "Geological Repositories: The Last Nuclear Frontier," statement, International Conference on Geological Repositories: Political and Technical Progress, Stockholm, Sweden, December 8–10, 2003, International Atomic Energy Agency, http://www.iaea.org/NewsCenter/ Statements/2003/ebsp2003n028.html.

326 **"The waste can be":** John M. Deutch and Ernest J. Moniz, "The Nuclear Option," *Scientific American* 295, no. 3 (September 2006): 76–83.

18 THE GIGANTIC CRYSTAL

330 **Deaths from natural gas:** World Nuclear Association, "Safety of Nuclear Power Reactors." See chart, Comparison of Accident Statistics in Primary Energy Production: http://www.world-nuclear.org/info/inf06.html.

332 **"Some oppose WIPP":** Wendell D. Weart, quoted by John German, "WIPP History Notable Quotes," *Sandia Lab News* 51, no. 7 (April 9, 1999), Sandia National Laboratories, http://www.sandia.gov/LabNews/LN04-09-99/wippside3 _story.html.

332 **Court decision against WIPP:** Chronology of WIPP, New Mexico Energy, Natural Resources, and Minerals Department: http://www.emnrd.state.nm.us/wipp/ chronolo.htm.

336 **"There is probably no single piece":** Margaret Chu, quoted by John German, ibid.

337 **"WIPP is not opening":** Don Hancock, quoted by John German, ibid.

349 **outside watchdogs:** John Fleck, "WIPP to Get Hot Waste; State Officials Set to Sign New Permit Today," *Albuquerque Journal*, October 16, 2006.

350 **Warning markers:** Kathleen M. Trauth, Stephen Curtis Hora, and Robert V. Guzowski, *Expert Judgment on Markers to Deter Inadvertent Human Intrusion into the Waste Isolation Pilot Plant* (Albuquerque, NM: Sandia National Laboratories, 1993), http://infoserve.sandia.gov/sand_doc/1992/921382.pdf.

Part 6 BORROWING FROM OUR CHILDREN

351 **epistle of Cabeza de Vaca:** Haniel Long, *The Power Within Us: Cabeza de Vaca's Relation of His Journey from Florida to the Pacific, 1528–1536* (New York: Duell, Sloan, and Pearce, 1944).

19 THE IRON CHAMBER

355 **Diagnostic medical radiation increase:** Roni Caryn Rabin, "With Rise in Radiation Exposure, Experts Urge Caution on Tests," *The New York Times*, June 14, 2007.

356 **Ambient radiation from U.S. Capitol Building:** U.S. Army Corps of Engineers,

"Radiation at FUSRAP Sites," FUSRAP fact sheet, April 1998, http://www
.lrb.usace.army.mil/fusrap/docs/fusrap-fs-radfusrap.pdf.

356 **scientists polled:** Carol L. Silva, Hank C. Jenkins-Smith, and Richard P. Barke,
"The Politics of Caution: The Bases of Scientists' Precautionary Choices," Bush
School of Government and Public Service, Texas A&M University, April 2005,
http://bush.tamu.edu/research/working_papers/csilva/PoliticsofCaution.pdf

357 **"The facility will allow":** Leo S. Gómez, David J. Brenner, and Otto G. Raabe, co-
chairs, "Executive Summary," *Ultra-Low-Level Radiation Effects Summit: Final
Report*, Carlsbad, New Mexico, January 15–16, 2006, p. 16, http://www.orionint
.com/ullre/report-2006.pdf.

358 **"Cancer is a fact":** Ibid.

20 "WATER THEM ANYWAY"

363 **Toshiba nuclear power plant:** Joel Gay, "Village invited to test cheap, clean
nuclear power," *Anchorage Daily News*, October 21, 2003, http://www.adn.com
/front/story/4214182p-4226215c.html.

363 **TXU's new nuclear plants:** Jack Uldrich, "TXU's Nuclear Ambitions," *The Motley
Fool* (April 11, 2007), http://www.fool.com/investing/high-growth/2007/04/11/
txus-nuclear-ambitions.aspx.

365 **recent independent studies:** World Nuclear Association, "New Economics of
Nuclear Power," December 2005, http://www.world-nuclear.org/reference/pdf/
economics.pdf.

365 **Nuclear plants economically competitive:** George S. Tolley and Donald W.
Jones, "The Economic Future of Nuclear Power, A Study Conducted at The University of Chicago," August 2004, Argonne National Laboratory, http://www.anl
.gov/Special_Reports/NuclEconSumAug04.pdf.

365 **Worldwide estimate of nuclear power subsidy:** HM Treasury's Stern Report, Part IV, http://www.hm-treasury.gov.uk/media/9A3/57/Ch_16_accelerating
_technological_innovation.pdf.

366 **Decision to expand nuclear power:** Mark Landler, "With Apologies, Nuclear
Power Gets a Second Look," *The New York Times*, January 28, 2007, http://www
.nytimes.com/2007/01/28/weekinreview/28land.html?ex=1327640400&en=7a4697
11e382f1d6&el=5088&partner-rssnyt&emc=rss.

366 **New reactors:** Dale Klein, interview by Jeff Nesmith and George Lobsenz,
The Newsmakers, C-Span, October 18, 2006, http://www.c-span.org/newsmakers/
klein.htm.

366 **Nuclear plant security:** Dale Klein, letter to the editor, *The New York Times*, January 17, 2007, http://select.nytimes.com/search/restricted/article?res=F00914FE3A
540C748DDDA80894DF404482.

371 **Uranium supply:** Colin MacDonald, "Uranium: Sustainable Resource or Limit to
Growth?" World Nuclear Association, Annual Symposium, 2003, http://www
.world-nuclear.org/sym/2003/macdonald.htm.

21 THE POWER WITHIN

372 **refuge for President Eisenhower:** GlobalSecurity, "Weapons of Mass Destruction
(WMD), Manzano," http://www.globalsecurity.org/wmd/facility/manzano.htm.

GLOSSARY

ABCC: Atomic Bomb Casualty Commission, founded after World War II to study effects of the atomic bombs dropped on Japan. See *Radiation Effects Research Foundation.*

Absorbed dose: The *energy* imparted to a unit mass of matter by *ionizing radiation.* Measured in *rads* or grays.

Absorbed dose rate: *Absorbed dose* divided by the time it takes to deliver that dose. High-dose rates are usually more damaging to humans and animals than low-dose rates. This is because repair of damage is more efficient when the *dose rate* is low.

Actinide: Any of a series of radioactive metallic *elements* ranging from *atomic number* 89 to 103, including *thorium* (90), *uranium* (92), and *plutonium* (94). Minor actinides, by-products of the *chain reaction,* are responsible for significant radiotoxicity in *spent nuclear fuel* but can be burned up as fuel in certain types of reactors.

Activation products: Atomic fragments produced in a reactor vessel by *radiation* from the *chain reaction.*

Activity: The rate at which radioactive material emits *radiation,* stated in terms of the number of nuclear disintegrations occurring over a unit of time. One disintegration per second is a *becquerel (Bq),* which has replaced the *curie (Ci)* as the standard unit of activity.

Acute exposure: A single *exposure* to a toxic substance. Acute exposures are usually characterized as lasting no longer than a day, as compared to longer, continuing, or chronic *exposures.*

Adaptive response: The ability of cells to respond to low *doses* of *radiation* with the induction of a series of genes and to reduce the level of *radiation*-induced damage when challenged with a subsequent high *dose* of *radiation.*

Added risk: The difference between the cancer *incidence* under the *exposure* condition and the background cancer *incidence* in the absence of *exposure.*

ALARA: "As Low As Reasonably Achievable." Government acronym for a goal for the minimal *exposure* of the public and workers to radiological materials. Every reasonable effort is to be made to keep *exposures* to *ionizing radiation* as far below the *dose limits* as practical, taking into account the state of technology and the economics of improvements in relation to the benefits to public health and safety, other societal and socioeconomic considerations, and utilization of *nuclear energy.*

Alpha decay: The *emission* of a *nucleus* of a helium *atom* from the *nucleus* of an *element,* generally of a heavy *element,* in the process of its *radioactive decay.*

Alpha particle: Positively charged particle emitted by certain radioactive material. An alpha particle is the *nucleus* of a helium *atom,* which is made up of two *neutrons* and two *protons.* Alpha particles are discharged by *radioactive decay* of many heavy *elements,* such as *uranium*-238 and *plutonium*-239.

Anthropogenic: Of human origin. Term often used to refer to human contributions to earth's burden of *greenhouse gases.*

Argonne National Laboratory (ANL): Located in Illinois. One of nine national laboratories under the aegis of the *Department of Energy* (DOE).

Argonne National Laboratory-West (ANL-W): Located in Idaho. In 2005 it merged with *Idaho National Engineering and Environmental Laboratory* (INEEL) to become *Idaho National Laboratory* (INL).

Atom: The smallest particle of an *element* that cannot be divided or broken up by chemical means. It consists of a central, positively charged core of *protons* and *neutrons* that make up the *nucleus* around which negatively charged *electrons* orbit.

Atomic energy: *Energy* released in nuclear reactions. Specifically, the *energy* released when a *neutron* initiates the breaking up of an *atom's nucleus* into smaller pieces (*fission*), or when two nuclei are joined together under millions of degrees of heat (*fusion*). It is more correctly called *nuclear energy*.

Atomic Energy Commission (AEC): Established in 1946 by Congress to oversee development of atomic science and technology. Broken up in 1974 into the Energy Research and Development Administration (ERDA) and the *Nuclear Regulatory Commission* (NRC). ERDA later became the *Department of Energy* (DOE).

Atomic mass: The number of *protons* and *neutrons* in an *atom*. *Uranium*-238 has an atomic mass of 238: 92 *protons* and 146 *neutrons*.

Atomic number: The number of *protons* in the *nucleus* of an *atom*. Each *element* has a different atomic number. All *uranium atoms*, for example, have an atomic number of 92, but *isotopes* of *uranium* differ in their *atomic mass*.

Atomic structure: The conceptualized concept of an *atom*, regarded as consisting of a central positively charged *nucleus* (*protons* and *neutrons*) around which negatively charged *electrons* revolve in various orbits.

Back end: The final stage of the nuclear *fuel cycle*.

Background radiation: Background radiation comes from the cosmos, the sun, radioactive *elements* in the upper atmosphere, the earth's crust, building materials, food, and the human body. Most natural background radiation comes from *radon* emanating from rocks and soil. Global fallout from the testing of nuclear explosive devices in the atmosphere is also included in estimates of background radiation. The typically quoted average individual *exposure* in the United States from background radiation is 360 *millirem* (3.6 milli*sieverts*) per year.

Basalt: Rock of volcanic origin. One of the geological formations recommended by the *National Academy of Sciences* (NAS) as suitable for deep disposal of nuclear waste.

Baseload power: *Power* reliably generated by a plant—*coal*-fired, nuclear, or hydroelectric—that provides a steady flow of *electricity* regardless of total *power* demand by the *grid*. These plants run constantly except in the case of repairs or scheduled maintenance. Baseload power plants are distinguished by their low-cost generation, high efficiency, and safety at set outputs.

Becquerel (Bq): The unit of *activity* equal to one disintegration per second.

Beta decay: The *emission* of *electrons* or positrons (particles identical to *electrons*, but with a positive electrical charge) from the *nucleus* of an *element* in the process of *radioactive decay* of the *element*.

Beta particle: A negatively charged particle emitted from a *nucleus* during *radioactive decay*. A beta particle has mass and charge equivalent to that of an *electron*. It can travel only a short distance through air and has a low ability to penetrate other materials. Thin sheets of metal or plastic may stop beta particles.

Beta radiation: *Radiation* consisting of *beta particles*.

Binding energy: The *energy* needed to separate a particle from a system of particles or to disperse all the particles of a system. To separate an atomic *nucleus* into its components, *protons* and *neutrons,* nuclear binding energy is required. It is also the *energy* that is released by combining individual *protons* and *neutrons* into a single *nucleus. Electron* binding energy, also called *ionization* potential, is the *energy* needed to separate an *electron* from an *atom, molecule,* or *ion,* and also the *energy* released when an *electron* joins an *atom, molecule,* or *ion.* The binding energy of a single *proton* or *neutron* in a *nucleus* is about a million times greater than that of a single *electron* in an *atom.*

Biofuels: Liquid fuels and blending components produced from *biomass* (plant) feedstocks, used primarily for transportation.

Biological Effects of Ionizing Radiation (BEIR): This refers to the subject of study by scientific committees that meet periodically under the auspices of the *National Research Council* (NRC). BEIR panels issue reports. For example, BEIR-VI defined the health effects of *radon.*

Biomass: Any organic (plant or animal) material that is available on a renewable basis, including agricultural crops and agricultural wastes and residues, wood and wood wastes and residues, animal wastes, municipal wastes, and aquatic plants.

Boiler: A tank in which water is heated to produce either hot water or steam that is circulated for the purpose of heating and *power.*

Boiling-water reactor (BWR): A nuclear reactor in which water is allowed to boil in the core. The resulting steam is used to drive a turbine generating *electric power.*

Breeder reactor: A nuclear reactor that "breeds" more *fissile material* during its operation than it consumes. Also called a "fast reactor" because of its use of *fast neutrons* for sustaining the *chain reaction.*

Carbon dioxide (CO_2): A colorless, odorless noncombustible gas, two parts oxygen to one part carbon. It is formed by the *combustion* of carbon and carbon compounds (such as *fossil fuels* and *biomass*) and by respiration, which is a slow *combustion* in animals and plants, and by the gradual oxidation of organic matter in the soil. Because humans have taken large quantities of hydrocarbons in the form of *fossil fuels* out of the ground and burned them, carbon dioxide has been released in excessive quantities into the atmosphere where, as a *greenhouse gas,* it traps heat.

Carbon equivalent: A metric measure used to compare the *emissions* from various *greenhouse gases* based upon their *global warming* potential (GWP). *Carbon dioxide* equivalents are commonly expressed as "million metric tons of *carbon dioxide* equivalents (MMTCDE)." The *carbon dioxide* equivalent for a gas is derived by multiplying the tons of the gas by the associated GWP. For example, the GWP for *methane* is 21 and for nitrous oxide 310. This means that *emissions* of a million *metric tons* of *methane* and nitrous oxide respectively is equivalent to *emissions* of 21 and 310 million *metric tons* of *carbon dioxide.*

Centrifuge enrichment: Different *isotopes* of gasified *uranium* (U-238 and U-235) are separated from one another by weight in a large number of rotating cylinders arranged in series. These series of centrifuge machines are interconnected to form cascades. As the cylinders spin, *uranium*-238, which is heavier than *uranium*-235, will tend to settle toward the outside. Centrifuge enrichment is generally fifty times more *energy* efficient than the older method of *gaseous diffusion enrichment.*

Cesium-137: A *beta-gamma*-emitting *fission product* that is made in reactors and in the environment has a *half-life* of 26.6 years. Cesium has metabolic properties similar to potassium. As a result it is rather uniformly distributed in the body. It clears from the body quickly, with a *half-life* of days and weeks, and therefore has a rather low effectiveness in increasing cancer *incidence.*

Chain reaction: A self-sustaining nuclear reaction that takes place during *fission*. A fissionable substance (e.g., *uranium*) absorbs a *neutron* and divides, releasing additional *neutrons* that are absorbed by other fissionable *nuclei*, releasing still more *neutrons*.

Chemical energy: *Energy* stored in a substance and released during a chemical reaction such as burning wood, *coal*, or oil. Chemical reactions always involve the *electrons* in an *atom*, as opposed to nuclear reactions, which always involve the atomic *nucleus*.

Cladding: The thin-walled metal jacket, usually made of a zirconium alloy, that encloses a fuel rod. The cladding prevents corrosion of the nuclear fuel and release of *fission products* into the coolant.

Cleanup: Actions taken to deal with a release or threat of release of a hazardous substance that could affect humans and the environment. The term *cleanup* is sometimes used interchangeably with the terms *environmental isolation, remedial action, removal action, response action*, and *corrective action*.

Climate change: A term used to refer to all forms of climatic inconsistency, but especially to significant change from one prevailing climatic condition to another.

Closed cycle (cooling): Refers to the systems in a nuclear plant that circulate reactor *coolant* in a closed loop, so that water that has been through the reactor does not enter any other circulating water and does not enter the environment.

Closed cycle (fuel): Refers to recycling of *spent nuclear fuel*.

Coal: A *fossil fuel* formed millions of years ago by the breakdown of organisms trapped underground without access to air.

Cohort study: An epidemiological study that observes subjects in differently exposed groups and compares the *incidence* of disease. Although ordinarily prospective in nature, such a study is sometimes carried out retrospectively, using historical data.

Collective dose: The sum of individual *doses* received in a given period of time by a specified population from *exposure* to a specified source of *radiation*. Its significance is speculative.

Combustion: Chemical oxidation accompanied by the generation of *light* and heat.

Coolant loop: Also known as the primary *cooling system*, coolant in a set of pipes circulates in a closed loop through the reactor and carries heat away from the fissioning fuel to a *heat exchanger*. There, the coolant transfers its heat to a secondary *cooling system* in a completely separate loop before returning to the reactor.

Cooling system: At nuclear and fossil-fuel *power plants*, the cooling system takes water directly from a large body of water, purifies it, uses it to absorb the waste heat from the burning *coal* or the *reactor core*, and then discharges the heated water back to its source. The heated purified water makes steam to drive turbines.

Cooling towers: The hourglass-shaped structures that rise above some nuclear or *coal*-fired plants. In a coolant circulation system separate from the primary one, they dissipate waste heat from the steam the plant is making by bringing warm water in contact with the air, to which heat and moisture are transferred. The cooled water is cycled back through the plant.

Criticality: A term used in reactor physics to describe the state when the number of *neutrons* released by *fission* is exactly balanced by the *neutrons* being absorbed (by the fuel or by other materials, such as boron) and escaping the *reactor core*. A reactor is said to be "critical" when it achieves a self-sustaining nuclear *chain reaction*, as when the reactor is operating.

Cumulative dose: The total *dose* resulting from repeated *exposures* of *ionizing radiation* to the same portion of the body or to the whole body, over a period of time. Often refers to an occupationally exposed worker.

Curie (Ci): A measure of *radioactivity* based on the observed *decay* rate of approximately

1 gram of radium. The curie was named in honor of Pierre and Marie Curie, pioneers in the study of radiation. One curie of radioactive material will have 37 billion atomic transformations (disintegrations) in one second, which is approximately the *activity* of 1 gram of radium. The *becquerel* has replaced the curie in the standard international system of nomenclature.

Daughter products: *Isotopes* that are formed by the *radioactive decay* of some other *isotope.*

Decay: The decrease in the amount of any radioactive material with the passage of time due to spontaneous *emission* from the atomic nuclei of either *alpha* or *beta* particles, often accompanied by *gamma radiation.* Every *radionuclide* has a definite *half-life.* See *radioactive decay.*

Department of Energy (DOE): The stated mission of the department is the advancement of national, economic, and *energy* security for the United States; promotion of scientific and technological innovation; and environmental *cleanup* of the national nuclear weapons complex.

Depleted uranium (DU): Natural *uranium* from which the radioactive *isotope* U-235 has been removed and which therefore has less than the 0.7 percent of U-235 found in natural *uranium.* DU is obtained from spent-fuel *elements,* from *uranium tailings,* and from residues of *uranium enrichment* and has about the same toxicity as lead. DU is less radioactive than *uranium* ore.

Dirty bomb: Colloquial name for a *radiological dispersal device* (RDD) consisting of conventional explosives laced with radioactive material.

DNA repair: The cell's ability to repair DNA damage and restore the original base sequences. This process can restore destruction produced by normal physiological processes, *ionizing radiation,* or chemicals. There are many forms of DNA repair, and many genes responsible for and involved in DNA repair have been identified.

Dose: The *absorbed dose,* given in *rads* (or, in International System units, grays), that represents the *energy* in ergs or *joules* absorbed from the *radiation* per unit mass of tissue. Furthermore, the biologically effective dose or dose equivalent, given in *millirem, rem,* or *sieverts* (an International System unit; 100,000 *millirem* equal 1 *sievert*), is a measure of the biological damage to living tissue from *radiation exposure.*

Dose limits: Recommendations of the *National Council on Radiation Protection and Measurements* (commonly referred to as NCRP) for *ionizing radiation exposure* that should not be exceeded for protection of health. The total *lifetime exposure* limit is the person's age multiplied by 1,000 *millirem.* For example, a fifty-year-old person would have a *lifetime exposure* limit of 50,000 *millirem.* The recommended annual *occupational-exposure* limit is 5,000 *millirem.* For pregnant women in occupational conditions, the limit is only 50 *millirem* to the fetus each month. For the public, the annual limit and continuous *exposure* limit is 500 *millirem.* These limits do not include natural background or medical *exposures.* The average *background radiation* each person in the United States receives each year is 360 *millirem.*

Dose rate: The quantity of *absorbed dose* delivered per unit of time.

Dose response: Correlation between a quantified *exposure* (*dose*) and the proportion of a population demonstrating a specific effect (response).

Dosimetry: The theory and application of the principles and techniques involved in the measurement and recording of *doses* of *ionizing radiation.*

Electric motor: A device that takes *electrical energy* and converts it into mechanical *energy* to turn a shaft.

Electric power: The amount of *energy* produced per second. The *power* produced by an electric current.

Electrical energy: The *energy* associated with electric charges and their movements.

Electricity: A form of *energy* characterized by the presence and motion of elementary charged particles generated by friction, induction, or chemical change.

Electricity generation: The process of producing *electric energy* or the amount of *electric energy* produced by transforming other forms of *energy*, commonly expressed in *kilowatt-hours* (*kWh*) or *megawatt-hours* (MWh).

Electromagnetic radiation: A traveling wave motion resulting from changing electric or magnetic fields. Familiar electromagnetic radiation ranges from *X-rays* and *gamma* rays of short wavelength through the ultraviolet, visible, and infrared regions to radar and radio waves of relatively long wavelength.

Electron: An elementary particle carrying one unit of negative electric charge. Its mass is $\frac{1}{1837}$ that of a *proton*. *Electrons* form part of an *atom* and move around its *nucleus*.

Element: One of the 103 known chemical substances that cannot be broken down farther without changing its chemical properties. Each element has a different *atomic number* on the *periodic table*. All matter is composed of elements.

Emission: A discharge or something that is given off; generally used in regard to discharges into the air. Or, releases of gases to the atmosphere from some type of human activity, chiefly the extraction and *combustion* of *fossil fuels*. Additional sources: the burning of *biomass*, cattle ranching, rice cultivation, garbage landfills. In the context of *global climate change*, the emissions consist of *greenhouse gases* (e.g., the release of *carbon dioxide* during fuel *combustion*).

Energy: The ability to do work or the ability to move an object. *Electrical energy* is usually measured in *kilowatt-hours* (*kWh*), while heat energy is usually measured in British thermal units (Btu).

Energy consumption: The use of *energy* as a source of heat or *power* or as a raw material input to a manufacturing process.

Energy efficiency: Refers to activities that are aimed at extracting more *energy* out of existing resources by substituting technically more advanced equipment, typically without affecting the services provided. Examples include high-efficiency (Energy Star) appliances; efficient lighting programs; high-efficiency heating, ventilating, and air-conditioning (HVAC) systems or control modifications; efficient building design; advanced *electric motor* drives; and heat recovery systems.

Energy Information Administration (EIA): Part of the U.S. *Department of Energy* (DOE), EIA provides *energy* statistics gathered by the federal government.

Enrichment: The process of increasing the proportion of a particular *isotope* in, usually, *uranium*. Naturally low in the *fissile isotope* U-235 (0.7 percent), *uranium* is enriched to increase the proportion to around 3–6 percent to make reactor fuel. See *centrifuge enrichment*.

Environmental Protection Agency (EPA): Created in 1970, the U.S. EPA is responsible for working with state and local governments to set standards that help control and prevent pollution and minimize the potential health effects of solid and hazardous waste and toxic and radioactive substances.

Epidemiology: The study of the distribution and dynamics of diseases and injuries in human populations. The two main types of epidemiological studies of chronic disease are *cohort* (follow-up) *studies* and case-control (retrospective) studies.

Exempt waste: Waste that contains amounts of *radioactivity* comparable to that found in household garbage.

Experimental Breeder Reactor-I (EBR-I): The first reactor in the United States to use *nuclear energy* to make *electricity*, at the *National Reactor Testing Station* (*NRTS*) in Idaho in 1951.

Experimental Breeder Reactor-II (EBR-II): An early reactor prototype designed to demonstrate a complete breeder-reactor *power plant* with on-site *reprocessing* of metallic fuel. EBR-II operated successfully in this capacity from 1964 to 1969 in Idaho. It was later used for the *Integral Fast Reactor* (IFR) program.

Exposure: Contact of an organism with a chemical, radiological, or physical agent.

External radiation dose: The *dose* from sources of *radiation* located outside the body.

Fast breeder reactor: *Fission* reactor that uses only *fast neutrons* to sustain the *chain reaction*. It is configured so that these *neutrons* breed more *fissile material* in a *fertile* blanket than is consumed in the core as it makes *energy*.

Fast neutrons: *Neutrons* with an *energy* on the order of millions of *electron volts*. Fast neutrons in the reactor are released directly from fissioning nuclei and *neutron-rich fission products*; they travel at a much higher speed than *neutrons* in a water-moderated reactor.

Fertile material: Material that is not *fissile* (fissionable by thermal *neutrons*) but can be converted into a *fissile material* by irradiation in a reactor.

Fissile material: Material consisting of *atoms* whose nuclei can be split when irradiated with low-*energy neutrons*.

Fission: A nuclear process in which a heavy, unstable *nucleus* of an *atom*, such as *uranium-235* or *plutonium-239*, splits into two daughter nuclei, releasing *neutrons* and a large amount of *energy*. The fission is usually caused by the absorption of a *neutron*. Nuclear *power plants* split the nuclei of *uranium atoms*.

Fission product: Any *atom* created by the *fission* of a heavy *element*. Fission products are usually radioactive.

Fossil fuels: Fuels (*coal*, oil, natural gas) that are the result of the compression of ancient plant and animal life over millions of years.

Free radical: An unstable and highly reactive *molecule*, bearing an atom with an unpaired *electron*, that nonspecifically reacts with a variety of organic structures such as DNA. The interaction of *ionizing radiation* with water can generate free radicals in the form of hydroxyl and hydroperoxyl groups that are potent oxidizing agents. Biologically, free radicals play essential roles in life processes. However, they can also produce cell damage and harm DNA and lipids, possibly causing cancer and other diseases as well as aging.

Fuel cell: An electrochemical conversion device. It converts *hydrogen* and oxygen into water and, in doing that, makes *electricity*.

Fuel cycle: The entire set of stages involved in the utilization of nuclear fuel, including extraction, transformation, transportation, and *combustion*. The "front end" refers to the extraction of any raw materials, any chemical or physical processing they undergo, *enrichment*, and fabrication. The "*back end*" refers to processing and disposal of irradiated, spent fuel after it is removed from a reactor.

Fusion: The phenomenon that occurs when two light nuclei of *atoms* are combined or "fused" together, as when the sun combines the nuclei of *hydrogen atoms* into helium *atoms*, releasing an enormous amount of *energy*.

Gamma radiation: High-*energy*, short-wavelength *electromagnetic radiation* emitted from the *nucleus* of an *atom*. Gamma rays are very penetrating and are stopped by dense materials such as lead and thick concrete.

Gaseous diffusion enrichment: An older method of *uranium enrichment* that employs the diffusion of gasified *uranium* through a porous membrane to separate out *uranium-*

235. Lighter than *uranium*-238, U-235 is more likely to diffuse through the membrane than U-238.

Geiger counter: Device to measure *radioactivity*. It consists of a Geiger-Müller tube that sparks whenever nuclear *radiation* enters it, and a counter that counts every time the tube has a spark.

Generating capacity: The amount of *electric power* a *power plant* can produce.

Generation I: Early commercial nuclear reactors, built in the 1950s and 1960s.

Generation II: Large-scale, economic commercial nuclear reactors built in the 1960s, 1970s, and 1980s. Most of the reactors currently operating in the West are Generation II.

Generation III: Evolution of Generation II reactors, built in the 1990s with improvements in all areas, including waste reduction, economics, and safety.

Generation III+: Further evolution of the Generation III. Generation III+ reactors are currently being licensed around the world.

Generation IV: Revolutionary new designs of nuclear reactors. They are expected to be installed in *power plants* over the next several decades. Many of them are *fast-neutron* reactors and incorporate recycling as one of their features.

Generator: A device that turns mechanical *energy* into *electrical energy*. The mechanical *energy* is sometimes provided by an engine or turbine.

Geothermal energy: The heat *energy* that is produced by natural processes inside the earth. It can be taken from hot springs or by reservoirs of hot water deep below the ground, or by breaking open the rock itself.

Gigawatt: One billion *watts*.

Global climate change (global warming): Usually refers to an increase in the near surface temperature of the earth that is causing *climate change* on a worldwide scale. Global warming has occurred in the distant past as the result of natural influences, but the term is today most often used to refer to the temperature increase presently occurring as a result of increased *emissions* of *greenhouse gases*, mostly from human activity. See *climate change*.

Greenhouse effect: Certain gases in the earth's upper atmosphere have the capacity to trap heat from the sun rather than to permit it to be reflected back into space. As more gases accumulate, the atmosphere heats up more, becoming like a greenhouse. This process affects climate on a global scale.

Greenhouse gases: Gases that trap the heat of the sun in the earth's atmosphere, producing the *greenhouse effect*. The two major greenhouse gases are water vapor and *carbon dioxide*. Lesser greenhouse gases include *methane*, ozone, chlorofluorocarbons, and nitrogen oxides. Human activities as well as natural processes emit greenhouse gases. The significant increase in greenhouse gases since the Industrial Revolution indicates that human activity is the chief cause in the relatively rapid rise in global temperature in the past two hundred years.

Grid: The layout of an electrical distribution system.

Half-life: The time in which half the *atoms* of a radioactive substance will have disintegrated, leaving half the original amount. Half the residue will disintegrate in another equal period of time.

Health physics: The science concerned with the recognition, evaluation, and control of health hazards that may arise from accidents or applications that result in *exposure* to *ionizing radiation*.

Heat exchanger: Any device that transfers heat from one fluid (liquid or gas) to another or to the environment.

Heavy water: Water that has deuterium (H_2, an *isotope* of *hydrogen* known as heavy hydrogen) in place of normal hydrogen. Heavy water behaves chemically in the same way as water.

Heavy water reactor (HWR): An older design that uses *heavy water* for a *moderator* instead of normal (light) water. The CANDU reactors in Canada use heavy water, which slows down the *neutrons* so that they can react with natural, rather than enriched, *uranium* to sustain a *chain reaction.*

High temperature reactor (HTR): This thermodynamically efficient design uses helium as the *coolant* and graphite as the *moderator* in order to generate much higher temperatures. One application would be to generate hydrogen. The *pebble-bed modular reactor* (PBMR) is an HTR.

High-level waste (HLW): Category of *radioactive waste*, usually *spent nuclear fuel* and *fission products*, that requires shielding and cooling. It makes up only 3 percent of the total volume of *radioactive waste* but has 95 percent of the total *radioactivity.*

Highly enriched uranium (HEU): *Uranium* enriched to more than 20 percent U-235, the rest being U-238. Used in submarine reactor fuel and in nuclear weapons. Also blended with *uranium* oxide to make low-enriched fuel for power reactors. See *mixed oxide fuel (MOX).*

High-to-low-dose extrapolation: The process of predicting human *risks* from low *radiation exposures* using either human or animal data on *risks* derived from high levels of *exposure.*

Homeostasis: The ability of an organism to maintain stability.

Hormesis: The biological theory that organisms are made more resilient by low-level *exposure* to a substance that is toxic in larger *doses.*

Hydroelectric plant: A *power plant* that uses moving water to power a turbine *generator* to produce *electricity.*

Hydrogen (H): A colorless, odorless, highly flammable gaseous *element*. It is the lightest of all gases and the most abundant *element* in the universe, occurring chiefly in combination with oxygen in water and also in acids, bases, alcohols, *petroleum*, and other hydrocarbons. Pure hydrogen can be used to store *energy* but is not a source of it.

Idaho National Engineering and Environmental Laboratory (INEEL): Former name of *Idaho National Laboratory* (INL). Originally the *National Reactor Testing Station (NRTS).*

Idaho National Laboratory (INL): Laboratory where nuclear power reactors were pioneered. Since 2005, the name of the combined *Idaho National Engineering and Environmental Laboratory* (INEEL) and *Argonne National Laboratory-West.*

Incidence: The number of new cases of a disease in a population over a period of time.

Individual risk: The *risk* to an individual based on the average *risk* to a population with similar *exposures.*

Integral Fast Reactor (IFR): A liquid-sodium-cooled fast reactor developed by Argonne National Laboratory using an integral (or closed) cycle to recycle spent fuel on-site and render it proliferation-resistant. The IFR was designed to be inherently safe, and to produce electricity economically by more efficiently exploiting the energy in the fuel.

Intermediate waste: Nuclear waste radioactive enough to be shielded by lead, concrete, or water. If it has a short *half-life*, it's buried in its shielding. If long-lived, it must be isolated in a deep geological *repository.*

Internal radiation dose: The *dose* to organs of the body from radioactive materials deposited and retained inside the body. It may consist of any combination of *alpha,* *beta*, and *gamma radiation*. Some organs can withstand large *doses* that are administered for therapeutic purposes.

International Atomic Energy Agency (IAEA): Created in 1957 by the United Nations as the "Atoms for Peace" world organization. Its mission is to work with its member states and multiple partners worldwide to promote safe, secure, and peaceful nuclear technologies.

International Commission on Radiation Protection (ICRP): An independent advisory group established in 1928 to advance, for the public benefit, the science of radiological protection, in particular by providing recommendations and guidance on all aspects of protection against *ionizing radiation*. ICRP provides an overview of *radiation* standards and regulations as well as information to help standardize them.

International Energy Agency (IEA): Created by the *Organization for Economic Cooperation and Development* (OECD) after the 1974 oil crisis to coordinate *energy* policies centered around the need for an emergency sharing system in case of *energy* disruptions. Today the twenty-six-member IEA's mission is to consider means of improving the supply and efficient use of *energy;* integrating *energy* and environmental policies; and opening dialogue between *energy* producers and consumers.

Iodine-131 (I-131): A short-lived (8.1 days) *isotope* made in reactors and in atomic explosions. If accidentally released, radioiodine can, in some exposed individuals, become concentrated in the thyroid gland, irradiating it. I-131 is routinely used to diagnose and treat thyroid disease. The deposition, retention, and *radiation dose* from this *isotope* can be modified by taking potassium iodide pills.

Ion: (1) An *atom* that has too many or too few *electrons*, causing it to have an electrical charge and, therefore, to be chemically active. (2) An *electron* that is not associated (in orbit) with a *nucleus*.

Ionization: The process of adding to or removing one or more *electrons* from *atoms* or *molecules*, thereby creating *ions* and *free radicals*. High temperatures, metabolic processes, electrical discharges, and *radiation* can cause ionization. If enough ionizations occur in human tissue, the resulting *free radicals* can cause damage to DNA.

Ionize: To split off one or more *electrons* from an *atom*, thus leaving it with a positive electric charge. The *electrons* usually attach to other *atoms* or *molecules*, giving them a negative charge.

Ionizing radiation: Any radiation capable of displacing *electrons* from *atoms* or *molecules*, thereby producing *ions*. Some examples are *alpha, beta,* and *gamma* particles, *X-rays, neutrons,* and ultraviolet light. Examples of non-ionizing radiation: microwaves and radio waves.

Isotope: *Atoms* of the same *element* that have an equal number of *protons* (and hence the same chemical properties) but a different number of *neutrons* and, therefore, different atomic weights. Although chemical properties are the same, radioactive and nuclear (*radioactive decay*) properties may be quite different for each isotope of an *element*.

Joule: Measure of *energy*. Deposition of 1 joule/kg is equal to 100 *rads*.

Kilowatt: A unit of *power,* usually used for *electric power* or *energy consumption*. A kilowatt equals 1,000 *watts*.

Kilowatt-hour (kWh): A measure of *electricity* defined as a unit of work or *energy*, measured as 1 *kilowatt* (1,000 *watts*) of *power* expended for one hour. One kWh is equivalent to 3,412 Btu or 3.6 million *joules*. The *energy* of 1 kilowatt-hour will keep a 40–*watt* bulb lit for a full day. Per capita kWh consumption in the United States in 2003: 14,057.

Japan: 8,701. Equatorial Guinea: 51. Cambodia: 9 (source: United Nations Development Programme, http://hdr.undp.org/hdr2006/statistics/indicators/198.html).

Latency period: The average period of time between *exposure* to an agent and the onset of a health effect.

Lethal dose (LD 50/30): The *dose* of radiation expected to cause the death of half of the exposed population within thirty days. For single whole-body acute *radiation exposure*, the LD 50/30 ranges from 400 to 500 *rem* (400,000 to 500,000 *millirem*).

Lifetime exposure: Total calculated *exposure* to *radiation* or a chemical that a human would receive in a lifetime (usually assumed to be seventy years).

Light: Radiant electromagnetic *energy* in the visible part of the spectrum.

Linear nonthreshold hypothesis (LNT or LNTH): The mathematical hypothesis that a toxic substance is harmful at all dosages, whether large or small, and that the harmful effects are directly proportional to the dosage. This model states that any amount of *radiation dose*, no matter how small, results in increased *radiation risk* and that for every unit of *dose*, there is an increase in *risk*. No evidence for this exists in regard to low-dose radiation.

Load: The *power* and *energy* requirements of users on the *electric power* system in a certain area or the amount of *power* delivered to a certain point.

Los Alamos National Laboratory (LANL): Located in New Mexico, LANL was established during World War II to work on the Manhattan Project: the making of the world's first atomic bomb. Today research on nuclear weapons design continues at LANL, along with a variety of other scientific studies. The laboratory is one of the largest multidisciplinary institutions in the world.

Low-enriched uranium (LEU): *Uranium* enriched in the *isotope* U-235 above the natural level of 0.7 percent but to below 20 percent, the rest being *uranium-238*. *Uranium* for use in *power plants* is low-enriched, usually 3–6 percent.

Low-level waste: A category for nuclear waste containing small amounts of *radionuclides* that may have short or long *half-lives* and do not require shielding when being handled.

Materials testing reactor (MTR): Used at *Idaho National Laboratory* (INL) to irradiate various materials to determine their capacity to be used in reactor components and to withstand other thermally hot and highly radioactive environments.

Megawatt: A unit of electrical *power* equal to 1,000 *kilowatts* or 1 million *watts*. Output of *power plants* is usually measured in megawatts.

Methane: A colorless, flammable, odorless hydrocarbon gas that is the major component of natural gas. Also an important source of *hydrogen* in various industrial processes. Methane has natural as well as man-made sources and is a *greenhouse gas*.

Metric ton: A unit of weight equal to 1,000 kilograms, or 2,204.6 pounds.

Millirem: A unit of *absorbed radiation dose* amounting to 0.001 *rem*. Eating a single banana will give you a *dose* of 0.01 *millirem*. Living near a nuclear plant is estimated to give you an annual *dose* of 0.009 *millirem*.

Mixed oxide fuel (MOX): A reactor fuel made by blending a small amount of *plutonium* or *highly enriched uranium* with natural *uranium*.

Moderator: Any material that can slow down *neutrons*. In a thermal *fission* reactor, a moderator is used to reduce the speed of *neutrons* so that they can cause *fission* in the *fissile* nuclei of the fuel. Water is typically the moderator in most power reactors.

Molecule: Particles that normally consist of two or more *atoms* joined together. A water molecule is made up of two *hydrogen atoms* and one oxygen *atom*.

Mutation: Any heritable change in DNA sequence. A mutation can be induced by changes at the chromosome, gene, or DNA level.

National Academy of Sciences (NAS): A private, nonprofit, self-perpetuating American honorific society of distinguished scholars. The purpose of NAS is to engage in scientific and engineering research with the aim of advancing science and technology and enhancing their use for the general welfare. Since the body's inception in 1863, the nation's leaders have relied upon the National Academies—consisting of not only the *National Academy of Sciences* but also the *National Research Council* (*NRC*), the National Academy of Engineering, and the Institute of Medicine—to conduct scientific and technological studies and offer objective advice on issues that bear on policy decisions.

National Council on Radiation Protection and Measurements (NCRP): A nonprofit corporation chartered in 1964 by Congress to make recommendations on *radiation* measurements, quantities, and units. NCRP's stated mission is "to formulate and widely disseminate information, guidance and recommendations on radiation protection and measurements which represent the consensus of leading scientific thinking."

National Institute for Occupational Safety and Health (NIOSH): A federal agency that, among other activities, recommends *occupational exposure* limits for various substances, and assists in occupational safety and health investigations and research.

National Reactor Testing Station (NRTS): Established in 1949 in Idaho, NRTS eventually became the *Idaho National Laboratory* (*INL*).

National Research Council (NRC): As part of the *National Academy of Sciences* (*NAS*), the National Research Council is designed to associate the broad community of science and technology with the needs of the government. The council is the operating agency for *NAS*.

Neutron: A particle slightly heavier than a *proton* and with no electrical charge that helps to make up the *nucleus*. In the proper conditions, neutrons, depending on their *energy*, can be absorbed by susceptible nuclei and can cause *fissile* nuclei to *fission*.

Nevada Test Site (NTS): Federal reservation where 928 atomic weapons tests were once conducted, 100 of them atmospheric. Congress chose Yucca Mountain at NTS as the site for a national *repository* for *high-level nuclear waste*.

Nonlinear threshold hypothesis: Model of *exposure* to a toxic substance that states that damaging effects only occur above a threshold *dose*, as yet to be determined, and that low-level *exposures*—that is, below that threshold—do not cause harm. See *linear nonthreshold hypothesis*.

Nuclear energy: *Energy* that comes from splitting *atoms* of radioactive materials, specifically the *energy* bound up in the *nucleus* of an *atom*.

Nuclear Energy Agency (NEA): An international agency within the *Organization for Economic and Cooperative Development* (OECD). According to NEA, "It helps its twenty-eight members to maintain and develop, through international co-operation, the scientific, technological and legal bases required for the safe, environmentally friendly and economical use of *nuclear energy* for peaceful purposes. Its studies and co-operative activities address nuclear safety and regulation; *radioactive waste* management; radiological protection and public health; nuclear science; economics, resources and technology; and legal affairs."

Nuclear Nonproliferation Treaty (NPT): International agreement, first signed in 1968 and since developed and expanded, to stop the spread of nuclear weapons by restricting the trade of all nuclear technology and materials to signatory nations, which agree to full compliance with international safeguards.

Nuclear propulsion: A concept that involves propelling a spacecraft by detonating nuclear explosives behind a shield at the base of the spacecraft.

Nuclear Regulatory Commission (NRC): An independent agency created from the U.S. *Atomic Energy Commission (AEC)* in 1975 to regulate civilian uses of nuclear material. Specifically, NRC is responsible for ensuring that activities associated with the operation of nuclear *power* and *fuel cycle* plants and the use of radioactive materials in medical, industrial, and research applications are carried out with adequate protection of public health and safety, the environment, and national security.

Nucleus: Composed predominantly of *protons* and *neutrons,* the nucleus is the central component of the *atom.* It is about 1 percent the size of the *atom* but accounts for almost all of its mass.

Occupational exposure: *Radiation exposure* attributable to people's jobs.

Occupational Safety and Health Administration (OSHA): A U.S. Department of Labor agency with safety and health regulatory and enforcement authority for most U.S. industry and business.

Office of Civilian Radioactive Waste Management (OCRWM): U.S. *Department of Energy (DOE)* body in charge of disposal of commercial nuclear waste.

Open fuel cycle: Refers to the complete cycle of nuclear fuel from origin to one trip through the reactor to disposal. Also called once-through. See *closed cycle.*

Organization for Economic Cooperation and Development (OECD): A global entity made up of thirty member countries. It has active relationships with some seventy other countries, economies, nongovernmental organizations, and civil societies. OECD is made up of multiple bodies, including the *International Energy Agency (IEA)* and the *Nuclear Energy Agency (NEA),* and is known for its research, publications, and statistical analyses.

Particulates: Fine liquid or solid particles such as dust, smoke, mist, fumes, and smog, found in the air or *emissions.* Sometimes "inhalable" or "respirable" is used to describe particles (less than 2 microns) that can be inhaled through the nose and enter the lungs.

Peak-load plant: An electrical plant usually housing old, low-efficiency steam units, gas turbines, diesels, or pumped-storage hydroelectric equipment normally used during the peak-load periods.

Pebble-bed modular reactor (PBMR): A *high temperature reactor (HTR)* that uses fuel encased in spheres of graphite and silicon carbide to ensure *fission product* containment and meltdown resistance.

Periodic table: A chart of all known *elements* arranged in a meaningful, orderly pattern according to their respective *atomic numbers* and configurations of *electrons.*

Petrochemicals: Organic and inorganic *petroleum* compounds and mixtures that include but are not limited to organic chemicals, cyclic intermediates, plastics and resins, synthetic fibers, elastomers, organic dyes, organic pigments, detergents, surface active agents, carbon black, and ammonia.

Petroleum: Generally refers to crude oil or the refined products obtained from the processing of crude oil (including gasoline, diesel fuel, jet fuel, heating oil). Petroleum also includes natural gas and natural gas liquids.

pH: A measure of the acidity or alkalinity of a material, either liquid or solid (pH is represented on a scale of 0 to 14 with 7 representing a neutral state, 0 representing the most acid, and 14 the most alkaline). Relatively rapid absorption by the ocean of increased *carbon dioxide* from *emissions* changes the pH of seawater, acidifying it.

Photovoltaic cell (PVC): A device, usually made from silicon, which converts some of the *energy* from *light* into *electrical energy.*

Plutonium (Pu): A heavy metal *element* that is an *alpha*-emitter. Plutonium is bred in nuclear reactors from the *chain reaction* and can be used as reactor fuel or to make nuclear weapons. In most of its chemical forms, plutonium is biologically inert. Only a very small fraction of any ingested plutonium is taken up by tissues; most is excreted. However, if it is inhaled as an aerosol of small particles, it can become imbedded in the lungs and remain for long periods, resulting in chronic irradiation of the cells. A fraction of the total deposited material can move from the lung to bone surfaces and the liver. Experimental animal studies show an increase in lung, bone, and liver cancer following high levels of deposition of this *radioisotope.*

Power: The rate at which *energy* is transferred. *Electrical energy* is usually measured in *watts.* Also used for a measurement of capacity.

Power degradation: The loss of *power* when *electricity* is sent over long distances. For this reason *power plants* are sited near population centers.

Power plant: A facility where *power,* especially *electricity,* is generated.

Power-generating efficiency: The percentage of the total *energy* content of a *power plant's* fuel that is converted into *electrical energy.* The remaining *energy* is lost to the environment as heat.

Pressurized-water reactor (PWR): The most common power-reactor design. It uses water kept to serve as both coolant and *moderator.* In the closed primary *coolant system,* the water, heated by thermal *energy* from the *chain reaction* while passing through the core, is kept under high pressure. This prevents the water from boiling into steam, so that it remains liquid even at high temperatures. Steam is then generated in a secondary *coolant loop* entwined around the primary one.

Probabilistic risk assessment (PRA): A systematic process for examining how engineered systems—built and operated according to regulatory requirements and practices—and human interactions with these systems work together to ensure plant safety. This process is quantitative, in that probabilities of events with potential public health consequences are calculated, as are the magnitudes of these potential health consequences. The *risk* of such events is the product of the event probabilities and their consequences.

Probability: The mathematical chance that a particular event will occur, given the spectrum of all possible events in a specific situation. See *risk.*

Proton: An elementary particle with a positive electric charge.

Quality factor (QF): A measure of the effectiveness of a particular radiation (for example, *X-rays* have a QF of 1, *alpha particles* a QF of 20) at producing injury in a biological system. Relates the physical effects of a particular radiation or mixture of radiations to that of an *electromagnetic radiation.*

Rad: Acronym for *radiation absorbed dose.* Approximately comparable to a *rem* or a *roentgen.* A rad is a unit of *absorbed dose* of *radiation* defined as deposition of 100 ergs of *energy* per gram of tissue. A rad times a *quality factor* is equal to a *rem.* A rad is approximately equal to a *roentgen.* In more current terminology, the rad has been replaced by the gray, which is equal to 100 rads.

Radiant energy: Any form of *energy* radiating from a source in waves. The sun produces radiant *energy.*

Radiation: Any high-speed *transmission* of *energy* in the form of particles or electromagnetic waves.

Radiation dose: The quantity of *radiation* or *energy* that is absorbed. See *absorbed dose* and *radiation exposure*.

Radiation Effects Research Foundation (RERF): A cooperative organization jointly sponsored by Japan and the United States that, since 1975, has been conducting research and studies on the effects of *radiation exposure* on humans in order to help maintain the health and welfare of atomic bomb survivors as well as to contribute to the enhancement of the health of all people.

Radiation exposure: Generally, a term relating to the amount of *ionizing radiation* that strikes a living being or inanimate material.

Radiation hormesis: The theory that small *doses* of *radiation* can induce beneficial biological processes and are healthful.

Radiation shielding: Reduction of *radiation* by interposing a shield of absorbing material between any radioactive source and a person, work area, or radiation-sensitive device.

Radiation sickness: The complex of symptoms characterizing the disease known as *radiation* injury, resulting from excessive *exposure* (greater than 200 *rads* or 2 gray) of the whole body (or large part) to *ionizing radiation*. The earliest of these symptoms are nausea, fatigue, vomiting, and diarrhea, which may be followed by loss of hair, hemorrhage, inflammation of the mouth and throat, and general weakness. In severe cases, where the *radiation exposure* has been around 1,000 *rad* (10 gray) or more, death may occur within two to four weeks. Those who survive six weeks after receiving a single large *dose* of radiation to the whole body may generally be expected to recover.

Radioactive decay: Property of undergoing spontaneous nuclear transformation in which nuclear particles or electromagnetic *energy* is emitted. All radioactive material decays over time into stable *elements*.

Radioactive waste: Materials left over from making *nuclear energy* as well as from other uses of nuclear materials in industry and therapeutic and diagnostic medicine.

Radioactivity: The spontaneous discharge of *radiation* from atomic nuclei of certain *elements*, usually in the form of *beta* or *alpha* radiation, together with *gamma radiation*. *Beta* or *alpha emission* results in transformation of the *atom* into a different *element*.

Radioisotope: A radioactive *isotope*. An unstable *isotope* of an *element* that decays or disintegrates spontaneously, emitting *radiation*. More than thirteen hundred natural and artificial radioisotopes have been identified.

Radiological dispersal device (RDD): A weapon made of conventional explosives to which radioactive material has been added. See *dirty bomb*.

Radionuclides: *Radioactive elements*. These may be either natural radionuclides such as *radium* or *uranium*, which are normally present in the earth, or artificial radionuclides, which are not normally present (or normally present in very small amounts) and are produced by nuclear *fission*.

Radium (Ra): A radioactive metallic *element*. It occurs in minute quantities associated with *uranium* in pitchblende and other minerals.

Radon (Rn): A radioactive *element* that is one of the heaviest gases known. It is a daughter of radium. Most of our *exposure* to radiation comes from *radon* in soil and rock.

Reactor core: The structure inside which *fission* occurs in *uranium* fuel, in millions of atomic nuclei, producing huge amounts of heat *energy*.

Reactor vessel head: The removable top section of the reactor pressure vessel, bolted into place during *power* operation and removed to permit refueling.

Refinery: An industrial plant that heats crude oil (*petroleum*) so that it separates into chemical components, which are then made into more useful substances.

Relative risk: Ratio of the disease rate (usually *incidence* or mortality) among those exposed to the rate among those not exposed.

Rem: Acronym for "*roentgen* equivalent man." A unit of equivalent *absorbed dose* of *radiation*, taking account of the *quality factor* of the particular *radiation*. A *millirem* is one-thousandth of a *rem*.

Renewable energy sources (renewables): Fuels that can be easily made or "renewed" and are never depleted, for example, hydropower (water), solar, wind, geothermal, and *biomass*. Nonrenewable fuels: oil, gas, *coal*.

Repair processes: Metabolic processes within a cell that can repair *radiation* damage before it is expressed as a biological effect such as cell killing.

Repository: A place to isolate and shield *radioactive wastes*, either as an interim measure or for the long term. The radioactive wastes will ultimately *decay* to natural *radioactivity* levels.

Reprocessing: Chemical treatment of *spent nuclear fuel* that separates unused *uranium* and *plutonium* from radioactive *fission product* wastes. This allows the reuse of valuable fuel material and greatly reduces the volume of *high-level waste* materials.

Risk: The product of severity (consequence) impact and likelihood (*probability*) impact. Specifically for carcinogenic effects, risk is estimated as the incremental *probability* of an individual's developing cancer over a lifetime as a result of *exposure* to a potential carcinogen. For noncarcinogenic (systemic) effects, risk is not expressed as a *probability* but rather is evaluated by comparing an *exposure* level over a period of time to a reference *dose* for a similar *exposure* period. Specifically for reactors, the *Nuclear Regulatory Commission* (*NRC*) uses three questions to define risk: What can go wrong? How likely is it? What are the consequences? The *NRC* identifies important scenarios from such an assessment.

Risk analysis: A detailed examination including *risk assessment, risk evaluation,* and *risk management* alternatives. Risk analysis is performed to determine the nature of unwanted negative consequences to human life, health, property, or the environment. Risk analysis can also provide information regarding undesirable events. The term is also applied to the process of quantifying the probabilities and expected consequences of identified *risks.*

Risk assessment: The process of establishing information regarding acceptable levels of *risk* for individuals, groups, society, or the environment. See *probabilistic risk assessment* (*PRA*).

Risk characterization: The last step in *risk assessment.* This process characterizes the potential for adverse health effects and evaluates the degree of uncertainty involved.

Risk estimate: Description of the *probability* that organisms exposed to a specific *dose* of a chemical or other pollutant will develop an adverse response.

Risk estimation: The scientific determination of *risk* characteristics, usually in as quantitative a way as possible. These include the magnitude, spatial scale, duration, and intensity of adverse consequences and their associated probabilities as well as a description of the cause-and-effect links.

Risk evaluation: A component of *risk assessment* in which judgments are made about the significance and acceptability of *risk.*

Risk factor: Characteristic (e.g., race, sex, age, obesity) or variable (e.g., smoking, *occupational exposure* level) associated with increased *probability* of a toxic effect.

Risk identification: Recognizing that a hazard exists and trying to define its characteristics. Often *risks* exist and are even measured for some time before their adverse consequences are recognized. In other cases, risk identification is a deliberate procedure to review and, it is hoped, anticipate possible hazards.

Risk management: The process of evaluating and selecting alternative regulatory and nonregulatory responses to *risk*. The selection process necessarily requires the consideration of legal, economic, and behavioral factors.

Risk-specific dose: The *dose* associated with a specified *risk* level.

Roentgen: A unit of *gamma radiation* measured by the amount of *ionization* per unit volume in air. In nonbony biological tissue 1 roentgen delivers a *dose* approximately equal to 1 *rad*.

Sandia National Laboratories (SNL): Established in New Mexico in 1949 to address the engineering aspects of nuclear weapons design and testing, SNL does scientific and technological research as well as studies reduction of nuclear proliferation, the threat of nuclear accidents, and their damage to the environment. The lab experiments with alternative *energy* sources, in particular wind and solar *power*, and also works with the Defense Department and the Department of Homeland Security in matters of protection and security. It is the lead laboratory at the *Waste Isolation Pilot Plant* (*WIPP*), and since 2006 has become the lead laboratory for the *Yucca Mountain Project* (*YMP*).

Scram: Automatic shutdown of a nuclear reactor.

Semiconductor: Any material that has a limited capacity for conducting an electric current. Semiconductors are crystalline solids, such as silicon, that have an electrical conductivity between that of a conductor and an insulator. *Photovoltaic cells* are semiconductors.

Shelter in place: Usually the most prudent course of action in the event of a nuclear accident or an attack involving dispersal of radioactive materials. Go indoors wherever you happen to find yourself. The building will protect you from *exposure*. Await instructions about whether you need to evacuate. Often it is safer to stay where you are than to go outdoors. Close doors and windows, take a shower, and put on clean clothing.

Sievert (Sv): Unit of *radiation* measurement. One hundred *rem* is equivalent to 1 Sv.

Solar cell: An electric cell that changes *radiant energy* from the sun into *electrical energy* by the *photovoltaic* process.

Solar energy: The *radiant energy* of the sun, which can be converted into other forms of *energy*, such as heat or *electricity*.

Sorb: Chemical term meaning "to take up" or "to hold," as in absorption or adsorption.

Spent nuclear fuel: Irradiated fuel that is permanently removed from a nuclear reactor when no longer capable of sustaining a *chain reaction*. Except for possible *reprocessing*, this fuel must eventually be taken from its temporary storage location at the reactor site to an interim or permanent *repository*.

Steam generator: A *generator* in which the prime movers (turbines) are powered by steam.

Tailings: Waste rock from mining operations that contains concentrations of mineral ore that are too low to be economically extracted.

Terawatt-hour (tWh): One trillion *watts*.

Thorium: Heavy metal, three times more abundant than *uranium*, and likely in the future to be used as a reactor fuel. Brazil and India have especially rich deposits.

Threshold hypothesis: See *non-linear threshold hypothesis*.

Transformer: A simple electrical device that transfers *energy* from one circuit to another without relying on moving parts. At *power plants*, transformers convert the *generator's* low-*voltage electricity* to higher-*voltage* levels for *transmission* to the *load* center, such as a city or factory.

Transmission (electric): The movement or transfer of *electric energy* over an interconnected group of lines and associated equipment between points of supply and points at which it is transformed for delivery to consumers or is delivered to other electric systems. Transmission is considered to end when the *energy* is transformed for distribution to the consumer.

Transmission line: A set of conductors, insulators, supporting structures, and associated equipment used to move large quantities of *power* at high *voltage*, usually over long distances between a generating or receiving point and major substations or delivery points.

Transuranic: An artificially made, radioactive *element* that has an *atomic number* higher than *uranium* in the *periodic table* of *elements*. Examples: neptunium, plutonium, americium. The *Department of Energy (DOE)* distinguishes transuranic waste from other forms, like *spent nuclear fuel*, and stores it at the *Waste Isolation Pilot Plant (WIPP)*.

United Nations Scientific Committee on the Effects of Atomic Radiation (UNSCEAR): Established by the General Assembly of the United Nations in 1955 to assess and report levels and effects of *exposure* to *ionizing radiation*. Governments and organizations throughout the world rely on the committee's estimates as the scientific basis for evaluating *radiation risk* and for establishing protective measures.

United States Enrichment Corporation (USEC): USEC supplies low-enriched reactor fuel to commercial *power plants*, about half of it made from *uranium* from dismantled Soviet warheads.

Uranium: A naturally occurring, silvery heavy metal that occurs in abundance in the earth's crust and oceans and serves as the primary fuel in current nuclear reactors.

Uranium fuel cycle: The series of steps involved in supplying fuel for nuclear power reactors. It includes mining, refining, the making of fuel elements, their use in a reactor, chemical processing to recover spent fuel, re*enrichment* of the fuel material, and fabrication into new fuel elements.

Vitrification: The process of incorporating hazardous material in glass, where it will remain immobile and will be shielded from interaction with the environment.

Volt (V): The volt is the International System of Units measure of electric potential or electromotive force.

Voltage: The difference in electrical potential between any two conductors or between a conductor and ground. It is a measure of the *electric energy* per *electron* that *electrons* can acquire and/or give up as they move between the two conductors.

Waste Isolation Pilot Plant (WIPP): The world's first successfully operating long-term deep geologic *repository* for nuclear waste. Located near Carlsbad, New Mexico.

Watt: A metric unit of *power*, usually used in electric measurements, which gives the rate at which work is done or *energy* used. A watt is equal to one *joule* per second. A person climbing stairs is doing work at the rate of 200 watts.

X-ray: Penetrating *electromagnetic radiation* having a wavelength that is much shorter than that of visible *light*. These rays are usually produced by excitation of the *electron* field around certain nuclei.

Yellowcake: A natural *uranium* concentrate in powder form, it is produced from the milling and leaching of *uranium* ore. Yellowcake, which takes its name from its color and which typically contains 70 to 90 percent U_3O_8 (*uranium* oxide) by weight, is used as feedstock for *uranium* fuel *enrichment* and fuel pellet fabrication.

Yucca Mountain Project (YMP): Its mission has been to explore the feasibility of the establishment of a *repository* for high-level nuclear waste, mainly *spent nuclear fuel*, at the *Nevada Test Site* (*NTS*). It is presently preparing a licensing application for the *repository* to be submitted to the *Nuclear Regulatory Commission* (*NRC*).

Sources

Nuclear Energy Agency
Nuclear Energy Institute
UN International Atomic Energy Agency
U.S. Department of Energy
U.S. Environmental Protection Program
World Nuclear Association

ACKNOWLEDGMENTS

During the years of my nuclear tour, many people patiently and generously shared their expertise and practical assistance, starting with Rip Anderson, Marcia Fernández, and Leo S. Gómez. Henry Beard, my ever-supportive partner and titleist, insisted I write this book. The advice I received from Timothy Crouse and Chris Crawford was invaluable. Elizabeth Bruce, Kenneth Christian, Raymond Guilmette, Peter S. Heller, William L. Hurt, Paul Lorenzini, Jonathan Schell, and Mark Stumpf also gave me many helpful suggestions. I am deeply grateful to my mentor, Richard Rhodes; to my agent, Mary Evans; to my editor, Jonathan Segal; to my research assistant, Georgia Cravey; and to my son-in-law Erik Satre for his illustrations. Without the affection, understanding, and encouragement provided by my daughter Astrid Cravens as well as Lew and Lili Critchfield and many other dear friends, this project could not have come to fruition. Tom and Rhonda Dimperio, Janet Gutierrez, Erica Rominger, and the staff of Stardust House in Albuquerque gave me the peace of mind to write by giving superb care to my father.

For guiding me through the nuclear world and elucidating its mysteries, I am indebted to George Basabilvazo, Robert Benedict, John D. Boice, Jr., Evaristo J. Bonano, Barbara Critchfield Chang, Douglas Critchfield, Robert Critchfield, Kevin Davis, Alberto Delgado, Evan Douple, Terry Fletcher, Jeffrey Hahn, Frank D. Hansen, Hank Jenkins-Smith, John E. Kelly, Charles Key, Michael Lineberry, Melvin G. Marietta, Harold F. McFarlane, Navy Commander P. G. McLaughlin, Fred A. Mettler, Edward L. Miles, Andrew Orrell, Paul Pugmire, Theodore Rockwell, Navy Lt. Philip Rosi, David Schoep, Les Shephard, Tom P. Shiel, Rita Sipe, Ruth Sponsler, Susan M. Stacy, Dayle Stewart, Harlan Summers, Laurie Wiggs, and Aristides A. Yayanos.

INDEX

A NOTE ABOUT THE AUTHOR

Gwyneth Cravens has published five novels. Her fiction and nonfiction have appeared in *The New Yorker*, where she also worked as a fiction editor, and in *Harper's Magazine*, where she was an associate editor. Her articles and op-eds on science and other topics have appeared in *Harper's Magazine*, *The New York Times*, *The Brookings Review*, and *The Washington Post*. She grew up in New Mexico and now lives on eastern Long Island.

A NOTE ON THE TYPE

This book was set in Janson, a typeface long thought to have been made by the Dutchman Anton Janson, who was a practicing typefounder in Leipzig during the years 1668–1687. However, it has been conclusively demonstrated that these types are actually the work of Nicholas Kis (1650–1702), a Hungarian, who most probably learned his trade from the master Dutch typefounder Dirk Voskens. The type is an excellent example of the influential and sturdy Dutch types that prevailed in England up to the time William Caslon (1692–1766) developed his own incomparable designs from them.

Composed by North Market Street Graphics,
Lancaster, Pennsylvania

Printed and bound by Berryville Graphics,
Berryville, Virginia

Designed by M. Kristen Bearse